Boundary Elements and other Mesh Reduction Methods XLII

WIT*PRESS*

WIT Press publishes leading books in Science and Technology.
Visit our website for the current list of titles.
www.witpress.com

WIT*eLibrary*

Home of the Transactions of the Wessex Institute.
Papers published in this volume are archived in the WIT elibrary in volume 126 of WIT Transactions on Engineering Sciences (ISSN 1743-3533).
The WIT electronic-library provides the international scientific community with immediate and permanent access to individual papers presented at WIT conferences.
http://library.witpress.com

Forty Second International Conference on Boundary Elements and other Mesh Reduction Methods

BEM/MRM 2019

Conference Chairmen

A.H.D. Cheng
University of Mississippi, USA
Member of WIT Board of Directors

A. Tadeu
ITeCons, University of Coimbra, Portugal
Member of WIT Board of Directors

International Scientific Advisory Committee

Organised by

Wessex Institute, UK
ITeCons, University of Coimbra, Portugal
University of Mississippi, USA

Sponsored by

Elsevier
WIT Transactions on Engineering Sciences
International Journal of Computational Methods and Experimental Measurements

WIT Transactions

Wessex Institute
Ashurst Lodge, Ashurst
Southampton SO40 7AA, UK

Editorial Board

Boundary Elements and other Mesh Reduction Methods XLII

Editors

A.H.D. Cheng
University of Mississippi, USA
Member of WIT Board of Directors

A. Tadeu
ITeCons, University of Coimbra, Portugal
Member of WIT Board of Directors

WIT PRESS Southampton, Boston

Editors:

A.H.D. Cheng
University of Mississippi, USA
Member of WIT Board of Directors

A. Tadeu
ITeCons, University of Coimbra, Portugal
Member of WIT Board of Directors

Published by

WIT Press
Ashurst Lodge, Ashurst, Southampton, SO40 7AA, UK
Tel: 44 (0) 238 029 3223; Fax: 44 (0) 238 029 2853
E-Mail: witpress@witpress.com
http://www.witpress.com

For USA, Canada and Mexico

Computational Mechanics International Inc
25 Bridge Street, Billerica, MA 01821, USA
Tel: 978 667 5841; Fax: 978 667 7582
E-Mail: infousa@witpress.com
http://www.witpress.com

British Library Cataloguing-in-Publication Data

A Catalogue record for this book is available
from the British Library

ISBN: 978-1-78466-341-4
eISBN: 978-1-78466-342-1
ISSN: (print) 1746-4471
ISSN: (on-line) 1743-3533

The texts of the papers in this volume were set individually by the authors or under their supervision. Only minor corrections to the text may have been carried out by the publisher.

Preface

This issue contains papers selected from the 42nd International Conference on Boundary Elements and other Mesh Reduction Methods, held in the beautiful medieval town of Coimbra, Portugal, organised by the Wessex Institute, and co-sponsored by the local host ITeCons – Institute for Research and Technological Development in Construction, Energy, Environment and Sustainability.

The annual conference on Boundary Elements and other Mesh Reduction Methods (BEM/MRM) started in 1978 and is now in its 42nd version. It is well established as the recognised international forum for the latest advances in those techniques and their application in science and engineering.

The conference series was founded by Professor Carlos Brebbia, with its first meeting held in Southampton. For the next 42 years, scientists and engineers have used this gathering to exchange the progresses made in the field. The continued success of the meeting is a result of the strength of the research on boundary elements and mesh reduction techniques being carried out all over the world.

Unfortunately, Professor Brebbia passed away on March 3, 2018. The conference lost its founder and the world lost a giant. Professor Brebbia's legacy, however, continues. The field that he founded and stimulated is as vibrant as ever. This is evident by the continuing demand of the BEM/MRM conference and the increasing number of participants.

The objective of the research papers presented at the meetings is the further development of techniques that reduce or eliminate the type of meshes required by first generation computational methods, such as finite differences or finite elements. This has steadily been achieved through the development of BEM as a computational tool and continues through more recent research into advanced techniques, leading to further mesh reduction aiming to produce truly meshless methods.

Also included are papers on the use of BEM for various engineering and scientific applications.

The Editors would like to thank all authors for the quality of their papers and other colleagues for their help in reviewing the material.

The Editors
2019

Contents

ANALYTICAL 3D BOUNDARY ELEMENT IMPLEMENTATION OF FLAT TRIANGLE AND QUADRILATERAL ELEMENTS FOR POTENTIAL AND LINEAR ELASTICITY PROBLEMS

NEY AUGUSTO DUMONT & TATIANA GALVÃO KURZ
Department of Civil and Environmental Engineering,
Pontifical Catholic University of Rio de Janeiro, Brazil

ABSTRACT

This paper introduces a formulation for 3D potential and linear elasticity problems that end up with the analytical handling of all regular, improper, quasi-singular, singular and hypersingular integrals of an implementation using linear triangle (T3) elements. The extension to flat Q4 and T6 elements is almost straightforward. Results at arbitrarily located internal points are also given analytically. The formulation is based on a generalized transformation to subtriangle coordinates that simplifies the problem's description and enables the adequate interpretation of all relevant geometric features of a discretized boundary segment, so that it becomes possible to arrive at manageable analytical expressions of all integrals. The paper outlines the main concepts and computational features of the proposed formulation, based on an array with all pre-evaluated integrals required in an implementation. An example of 3D potential problems illustrates all particular cases and the most challenging topological configurations one might deal with in practical applications. The procedure may be easily implemented in a general boundary element code, as the usual numerical quadrature schemes for source points sufficiently far from the integration field remain applicable. There is a work in progress for the implementation of the procedure in the frame of a fast multipole algorithm.

Keywords: collocation boundary element method, numerical integration, analytical integration, 3D problems.

1 INTRODUCTION

The collocation boundary element method (BEM) [1] has been recently reconceptualized by the first author for general 2D and 3D problems [2], with the proposition of a general, simple and unified procedure for the machine-precision treatment of all kinds of singularities that may occur in 2D steady-state potential and elastostatics problems, as an extension of [3], [4].

The numeric issues for 3D problems are of a completely different nature [5]. The triangle-to-square coordinate transformation for the evaluation of improper integrals seems to have been firstly suggested for BEM applications in [6].

Simple analytical results are arrived at for all kinds of integrals required for 3D steady-state potential and elastostatics problems – for the boundary discretized with flat segments. This includes the evaluation of results at arbitrarily placed internal points, with hypersingularities also dealt with analytically. The case of a generally curved boundary segment is still not considered by the authors – and may not be mastered in the present framework. Although the proposed developments apply directly to elastostatics problems, space restrictions force us to draw attention only to potential problems.

2 PROBLEM FORMULATION

The following developments, as outlined in [5], apply to the adequate evaluation of integrals that embed improper, singular or quasi-singularities. The source point 0 is

WIT Transactions on Engineering Sciences, Vol 126, © 2019 WIT Press
www.witpress.com, ISSN 1743-3533 (on-line)
doi:10.2495/BE420011

indicated in Fig. 1(c), where $r=\sqrt{(x-x_0)^2+(y-y_0)^2+(z-z_0)^2}$ is the distance from the singularity (0 or source) point $(x_0,y_0,z_0)\equiv(x(\xi_s,\eta_s),\,y(\xi_s,\eta_s),\,z(\xi_s,\eta_s))$ to a generic (F or field) point $(x,y,z)\equiv(x(\xi,\eta),\,y(\xi,\eta),\,z(\xi,\eta))$ in Cartesian coordinates, expressed in terms of the indicated natural triangle coordinates $(\xi\in[0,1-\eta],\eta\in[0,1])$ in Fig. 1(a) as well as $(\xi,\eta\in[0,1])$ for the quadrilateral in the centre.

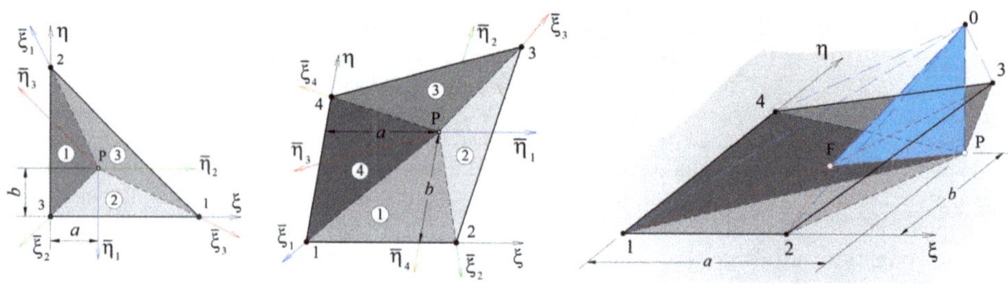

Figure 1: (a) Triangle; and (b) Quadrilateral elements described with two different systems of natural coordinates, (ξ,η) and $(\bar{\xi}_l,\bar{\eta}_l)$, for an in-plane singularity $\mathrm{P}(\xi=a,\eta=b)$; (c) Shows the general problem in the Cartesian space for a quadrilateral element with a quasi-singular source 0.

According to this proposition, the coordinates (ξ,η) are linear transformations of $(\bar{\xi}_l,\bar{\eta}_l)$, as for a triangle element [5],

$$[\xi,\eta]_1=\left[a(1-\bar{\xi}_1),\bar{\xi}_1(1-b-\bar{\eta}_1)+b\right];\quad |J|_1=\bar{\xi}_1 a \quad \text{for } \Delta_1, \tag{1}$$

$$[\xi,\eta]_2=\left[\bar{\xi}_2(\bar{\eta}_2-a)+a,b(1-\bar{\xi}_2)\right];\quad |J|_2=\bar{\xi}_2 b \quad \text{for } \Delta_2, \tag{2}$$

$$[\xi,\eta]_3=\left[\bar{\xi}_3(1-a-\bar{\eta}_3)+a,\bar{\xi}_3(\bar{\eta}_3-b)+b\right];\quad |J|_3=\bar{\xi}_3(1-a-b) \quad \text{for } \Delta_3, \tag{3}$$

as well as for a quadrilateral element (subtriangles referred to as Δ_l),

$$[\xi,\eta]_1=\left[\bar{\xi}_1(\bar{\eta}_1-a)+a,b(1-\bar{\xi}_1)\right];\quad |J|_1=\bar{\xi}_1 a \quad \text{for } \Delta_1, \tag{4}$$

$$[\xi,\eta]_2=\left[\bar{\xi}_2(1-a)+a,\bar{\xi}_2(\bar{\eta}_2-b)+b\right];\quad |J|_2=\bar{\xi}_2(1-a) \quad \text{for } \Delta_2, \tag{5}$$

$$[\xi,\eta]_3=\left[\bar{\xi}_3(1-a-\bar{\eta}_3)+a,\bar{\xi}_3(1-b)+b\right];\quad |J|_3=\bar{\xi}_3(1-b) \quad \text{for } \Delta_3, \tag{6}$$

$$[\xi,\eta]_4=\left[a(1-\bar{\xi}_4),\bar{\xi}_4(1-b-\bar{\eta}_4)+b\right];\quad |J|_4=\bar{\xi}_4 a \quad \text{for } \Delta_4. \tag{7}$$

The ratio $|J|_l/r(\bar{\xi}_l,\bar{\eta}_l)$ is always finite, which justifies the classically proposed coordinate transformation, but with the added advantage of enabling the analytical evaluation of the problem's key integrals [5]. As implemented, transformations end up carried out directly from Cartesian to subtriangle coordinates.

These developments are – to the authors' best knowledge – a novel generalization of a transformation classically proposed in the literature on finite and boundary element methods that uses a square-to-triangle degeneration in order to deal with a singularity [6]. The singularity of concern for the transformation is located at a point $\mathrm{P}(\xi = a, \eta = b)$, as in Fig. 1. For an analytical function $f(\xi,\eta)$ and the T3 triangle, integration is carried out over transformed spaces corresponding to the indicated subtriangles, $\widehat{\mathrm{P12}}$, $\widehat{\mathrm{P23}}$ and $\widehat{\mathrm{P31}}$,

$$\int_0^1\int_0^{1-\eta} f(\xi,\eta)\,\mathrm{d}\xi\,\mathrm{d}\eta = \sum_{l=1}^{3}\int_0^1\int_0^1 f\left(\xi(\bar{\xi}_l,\bar{\eta}_l),\eta(\bar{\xi}_l,\bar{\eta}_l)\right)\left|J_l(\bar{\xi}_l,\bar{\eta}_l)\right|\mathrm{d}\bar{\xi}_l\,\mathrm{d}\bar{\eta}_l\,, \tag{8}$$

considered as degenerated square subspaces $\left(\bar{\xi}_l \in [0,1], \bar{\eta}_l \in [0,1]\right)$, $l = 1,2,3$, where one of the nodal points of the unit square collapses in order to coincide with the point of singularity P. The auxiliary coordinate systems $(\bar{\xi}_l,\bar{\eta}_l)$ have origin at node P and are oriented as indicated in Fig. 1. Depending on the depicted values of a and b one or two of these triangles may be void or correspond to negative areas, so that both cases of singularity and quasi-singularity are dealt with in the same algorithm. In the presently proposed general approach, the singularity may be located even outside the (ξ,η) plane, as shown in Fig. 1(c) for a quadrilateral element represented in the Cartesian space and a singularity (source) point 0. The distance from the source 0 to a field point F on the element is the hypotenuse $\overline{\mathrm{0F}}$ of the right triangle whose legs are the distances $\overline{\mathrm{0P}}$ from the source point to the plane and $\overline{\mathrm{PF}}$.

2.1 Development for the case of a real quasi-singularity

Although eqn (8) can be used in a straightforward way for improper integrals, the integral cannot be directly obtained for $0 \equiv \mathrm{P}$ when embedding a real singularity or hypersingularity. When $0 \equiv \mathrm{P}$ is outside the triangle segment, as for results at internal points that are close to the boundary and on the (ξ,η) plane, one should better exclude the integration areas that are outside the actual integration interval, with eqn (8) evaluated according to one of the following particular cases (function arguments omitted):

$$\int_0^1\int_0^{1-\eta} f\,\mathrm{d}\xi\,\mathrm{d}\eta = \begin{cases} \left|\tilde{J}\right|_1 \displaystyle\int_0^1\left(\int_{\frac{b}{b+\bar{\eta}_1-1}}^1 \left(f\bar{\xi}_1\right)\mathrm{d}\bar{\xi}_1\right)\mathrm{d}\bar{\eta}_1 + \left|\tilde{J}\right|_3 \int_0^1\left(\int_{\frac{b}{b-\bar{\eta}_3}}^1 \left(f\bar{\xi}_3\right)\mathrm{d}\bar{\xi}_3\right)\mathrm{d}\bar{\eta}_3 & \text{for } b<0 \\ \left|\tilde{J}\right|_2 \displaystyle\int_0^1\left(\int_{\frac{a}{a-\bar{\eta}_2}}^1 \left(f\bar{\xi}_2\right)\mathrm{d}\bar{\xi}_2\right)\mathrm{d}\bar{\eta}_2 + \left|\tilde{J}\right|_3 \int_0^1\left(\int_{\frac{a}{a+\bar{\eta}_2-1}}^1 \left(f\bar{\xi}_3\right)\mathrm{d}\bar{\xi}_3\right)\mathrm{d}\bar{\eta}_3 & \text{for } a<0 \\ \left|\tilde{J}\right|_1 \displaystyle\int_0^1\left(\int_{\frac{a+b-1}{a+b+\bar{\eta}_1-1}}^1 \left(f\bar{\xi}_1\right)\mathrm{d}\bar{\xi}_1\right)\mathrm{d}\bar{\eta}_1 + \left|\tilde{J}\right|_2 \int_0^1\left(\int_{\frac{a+b-1}{a+b-\bar{\eta}_2}}^1 \left(f\bar{\xi}_2\right)\mathrm{d}\bar{\xi}_2\right)\mathrm{d}\bar{\eta}_2 & \text{for } a,b\geq 0. \end{cases} \tag{9}$$

Up to three cases are given in the first and third rows of expressions on the right-hand side: $a<0$ or $a=0$ or $a>0$ for $b<0$ as well as $(a=0,b>0)$ or $a,b>0$ or $(a>0,b=0)$ for $a<0$. One or two cases are comprised in the second row: $b=0$ or $b>0$ for $a<0$. Depending on $\left|\tilde{J}\right|_l$, as given in eqns (1)–(7), some integrals may be void or negative. The integration limit $\bar{\xi}_l \to 0$ is never reached.

Eqns (8) – for improper integrals and weak quasi-singularities – and (9) are compact sets of expressions directly needed in code implementations for potential and elasticity problems, provided that the integrals with integrands $\left(f\bar{\xi}_l\right)$ can be obtained analytically, which has been shown to be the case [5].

3 GENERAL FORMULATION OF THE 3D PROBLEM

For a linear triangle or a flat quadrilateral element, the evaluation of the coordinates $(a,b)\equiv(\xi_{\mathrm{P}},\eta_{\mathrm{P}})$ of Fig. 1 is a simple geometry matter. After a tedious development that uses the triangle to subtriangle transformations given in eqns (1)–(7), the distance square $\Delta r^2 \equiv \overline{0\mathrm{F}}^2 = \overline{0\mathrm{P}}^2 + \overline{\mathrm{PF}}^2$ of Fig. 1 (see [5] for triangles and more generally) becomes expressed for each subtriangle as

$$\Delta r_l^2 = \Delta r_0^2 + c_l^2\bar{\xi}_l^2 \equiv \Delta r_0^2 + \left(\tilde{a}_l\bar{\eta}_l^2 + \tilde{b}_l\bar{\eta}_l + \tilde{c}_l - \Delta r_0^2\right)\bar{\xi}_l^2, \tag{10}$$

where $\Delta r_0^2 \equiv \overline{0\mathrm{P}}^2$. One checks that, for a triangle whose area is not void, $\tilde{a}_l > 0$. Moreover, particular cases are $c_l^2 = \tilde{a}_l\bar{\eta}_l^2$, when the singularity is located at node 2,3 or 1 for $l = 1,2$ or 3, respectively, as for the triangle element. Also, $\tilde{c}_l \geq 0$. Moreover, for a subtriangle $l = 1,2,3$ and using a cyclic notation for $L(l) := 3,1,2,3,1$, one obtains according to Fig. 1(a):

$$\begin{aligned}
&\tilde{a}_l + \tilde{b}_l + \tilde{c}_l = \overline{0(L(l)-1)}^2 \geq 0 \\
&\tilde{a}_l + \tilde{b}_l + \tilde{c}_l - \Delta r_0^2 = \overline{\mathrm{P}(L(l)-1)}^2 \geq 0 \text{ for the triangle element} \\
&4\tilde{a}_l\left(\tilde{c}_l - \Delta r_0^2\right) - \tilde{b}_l^2 \geq 0.
\end{aligned} \tag{11}$$

In the case of a singularity or real quasi-singularity, $\Delta r_0^2 \equiv \overline{0\mathrm{P}}^2 = 0$ and the distance tends to zero with $\bar{\xi}_l \to 0$ independently from the respective coordinates $\bar{\eta}_l$. All particularities of subtriangles with void or negative areas (void or negative integrals) are taken care of by the Jacobian transformations given in eqns (1)–(7) as well as by developments according to eqns (8) and (9).

Eqn (10) – for flat triangle and quadrilateral elements (with straight edges, of course) – is part of a general outline for high order, distorted elements, with radial distances given as

$$\begin{aligned}
&\mathrm{T3:}\ \ \Delta r_l^2 = \Delta r_0^2 + \bar{\xi}_l^2\sum_{j=0}^{2} c_{0j}\bar{\eta}_l^j; \qquad \mathrm{Q4:}\ \ \Delta r_l^2 = \Delta r_0^2 + \bar{\xi}_l^2\sum_{i=0}^{2}\bar{\xi}_l^i\sum_{j=0}^{2} c_{ij}\bar{\eta}_l^j \\
&\mathrm{T6:}\ \ \Delta r_l^2 = \Delta r_0^2 + \bar{\xi}_l^2\sum_{i=0}^{2}\bar{\xi}_l^i\sum_{j=0}^{2+i} c_{ij}\bar{\eta}_l^j; \ \ \mathrm{Q8:}\ \ \Delta r_l^2 = \Delta r_0^2 + \bar{\xi}_l^2\sum_{i=0}^{4}\bar{\xi}_l^i\sum_{j=0}^{\min(2+i,4)} c_{ij}\bar{\eta}_l^j,
\end{aligned} \tag{12}$$

where c_{ij} are coefficients to be evaluated and adequately interpreted for their geometric representation, as done in eqn (11). It is still open whether such general problems may be dealt with in the present, analytical framework.

4 EVALUATIONS FOR POTENTIAL PROBLEMS

Owing to space restrictions, the explicit expressions for the single- and double-layer potential matrices **G** and **H** for 3D potential and elasticity are not given [1], [5]. The developments for arriving at such expressions are actually relevant to show the kind of integrals we must deal with as well as how important it is to take the properties outlined in Section 3 into account [5]. The results are briefly summarized in the following. All integrals could be evaluated analytically using the mathematical software Maple [7] and some ingenuity.

4.1 Single-layer potential matrix G

For the single-layer potential matrix $\mathbf{G}$ [1], [5], the integrals

$$\int_0^1\int_0^1 \frac{\langle 1 \quad \bar{\xi}_l \quad \bar{\xi}_l\bar{\eta}_l\rangle}{\sqrt{\Delta r_0^2 + c_l^2\bar{\xi}_l^2}}\bar{\xi}_l\,\mathrm{d}\bar{\xi}_l\mathrm{d}\bar{\eta}_l, \quad l=1,2,3 \quad \text{for } \Delta r_0^2 \geq 0, \tag{13}$$

are required, where the array $\langle 1 \quad \bar{\xi}_l \quad \bar{\xi}_l\bar{\eta}_l\rangle$ contains the three terms of a bivariate polynomial that are relevant to the present developments. The integrals for $\Delta r_0^2 = 0$ (when $\bar{\xi}_l$ cancels out in numerator and denominator) and for $\Delta r_0^2 > 0$ are actually dealt with separately in the computational implementation.

4.2 Double-layer potential matrix H

For matrix $\mathbf{H}$ and $\Delta r_0^2 > 0$, the analytical expressions of the three terms

$$\int_0^1\int_0^1 \frac{\langle 1 \quad \bar{\xi}_l \quad \bar{\xi}_l\bar{\eta}_l\rangle}{\left(\Delta r_0^2 + c_l^2\bar{\xi}_l^2\right)^{3/2}}\bar{\xi}_l\,\mathrm{d}\bar{\xi}_l\mathrm{d}\bar{\eta}_l, \quad l=1,2,3; \quad \Delta r_0^2 > 0, \tag{14}$$

are required. For $\Delta r_0^2 = 0$ (potential problems), these integrals are void. For a diagonal term of $\mathbf{H}$, the finite part of the integrals are also void (for potential problems, but not for elasticity), but discontinuous terms to locally go around the singularity must be evaluated. This in principle involves a complicated geometry description of all elements that share a source point but could actually be dealt with in the frame of a simple algorithm [5].

4.3 Results at internal points

The evaluation of potential results at internal points requires the same types of integrals above. For the quasi-singularity case of $\Delta r_0^2 > 0$, the evaluation of gradient results deals with the analytical expressions of 12 integrals, schematically represented for $p=3$ and $p=5$:

$$\int_0^1\int_0^1\begin{bmatrix}1 & \bar{\xi}_l & \bar{\xi}_l\bar{\eta}_l \\ \cdot & \bar{\xi}_l^2 & \bar{\xi}_l^2\bar{\eta}_l \\ \cdot & \cdot & \bar{\xi}_l^2\bar{\eta}_l^2\end{bmatrix}\frac{\bar{\xi}_l\,\mathrm{d}\bar{\xi}_l}{\left(\Delta r_0^2 + c_l^2\bar{\xi}_l^2\right)^{p/2}}\mathrm{d}\bar{\eta}_l,\ p = 3 \text{ or } 5,\ l = 1,2,3;\ \Delta r_0^2 > 0. \tag{15}$$

For a real quasi-singularity, that is, $\Delta r_0^2 = 0$ for the source point outside the boundary segment but on its plane, one evaluates the five integrals

$$\int\int\begin{bmatrix}1 & \bar{\eta}_l & \bar{\eta}_l^2 \\ \cdot & 1/\bar{\xi}_l & \bar{\eta}_l/\bar{\xi}_l\end{bmatrix}\frac{\mathrm{d}\bar{\xi}_l}{c_l^3}\mathrm{d}\bar{\eta}_l,\quad l = 1,2,3 \quad \text{for } \Delta r_0^2 = 0, \tag{16}$$

in the interval $[0,1]$, in principle, but making the adjustments indicated in eqn (9) for $\bar{\xi}_l$ in the denominator. For elasticity problems and triangle and quadrilateral flat elements, there are a few more polynomial terms in the numerators of eqns (13)–(16) and the denominators are affected by powers one order higher. However, all integrals may be evaluated analytically, as well [5].

5 A QUEST FOR SIMPLIFICATION OF THE TRIANGLE ELEMENT T3

Fig. 2(a) shows the element T3 for potential or displacement nodal values referred to the vertices, with degrees of freedom common to all elements that share a vertex (topological issues concerning fracture mechanics, for instance, not considered). Linear potential or displacement fields along the element are obtained by interpolation – exactly as in the isoparametric formulation of the displacement finite element method. On the other hand, the representation of normal gradients or traction forces, as for the evaluation of the single-layer potential matrix $\mathbf{G}$, requires a framework that has no parallel in the (traditional) finite element formulation. Normal gradients and traction forces are surface attributes and cannot be attached to nodes. If we geometrically refer such attributes to an element's boundary point, this must be interpreted as a limiting case, as illustrated in Fig. 2(a). In such a case, normal gradients or traction forces are also linearly interpolated along the element in terms of the given surface attributes. A numerical model implemented with such an element must exactly (within machine precision) represent constant and linear potential or displacement fields. Since for a flat surface the corresponding normal gradients (and traction forces) have respectively zero and constant values, one might ask whether the use of a single parameter for such quantities – geometrically referred to the element's middle point, as in Fig. 2(b) – would lead to reasonable numerical results. This will be illustratively assessed in the numerical simulation of Section 6. We also must keep in mind the total number of degrees of freedom involved in such numerical simulations.

Euler's theorem for a simply connected polyhedron is $V + F - E = 2$, where V is the number of vertices, F the number of faces and E the number of edges. It may also be shown that, in the special case of the simply connected polyhedron made up of triangular faces, $3F = 2E$. Then, we obtain for the T3 element and a simply connected domain that $F = 2V - 4$. If there are cavities in the domain, we must apply this formula to each separate cavity. On the other hand, if the domain is generally not convex, such as in a toroid, the relations above do not apply. Nevertheless, one may affirm that, when refining a 3D boundary with T3 triangle elements, the number of faces F, that is, elements, tends to be equal to twice the number of vertices V, that is, nodes: $F \to 2V$ as $V \to \infty$. A similar reasoning leads to the conclusion that, for a boundary discretization with Q4 quadrilateral

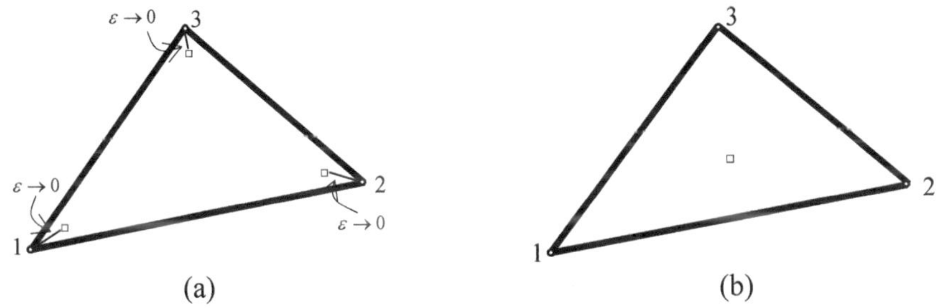

Figure 2: Element T3 with three nodal points for potentials (at the vertices) and surface attributes for normal gradients. (a) Three geometrically attached to points that tend to the vertices; and (b) One for a surface point referred to the triangle barycenter.

elements, $F \to V$ as $V \to \infty$. Then, in a numerical simulation of a potential 3D problem using T3 elements with three nodal parameters for potentials and three attributes for the normal gradients, the double-layer potential matrix $\mathbf{H}$ is always square and of order V, whereas the single-layer potential matrix $\mathbf{G}$ has V rows and $3F$ ($\sim 6V$) columns, in general – both matrices of rank $V-1$ for a finite domain, if the problem is consistently formulated [2], [3]. This is the case of boundary condition discontinuities of the discretized surfaces that have a vertex in common. Observe that this assertion is more general and conceptually more appropriate than just saying that we are dealing with locally non-smooth surfaces. If one is sure that there are no boundary discontinuities of the normal gradients on the surface around a vertex, one may consider – for computational economy – one normal gradient degree of freedom at the considered vertex. In the case of only one normal gradient attribute geometrically referred to the middle of the T3 element, the number of columns of $\mathbf{G}$ is just F and its rank is the same as before. (The dilemma *continuous* versus *discontinuous* elements is just a matter of misconception [2]).

6 A SIMPLE NUMERICAL ILLUSTRATION

Fig. 3 represents a very irregular 3D domain with a cavity (drawing rotated with angles $\vartheta = 105°, \varphi = 80°$), whose boundary is discretized with 16 linear triangle elements and 12 nodes. This is a development of the numerical example presented in [5]. The Cartesian coordinates of these nodes are

$$Coord = \left[\begin{array}{cccccccc} 1 & 2 & 3 & 4 & 5 & 6 & 7 & 8 \\ \hline 1 & 1 & 0.2 & -1 & -2 & -1 & 0.5 & -0.25 \\ -0.5 & 0.4 & 0 & 0.5 & 0 & -0.5 & -1 & \dfrac{-0.55}{3} \\ 0.5 & -0.5 & 1 & 1 & -0.2 & 0.5 & -1 & 0.05 \end{array}\right.$$
$$\left.\begin{array}{cccc|cc|cccc} 9 & 10 & 11 & 12 & \bar{1} & \bar{2} & \bar{3} & \bar{4} & \bar{5} & \bar{6} \\ \hline -0.3 & -0.5 & -0.4 & 0 & -0.035 & 0.285 & -0.075 & 0.16 & 1.04 & -0.3 \\ -0.2 & -0.1 & -0.3 & 0 & \dfrac{-0.275}{3} & -0.5 & \dfrac{-0.605}{3} & -0.02 & 0.42 & -0.15 \\ 0.3 & 0.4 & 0.5 & 0.6 & 0.475 & -0.05 & 1.055 & 1.075 & -0.575 & 0.45 \end{array}\right], \quad (17)$$

using the first row for the node numbering, in an array that also contains the coordinates of internal points $\overline{1},\overline{2}$ and external points $\overline{3},\overline{4},\overline{5},\overline{6}$ at which potential and gradient results are to be evaluated. The nodal incidence is

$$Inc = \left[\begin{array}{cccccccccccc|cccc} 1 & 2 & 3 & 4 & 5 & 6 & 7 & 8 & 9 & 10 & 11 & 12 & 13 & 14 & 15 & 16 \\ \hline 1 & 3 & 5 & 5 & 3 & 1 & 8 & 4 & 4 & 8 & 7 & 7 & 9 & 10 & 9 & 9 \\ 2 & 2 & 3 & 6 & 6 & 3 & 2 & 2 & 8 & 6 & 6 & 8 & 10 & 12 & 11 & 12 \\ 3 & 4 & 4 & 3 & 7 & 7 & 1 & 8 & 5 & 5 & 8 & 1 & 11 & 11 & 12 & 10 \end{array}\right]. \tag{18}$$

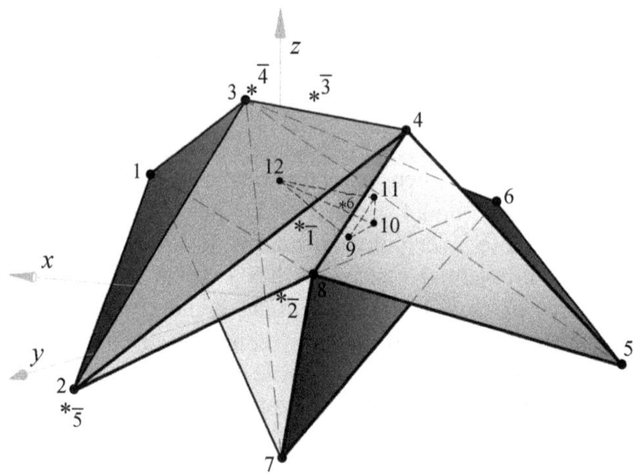

Figure 3: Very irregular domain with internal points $\overline{1},\overline{2}$ and external points $\overline{3},\overline{4},\overline{5},\overline{6}$ at which potential and gradient results are to be evaluated.

It is worth observing that elements 6 (nodes [1 3 7]) and 12 (nodes [7 8 1]) are almost coplanar. Nodes 9–12 correspond to a cavity built up with elements 13–16. The external point $\overline{6}$ is precisely in the middle of the cavity. All nodes and internal and external points are very close to the boundary segments. For such a small example (no far nodes, for which a numerical quadrature would be preferable in order to save computational costs), the diagonal terms of $\mathbf{H}$ are analytically evaluated by just imposing orthogonality to a constant potential field. Consistency check of the boundary element equation $\mathbf{Gt} = \mathbf{Hd}$ for 3 linearly varying potential fields holds for 19 digits in an implementation using Maple software [7] with 20 digits of precision. The internal point $\overline{1}$ and the external points $\overline{3},\overline{6}$ present complex quasi-singularities with respect to all 16 elements, with gradient results evaluated according to eqn (15). The internal point $\overline{2}$ is on the plane of element 6 and corresponding gradients must be evaluated for all three subtriangles as given in the second row on the right-hand side of eqn (9), for $a = -0.1, b = 0.55$, with negative area for subtriangle 1. The external points $\overline{4}$ and $\overline{5}$ are in the intersection of the planes given by elements 1 and 2. Gradient results at point $\overline{4}$ require evaluations for element 1 $(a = 0, b = -0.05)$ over subtriangles 2 (negative area) and 3, and for element 2

$(a=1.05, b=-0.05)$ over subtriangles 1 and 2 (the latter with negative area), both cases according to the first row of eqn (9). Gradient results at point $\overline{5}$ require evaluations for element 1 $(a=0, b=1.05)$ over subtriangles 2 and 3 (negative area) according to the third row of eqn (9), and for element 2 $(a=-0.05, b=1.05)$ over subtriangles 1 (negative area) and 2 according to the second row of eqn (9). Then, all possible cases of quasi-singularity and topology issues are numerically assessed. Accuracy for potential results at all six points is checked for about 18 digits. For gradient results round-off errors unavoidably occur and accuracy is checked for at least 16 digits in the case of points $\overline{1}, \overline{2}, \overline{6}$ and at least 13, 12 and 10 digits in the case of points $\overline{3}$, $\overline{4}$ and $\overline{5}$, respectively. (Results at external points $\overline{3}, \overline{4}, \overline{5}, \overline{6}$ obviously compare with zero.)

The domain of Fig. 3 is submitted to a total of 36 independent potential fields (solutions of the Laplace equation) corresponding to 1 constant, 3 linear, 5 quadratic, 7 cubic, 9 quartic and 11 quintic polynomials. The respective results measured on the boundary as nodal potentials and normal gradient attributes are applied to first check the consistency of the BEM equation $\mathbf{Gt}=\mathbf{Hd}$. This equation turns out to hold exactly (within machine precision) for the simplest four solutions. Fig. 4(a) shows the Euclidean norms of relative errors $|\mathbf{Gt}=\mathbf{Hd}|$ of this equation for the remaining 32 test solutions, vertical dash lines separating the polynomial orders. Four different kinds of results are grouped. The errors are in general very large, as the applied fields correspond to high degree polynomials and there are too many boundary irregularities for such a coarse mesh. The solid (black) and long dash (red) lines correspond respectively to the elements in Fig. 2(a) (3 parameters) and 2(b) (1 parameter) for the single-layer potential matrix $\mathbf{G}_{adm}$ evaluated in the admissible way preconized by Dumont [2]. The dash–dot (black) and dash (red) lines correspond to the same results above, but with $\mathbf{G}$ evaluated as traditionally (and conceptually wrongly). For the 5 quadratic potential fields the measured normal gradients are in balance and there is no difference between admissible and inadmissible evaluations. As communicated in [3], there is no means of telling whether admissible evaluations should lead to more accurate solutions – this is observed in the graphic, in general, but exceptions are also shown. In this example, the matrix $\mathbf{H}$ is of order 12 and rank 11. The most relevant conclusions concern the matrix $\mathbf{G}$, which has 12 rows, but 48 columns for the element with 3 parameters, and only 16 rows for the element with just 1 parameter for normal gradients – for no numerically sensible accuracy loss, according to the graphics. Since there are no approximations in the integral evaluations, the differences in results are solely attributable to the different simulation schemes. The comparatively good results for the discretization with one element middle node for normal gradients is not a surprise, as the middle point is the optimal location for a quadrature scheme.

Out of the many numerical assessment possibilities for evaluations at internal (and external) points, Fig 4(b) and 4(c) refer to the internal point $\overline{1}$ of Fig. 3, with the same type of simulations described in Fig. 4(a) – together with the analytical values for the sake of comparison. The conclusions arrived at for the first graphic are here corroborated.

7 CONCLUDING REMARKS

The developments of this paper apply to steady-state potential and elastostatics problems implemented for T3 and flat Q4 elements. Owing to space restrictions only the T3 implementation for potential problems is outlined and numerically assessed with some

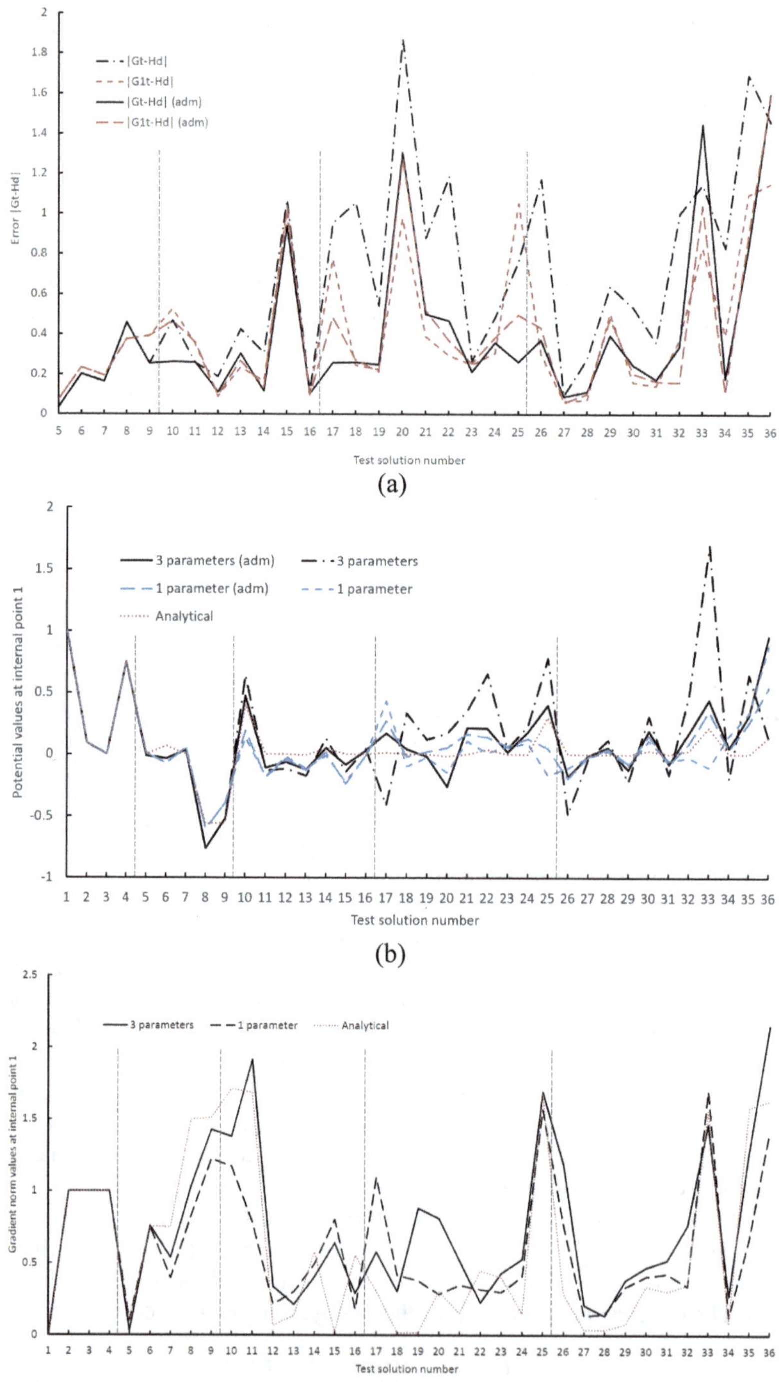

Figure 4: Some numerical results for the example of Fig. 3. (a) Consistency check of the equations; (b) Potential; and (c) Euclidean norm of gradient results at point $\overline{1}$.

detail. All integrations are carried out analytically. The arrays with such analytical results will be published opportunely, but may be made available on request. Since there are no unsolved numerical integration issues, no matter how close a source point is to an element surface and how particularly located, it is for the first time possible to numerically assess that the use of just one middle node parameter for the boundary representation of normal gradients with the T3 element is of definitive computational advantage. It is also for the first time numerically illustrated that the concept of an admissible single-layer potential matrix, as preconized by the first author already in 1998, leads to better and more consistent results also for 3D problems than as outlined in the classical literature on boundary element methods.

ACKNOWLEDGEMENTS

This work was supported by the Brazilian agencies CAPES, CNPq and FAPERJ.

REFERENCES

[1] Brebbia, C.A., Telles, J.C.F. & Wrobel, L.C., *Boundary Element Techniques*, Springer-Verlag: Berlin and New York, 1984.

[2] Dumont N.A., The collocation boundary element method revisited: perfect code for 2D problems. *International Journal of Computational Methods and Experimental Measurements*, **6**(6), pp. 965–975, 2018.

[3] Dumont, N.A., An assessment of the spectral properties of the matrix G used in the boundary element methods. *Computational Mechanics*, **22**(1), pp. 32–41, 1998.

[4] Dumont, N.A., On the efficient numerical evaluation of integrals with complex singularity poles. *Engineering Analysis with Boundary Elements*, **13**, pp. 155–168, 1994.

[5] Kurz, T.G., Analytical 3D Boundary element implementation of flat triangle and quadrilateral elements for potential and elasticity problems, PhD thesis, in progress, in Portuguese, PUC-Rio, Brazil, 2019.

[6] Lachat, J.C. & Watson, J.O., Effective numerical treatment of boundary integral equations: A formulation for three-dimensional elastostatics. *International Journal for Numerical Methods in Engineering*, **10**(5), pp. 991–1005, 1976.

[7] Maplesoft, a division of Waterloo Maple Inc., Maple 15, Waterloo, Ontario.

A BEM BASED ON THE BÉZIER/BERNSTEIN POLYNOMIAL FOR ACOUSTIC WAVEGUIDE MODELIZATION

ANTONIO ROMERO[1], PEDRO GALVÍN[1] & ANTÓNIO TADEU[2,3]
[1]Escuela Técnica Superior de Ingeniería, Universidad de Sevilla, Spain
[2]ITeCons – Institute for Research and Technological Development in Construction, Energy, Environment and Sustainability, Portugal
[3]ADAI – LAETA, Department of Civil Engineering, University of Coimbra, Portugal

ABSTRACT

This paper proposes a novel boundary element approach formulated on the Bézier–Bernstein basis to yield a geometry-independent field approximation. The proposed method is geometrically based on both computer aided design (CAD) and isogeometric analysis (IGA), but field variables are independently approximated from the geometry. This approach allows the appropriate approximation functions for the geometry and variable field to be chosen. We use the Bézier–Bernstein form of a polynomial as an approximation basis to represent both geometry and field variables. The solution of the element interpolation problem in the Bézier–Bernstein space defines generalised Lagrange interpolation functions that are used as element shape functions. The resulting Bernstein–Vandermonde matrix related to the Bézier–Bernstein interpolation problem is inverted using the Newton–Bernstein algorithm. The applicability of the proposed method is demonstrated by solving the Helmholtz equation over an unbounded region in a two-and-a-half dimensional (2.5D) domain.

Keywords: subparametric method, Bézier–Bernstein curve, Newton–Bernstein algorithm, computer-aided design, isogeometric analysis.

1 INTRODUCTION

We present a geometry-independent field approximation of the boundary element method (BEM). The proposed formulation shares some characteristics with the standard collocation methods and isogeometric analysis (IGA), exploring the advantages of each method. A similar concept was recently introduced by Atroshchenko and Bordas [1].

The geometry is represented by Bézier curves defined as a linear combination of Bernstein basis with control points. The Bernstein basis is widely used in CAD because of its good properties for building and modifying geometric shapes [2]. On the other hand, the use of orthogonal basis allows simple and efficient field approximation. To define an orthogonal basis for the field interpolation, therefore, a least-square problem in terms of Bernstein–Vandermonde matrix is generated by the interpolation constraints whose solution is a Lagrange interpolant relative to the Bernstein basis [3], [4]. This methodology is an efficient and robust approach that combines the Bernstein basis for geometry representation and the Lagrange interpolating polynomial relative to the Bernstein basis for the field approximation.

The proposed method is powerful because it includes the main characteristics of standard BEM formulation and isogeometric analysis: (i) the geometry is exactly defined as Bézier curves; (ii) field variables are represented at nodal points instead of control variables, which enables easy definition of the boundary conditions; (iii) h-refinement is performed directly on the geometry without CAD interaction, as it is performed by IGA; and (iv) p-refinement only affects the field approximation, rather than the geometry discretisation.

In this paper, the capability of the proposed method is demonstrated for solving the Helmholtz equation in longitudinally invariant problems. This situation is referred to as a two-and-a-half dimensional (2.5D) problem [5], in which the three-dimensional (3D) solution is

www.witpress.com, ISSN 1743-3533 (on-line)
doi:10.2495/BE420021

computed by means of a Fourier transform. The structure of the rest of the paper is as follows. First, the Bézier–Bernstein space is briefly described and the Lagrange interpolant relative to the Bernstein basis is derived. Next, the boundary element formulation in the Bézier–Bernstein space is presented, after which, the proposed methodology is verified by means of two benchmark problems and a numerical example is presented.

2 NUMERICAL MODEL

This approach is based on the Helmholtz integral equation for solving 3D interior, exterior and scattering problems in the frequency-wavenumber domain. The pressure field for the acoustic waveguide at point $\mathbf{x} = \mathbf{x}(x, y, z)$ is computed as the superposition of 2D solutions with different wavenumbers κ_z [6]:

$$p(\mathbf{x}, \omega) = \int_{-\infty}^{+\infty} \widetilde{p}(\widetilde{\mathbf{x}}, \kappa_z, \omega) e^{-\iota \kappa_z z} \, d\kappa_z, \tag{1}$$

where $\widetilde{p}(\widetilde{\mathbf{x}}, \kappa_z, \omega)$ is the frequency-wavenumber representation of the sound pressure, $\widetilde{\mathbf{x}} = \mathbf{x}(x, y)$ and $\iota = \sqrt{-1}$.

The boundary integral representation is derived from the Helmholtz equation. The fundamental solution $\widetilde{\Psi}(\widetilde{\mathbf{x}}, \kappa_z, \omega; \widetilde{\mathbf{x}}_i)$ for sound pressure at receiver position $\widetilde{\mathbf{x}}$ due to a source acting at $\widetilde{\mathbf{x}}_i$ is the solution to:

$$\nabla^2 \widetilde{\Psi}(\widetilde{\mathbf{x}}, \kappa_z, \omega; \widetilde{\mathbf{x}}_i) + \kappa^2 \widetilde{\Psi}(\widetilde{\mathbf{x}}, \kappa_z, \omega; \widetilde{\mathbf{x}}_i) = -\delta(r), \tag{2}$$

where $\kappa = \omega / c_f$ is the problem wavenumber, c_f is the sound propagation velocity, δ is the Dirac delta function and $r = ||\widetilde{\mathbf{x}}_i - \widetilde{\mathbf{x}}||$ is the distance from the source to the receiver. The solution to this equation defines the fundamental solution for an unbounded region in the frequency-wavenumber:

$$\widetilde{\Psi}(\widetilde{\mathbf{x}}, \kappa_z, \omega; \widetilde{\mathbf{x}}_i) = -\frac{\iota}{4} H_0^{(2)}(\kappa_\alpha r), \tag{3}$$

where $\kappa_\alpha = \sqrt{\kappa^2 - \kappa_z^2}$ and $H_0^{(2)}$ is the Hankel function.

The following sections deal with: (i) the Bézier–Bernstein space for geometry representation, (ii) the Lagrange interpolant relative to Bernstein basis for variable field approximation, and (iii) the boundary element formulation in the Bézier–Bernstein space.

2.1 The Bézier–Bernstein space

The Bernstein basis of degree n is defined over the interval $t \in [0, 1]$ as:

$$B_k^n(t) = \binom{n}{k} t^k (1 - t)^{n-k}, \quad k = 0, \ldots, n. \tag{4}$$

The change of variable to map $x \in [a, b]$ to $t \in [0, 1]$ allows the definition of the Bernstein polynomial related to a continuous function $f(t)$ as follows:

$$P_n(t) = \sum_{k=0}^{n} f\left(\frac{k}{n}\right) B_k^n(t). \tag{5}$$

Then, eqn (5) converges to $f(t)$ as n goes to infinity due to the Bernstein basis being non-negative and forming a partition unity, although the polynomial form does not interpolate the sample values $f(k/n)$. The Bernstein basis has partition unity property, is defined as non-negative and symmetric about the interval mid-point $(B_{n-k}^{n}(1-t) = B_{n}^{k}(t))$.

The application of polynomials in Bernstein form grows with the development of Bézier curves $\mathbf{r}_n(t)$ in computer-aided design:

$$\mathbf{r}_n(t) = \sum_{k=0}^{n} \mathbf{b}_k B_k^n(t), \tag{6}$$

where $\mathbf{b}_k$ are the control points used to approximate the geometry. A control polygon is obtained by connecting the control points and this is used to modify the curve geometry. The de Casteljau algorithm is often used for evaluating and splitting a Bézier curve $\mathbf{r}_n(t)$ at a given point t [2].

Although the de Casteljau algorithm allows an easy evaluation of a Bézier curve, it is computationally expensive. An efficient curve computation is achieved using the polar form (or blossom) of a Bézier curve $\mathbf{r}_n(t)$ [7], which defines a multi-affine transformation satisfying:

$$\mathbf{b}_k = \mathbf{R}(\underbrace{0, \ldots, 0}_{n-k}, \underbrace{1, \ldots, 1}_{k}), \tag{7}$$

where $\mathbf{R}(t_1, \ldots, t_n)$ is computed as:

$$\mathbf{R}(t_1, \ldots, t_n) = \sum_{\substack{I \cap J = \emptyset \\ I \cup J = \{1,2,\ldots,n\}}} \prod_{i \in I} (1 - t_i) \prod_{j \in J} t_j \mathbf{b}_{|J|}. \tag{8}$$

Thus, a polynomial in Bernstein form can be formulated in polar form substituting eqn (7) into eqn (6) as follows:

$$\mathbf{r}_n(t) = \sum_{k=0}^{n} \mathbf{R}(\underbrace{0, \ldots, 0}_{n-k}, \underbrace{1, \ldots, 1}_{k}) B_k^n(t) = \mathbf{R}(t, \ldots, t). \tag{9}$$

The polar form of a Bézier curve $\mathbf{r}_n(t)$ of degree n defines a multi-affine polynomial $\mathbf{R}(t_1, \ldots, t_n)$ with n variables that satisfy $\mathbf{r}_n(t) = \mathbf{R}(t, \ldots, t)$.

Cubic Bézier approximations are widely used to represent smooth patches in CAD. More complex shapes would be defined by either increasing the curve order or using a piecewise Bézier curve. Fig. 1 shows a third-order Bézier patch and its related control points represented in polar form.

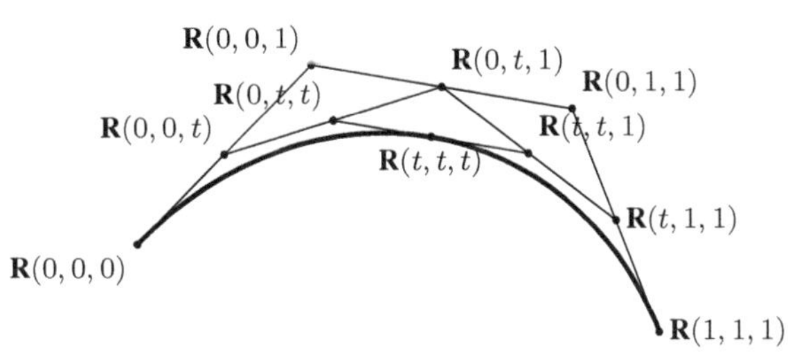

Figure 1: Third-order Bézier patch.

The geometry approximation presented in this section is used to formulate the BEM in the Bézier–Bernstein space. The boundary geometry is defined by patches with an arbitrary approximation degree, that depends on the geometry shape complexity. Then, each patch is subdivided into elements according to h-refinement.

2.2 Field approximation

The Bézier–Bernstein space allows the approximation of a curve geometry through control points. Moreover, the element shape function can be derived from the Lagrange interpolant relative to the Bernstein basis presented in the previous section. The Lagrange interpolant relative to the Bernstein basis has become useful for high-order element development [12]. Although a polynomial in Bernstein form satisfies $P_n(0) = f(0)$ and $P_n(1) = f(1)$, these properties do not remain valid at internal element points as has been stated. Therefore, it is necessary to properly modify the control points b_k to ensure the following condition:

$$P_n(x_i) = \sum_{k=0}^{n} b_k B_k^n(x_i) = f(x_i). \tag{10}$$

This condition is commonly expressed as a linear system of equations through the Bernstein–Vandermonde matrix $A_{ij} = B_i^j(x_i)$, for $i, j = 0, \dots, n$. The control points $\mathbf{b} = [b_0, \dots, b_n]^T$ are given by the solution of $\mathbf{Ab} = \mathbf{f}$, where the vector $\mathbf{f} = [f(x_0), \dots, f(x_n)]^T$ collects the interpolation data.

In this work, we have used the Newton–Bernstein algorithm proposed by Ainsworth and Sánchez [11] to compute the control points. The Newton–Bernstein algorithm [11] starts with the Newton form of the interpolant P_n at nodes x_i:

$$P_k(x) = \sum_{i=0}^{k} f[x_0, \dots, x_k] w_i(x), \quad k = 0, \dots, n, \tag{11}$$

where $f[x_0, \dots, x_k]$ is the divided difference defined by the following recursive formula:

$$f[x_j, \dots, x_k] = \frac{f[x_{j+1}, \dots, x_k] - f[x_j, \dots, x_{k-1}]}{x_k - x_j}, \quad k = j+1, \dots, n; \quad j = 0, \dots, n. \tag{12}$$

Thus, the following expressions for control points w_k and b_k, for $k = 0, \dots, n$ are defined:

$$w_k(x_j) = \frac{j}{k} w_{k-1}(x_{j-1})(1 - x_{k-1}) - \frac{k-j}{k} w_{k-1}(x_j) x_{k-1}, \tag{13}$$

$$b_k(x_j) = \frac{j}{k} b_{k-1}(x_{j-1}) + \frac{k-j}{k} b_{k-1}(x_j) + w_k(x_j) f[x_0, \dots, x_k], \tag{14}$$

and for $j = 0, \dots, k$ the latter expressions are $w_0(x_0) = 1$, $b_0(x_0) = f[x_0]$, $w_{-1}(x_{k-1}) = b_{-1}(x_{k-1}) = 0$ and $w_k(x_{k-1}) = b_k(x_{k-1}) = 0$.

The Lagrange polynomial P_n of order n defined through the Bernstein basis gives $n+1$ interpolation functions ϕ^j computed after imposing the constrains $P_n^j(x_k) = \phi^j(x_k) = \delta(x_k - x_j)$, for $j, k = 0, \dots, n$. Then, $\mathbf{b}$ becomes a matrix of unknown control points and $\mathbf{f} = \mathbf{I}$ is the identity matrix. The related interpolation functions are computed for a given point x as $[\phi^1(x), \dots, \phi^j(x), \dots, \phi^{n+1}(x)]^T = \mathbf{bA}(x)$. The evaluation of the shape functions also benefits from the computational advantages of using the polar form given in the previous section.

2.3 Boundary element formulation

Since the Bézier–Bernstein basis allowed the independent geometry and field variable approximation, the formulation of the BEM in Bézier–Bernstein space is quite straightforward, as in a classical formulation. This section describes the boundary element formulation for potential problems in acoustics [13]. The integral representation of the sound pressure in the frequency-wavenumber domain for a point $\widetilde{\mathbf{x}}_i$ located at the arbitrary boundary Γ can be written as:

$$
\begin{aligned}
&c_i(\widetilde{\mathbf{x}}_i)\widetilde{p}_i(\widetilde{\mathbf{x}}_i,\kappa_z,\omega)\\
&\quad = -\int_{\Sigma}\left(\iota\rho\omega\widetilde{v}_i(\widetilde{\mathbf{x}}_i,\kappa_z,\omega)\widetilde{\Psi}(\widetilde{\mathbf{x}},\kappa_z,\omega;\widetilde{\mathbf{x}}_i)+\widetilde{p}_i(\widetilde{\mathbf{x}}_i,\kappa_z,\omega)\frac{\partial\widetilde{\Psi}(\widetilde{\mathbf{x}},\kappa_z,\omega;\widetilde{\mathbf{x}}_i)}{\partial\mathbf{n}}\right)d\Sigma, \qquad (15)
\end{aligned}
$$

where $\widetilde{p}(\widetilde{\mathbf{x}}_i,\kappa_z,\omega)$ and $\widetilde{v}(\widetilde{\mathbf{x}}_i,\kappa_z,\omega)$ are the sound pressure and the particle normal velocity at the cross section Σ of the boundary Γ, respectively. $\widetilde{\Psi}(\widetilde{\mathbf{x}},\kappa_z,\omega;\widetilde{\mathbf{x}}_i)$ is the solution to the Helmholtz equation at point $\widetilde{\mathbf{x}}$ due to a point source located at $\widetilde{\mathbf{x}}_i$ (eqn (3)). The integral-free term $c_i(\widetilde{\mathbf{x}}_i)$ depends only on the boundary geometry at the collocation point $\widetilde{\mathbf{x}}_i$.

The boundary is discretised into N elements with $\Sigma=\bigcup_{j=1}^{N}\Sigma^j$, leading to a boundary approximation of the element field variables using the interpolation shape function ϕ of order p:

$$\sum_{i=1}^{p+1}\phi^i\widetilde{p}^i=\boldsymbol{\phi}\widetilde{\mathbf{p}}^e, \qquad (16)$$

$$\sum_{i=1}^{p+1}\phi^i\widetilde{v}^i=\boldsymbol{\phi}\widetilde{\mathbf{v}}^e. \qquad (17)$$

The proposed method employs the Lagrange interpolants relative to the Bernstein basis described in Section 2.2 as approximation shape functions. Thus, eqn (15) is written as:

$$c_i\widetilde{p}_i=-\sum_{j=1}^{Q}\left[\iota\rho\omega\left\{\int_{\Sigma^j}\phi^j\widetilde{\Psi}\,d\Sigma\right\}\widetilde{v}^j+\left\{\int_{\Sigma^j}\phi^j\frac{\partial\widetilde{\Psi}}{\partial\mathbf{n}}\,d\Sigma\right\}\widetilde{p}^j\right], \qquad (18)$$

where Q is the total number of nodes at the boundary and $\Sigma^j(\widetilde{\mathbf{x}})$ stands for the elements which contain the node j. The following integrals are defined for each collocation point $\widetilde{\mathbf{x}}_i$ and integration element Σ^j:

$$\mathbf{h}_i^j(\widetilde{\mathbf{x}},\kappa_z,\omega;\widetilde{\mathbf{x}}_i)=\int_{\Sigma^j}\phi^j\frac{\partial\widetilde{\Psi}}{\partial\mathbf{n}}\,d\Sigma, \qquad (19)$$

$$\mathbf{g}_i^j(\widetilde{\mathbf{x}},\kappa_z,\omega;\widetilde{\mathbf{x}}_i)=\int_{\Sigma^j}\phi^j\widetilde{\Psi}\,d\Sigma. \qquad (20)$$

Eqns (19) and (20) define the element matrices as in standard BEM formulation. Now, the Bézier–Bernstein space is used to describe the exact element geometry as $\Sigma^j(\widetilde{\mathbf{x}})=\mathbf{r}_n^j(t)$. Hence, the later integrals along a boundary element Σ^j are rewritten in the univariate basis $t\in[0,1]$ as:

$$\int_{\Sigma^j}\widetilde{f}(\widetilde{\mathbf{x}},\kappa_z,\omega;\widetilde{\mathbf{x}}_i)\,d\Sigma=\int_0^1 f(\widetilde{\mathbf{x}}(t),\kappa_z,\omega;\widetilde{\mathbf{x}}_i)\left|\frac{d}{dt}\mathbf{r}_n^j(t)\right|dt, \qquad (21)$$

where $\widetilde{f}(\widetilde{\mathbf{x}}, \kappa_z, \omega; \widetilde{\mathbf{x}}_i)$ represents the kernels in eqns (19) and (20). Thus, eqn (21) is transformed into a symmetric integration interval $\xi \in [-1, 1]$ to employ a Gauss–Legendre quadrature as:

$$\int_{\Sigma^j} \widetilde{f}(\widetilde{\mathbf{x}}, \kappa_z, \omega; \widetilde{\mathbf{x}}_i)\, d\Sigma = \int_{-1}^{1} \widetilde{f}(\widetilde{\mathbf{x}}(\xi), \kappa_z, \omega; \widetilde{\mathbf{x}}_i) \left| \frac{d}{dt}\mathbf{r}_n^j(t) \right| \frac{dt(\xi)}{d\xi}\, d\xi. \tag{22}$$

The relation $2t = \xi + 1$ defines the Jacobian of the transformation $dt/d\xi = 1/2$.

Finally, eqns (19) and (20) are expressed as follows:

$$\mathbf{h}_i^j(\widetilde{\mathbf{x}}, \kappa_z, \omega; \widetilde{\mathbf{x}}_i) = \frac{1}{2} \int_{-1}^{1} \phi^j \frac{\partial \widetilde{\Psi}}{\partial \mathbf{n}} \left| \frac{d}{dt}\mathbf{r}_n^j(t) \right| d\xi, \tag{23}$$

$$\mathbf{g}_i^j(\widetilde{\mathbf{x}}, \kappa_z, \omega; \widetilde{\mathbf{x}}_i) = \frac{1}{2} \int_{-1}^{1} \phi^j \widetilde{\Psi} \left| \frac{d}{dt}\mathbf{r}_n^j(t) \right| d\xi. \tag{24}$$

The system of equations for all the boundary elements becomes:

$$\widetilde{\mathbf{H}}(\widetilde{\mathbf{x}}, \kappa_z, \omega; \widetilde{\mathbf{x}}_i)\widetilde{p}(\widetilde{\mathbf{x}}, \kappa_z, \omega) = \widetilde{\mathbf{G}}(\widetilde{\mathbf{x}}, \kappa_z, \omega; \widetilde{\mathbf{x}}_i)\widetilde{v}(\widetilde{\mathbf{x}}, \kappa_z, \omega), \tag{25}$$

where $\widetilde{\mathbf{H}}(\widetilde{\mathbf{x}}, \kappa_z, \omega; \widetilde{\mathbf{x}}_i)$ and $\widetilde{\mathbf{G}}(\widetilde{\mathbf{x}}, \kappa_z, \omega; \widetilde{\mathbf{x}}_i)$ are the fully non-symmetrical boundary element system matrices, and $\widetilde{p}(\widetilde{\mathbf{x}}, \kappa_z, \omega)$ and $\widetilde{v}(\widetilde{\mathbf{x}}, \kappa_z, \omega)$ are the sound pressure and particle normal velocity at the boundary Σ, respectively.

The boundary integrals are computed using a standard Gauss–Legendre quadrature with $4(p+1)$ integration points whenever the collocation point is sufficiently distant from the integration element. Otherwise, the solution of singular or weakly singular integrals is numerically computed using a smoothing transformation by means of a Gauss–Legendre quadrature [14]. Fig. 2 shows a scheme for the treatment of singular and weakly singular integrals. This figure represents a collocation point $\widetilde{\mathbf{x}}_i$ and an integration element Σ^j. The projection point $t(\widetilde{\mathbf{x}}_i)$ is found from $\widetilde{\mathbf{x}}_i$ as the point that minimizes the distance $r(\widetilde{\mathbf{x}}_i)$ to the integration element. Obviously, if the collocation point belongs to the integration element, then $t(\widetilde{\mathbf{x}}_i)$ coincides with the coordinate of this node. Once the point $t(\widetilde{\mathbf{x}}_i)$ has been identified, the element integration defined by eqn (21) becomes:

$$\int_{\Sigma^j} \widetilde{f}(\widetilde{\mathbf{x}}, \kappa_z, \omega; \widetilde{\mathbf{x}}_i)\, d\Sigma$$
$$= \int_{0}^{t(\widetilde{\mathbf{x}}_i)} f(\widetilde{\mathbf{x}}(t), \kappa_z, \omega; \widetilde{\mathbf{x}}_i) \left| \frac{d}{dt}\mathbf{r}_n^j(t) \right| dt + \int_{t(\widetilde{\mathbf{x}}_i)}^{1} f(\widetilde{\mathbf{x}}(t), \kappa_z, \omega; \widetilde{\mathbf{x}}_i) \left| \frac{d}{dt}\mathbf{r}_n^j(t) \right| dt, \tag{26}$$

where the integration path is subdivided into two intervals according to $t(\widetilde{\mathbf{x}}_i)$. These integrals are numerically solved by a smoothing transformation of the form $\widetilde{\mathbf{x}} = \varphi_{l,r}(s)$ [14]:

$$\varphi_{l,r}(s) = \frac{(\alpha+\beta-1)!}{(\alpha-1)!(\beta-1)!} \int_0^s u^{\alpha-1}(1-u)^{\beta-1}\, du, \quad s \in [0,\, 1], \tag{27}$$

where

$$\Xi_{\alpha,\beta}(s) = \int_0^s u^{\alpha-1}(1-u)^{\beta-1}\, du \tag{28}$$

is computed using the recurrence expression:

$$\Xi_{\alpha,\beta}(s) = \frac{1}{\alpha+\beta-1}\left[-s^{\alpha-1}(1-s)^{\beta} + (\alpha-1)\Xi_{\alpha-1,\beta}(s)\right], \quad \alpha \geq 2,\, \beta \geq 1 \tag{29}$$

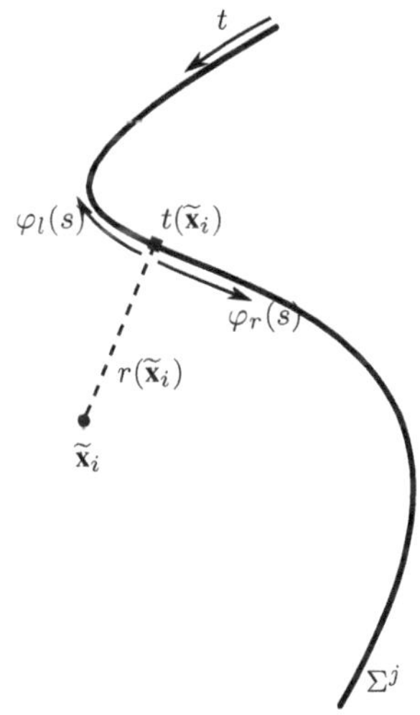

Figure 2: Treatment of singular and weakly singular integral.

with the initial value $\Xi_{1,1}(s) = s$. Accurate results for singular integrals are given in the next section, setting $\alpha = 3$ and $\beta = 1$, and using a number of integration points $10(p+1)$.

Therefore, eqn (26) becomes:

$$\int_{\Sigma^j} \widetilde{f}(\widetilde{\mathbf{x}}, \kappa_z, \omega; \widetilde{\mathbf{x}}_i)\, d\Sigma = -\int_0^1 f(\widetilde{\mathbf{x}}(\varphi_l(s)), \kappa_z, \omega; \widetilde{\mathbf{x}}_i) \left| \frac{d}{dt}\mathbf{r}_n^j(t) \right| \frac{dt}{d\varphi_l}\frac{d}{ds}\varphi_l(s)\, ds$$
$$+ \int_0^1 f(\widetilde{\mathbf{x}}(\varphi_r(s)), \kappa_z, \omega; \widetilde{\mathbf{x}}_i) \left| \frac{d}{dt}\mathbf{r}_n^j(t) \right| \frac{dt}{d\varphi_r}\frac{d}{ds}\varphi_r(s)\, ds, \quad (30)$$

where $dt/d\varphi_l = -t(\widetilde{\mathbf{x}}_i)$ and $dt/d\varphi_r = 1 - t(\widetilde{\mathbf{x}}_i)$.

The integration interval is then transformed to $\xi \in [-1,\, 1]$ to employ a Gauss–Legendre quadrature as described for regular integrals. The two integrals are reduced to only one integral if the projection point $t(\widetilde{\mathbf{x}}_i)$ lies at the element endpoints.

We chose a maximum radius r^* between the collocation point and the integration element for the identification of singular integrals. Otherwise, the collocation point is sufficiently distant from the integration element and the resulting integrals are regular. The asymptotic behaviour of the fundamental solution is accounted for by selecting the critical radius r^* when the integral becomes singular [15]:

$$\lim_{r\to 0} \widetilde{\Psi} = \frac{\iota}{4} + \frac{1}{2\pi}\log(\kappa_\alpha r), \quad (31)$$

$$\lim_{r\to 0} \frac{\partial \widetilde{\Psi}}{\partial n} = \frac{1}{2\pi r}\frac{\partial r}{\partial n}. \quad (32)$$

3 NUMERICAL EXAMPLE

In this section, the capability of the proposed method is shown with a numerical example. The problem solved herein models the three-dimensional wave propagation in a fluid channel with a fixed scatterer. The fluid channel is bounded by two reflection planes that were assumed to be rigid with null particle normal velocity. The total channel height was $H = 6\,\text{m}$ and fluid air properties were taken with sound wave propagation velocity $c_f = 340\,\text{m/s}$ and density $\rho = 1.225\,\text{kg/m}^3$.

A modified Green's function for the fluid channel was used to limit the discretisation to the scatterer boundary. The method of images was used to construct the modified Green's function [15]. The scatterer and the fluid channel were centred at the origin. The fluid channel and the scatterer were excited by a point source acting at location $\mathbf{x} = (-3, 0, 0)\,\text{m}$ with $\kappa = 20\,\text{rad/m}$. The source was defined by the modified Green's function. The problem solution was computed in the frequency-wavenumber domain for a longitudinal wavenumber varying over a range from zero to $\kappa_z = 2\kappa$ in 128 equidistant steps. The pressure field was evaluated over a grid of receivers with a density of six points per wavelength.

Two geometries were considered: (i) a cylindrical scatterer and (ii) the complex shaped scatterer represented in Fig. 3(a) directly imported from a CAD model. The geometry of the second example was defined by twelve C^1 cubic patches. Problem discretisation was given by $\kappa h = 6$ and $d_\lambda = 6$, which leads to an element order $p = 6$. The element nodes were located at the Chebyshev points of the second kind. Once the boundary discretisation had been defined, the geometry patches were subdivided as shown in Fig. 3(b). The cylindrical scatterer was discretised following a similar procedure.

Fig. 4 shows the incident wavefield and the scattered wave. The representation is limited to a half model, according to the problem symmetry. The incident pressure field shows a spherical distribution that was gradually lost as the distance to the source point increased due to channel wave refraction (Fig. 4(a)). The amplitude decreased with the distance to the excitation point, according to the radiation condition. The pattern of the incident wave changed under the effect of the scatterers. A complex pattern was found, caused by waves reflected from both the channel borders and the scatterer boundary. A shadow region was found behind the scatterers where pressure amplitude was considerably lower.

This example has shown that the proposed method is, therefore, a useful tool for studying boundary problems with exact CAD geometry. An independent geometry and variable field approximation allowed efficient modelling that took both the complexity of the boundary shape and the studied wavenumber into account. In this case, it was enough to use cubic

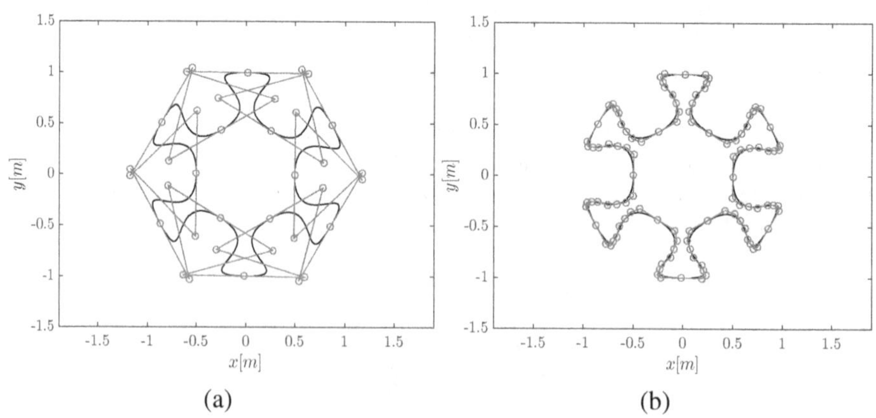

Figure 3: (a) Complex geometry imported from CAD model; (b) Related discretisation for $\kappa h = 6$. Control points and their related control polygons are represented by grey circles and grey lines.

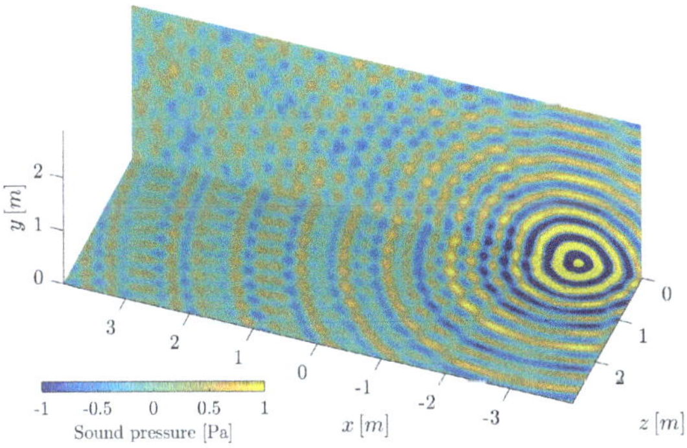

(a)

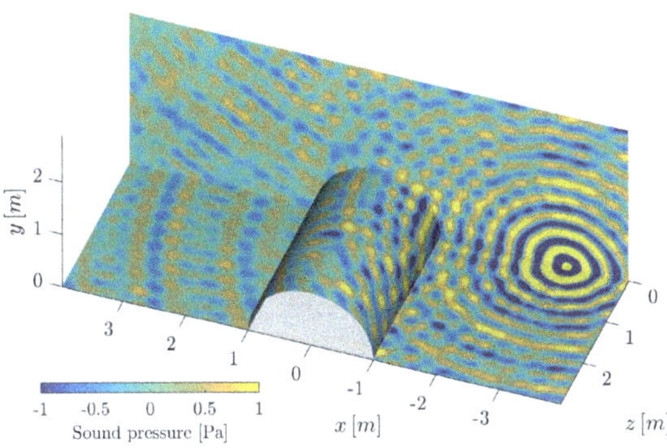

(b)

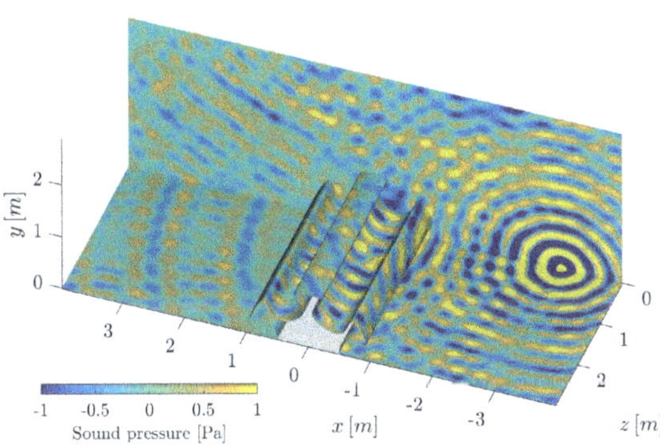

(c)

Figure 4: Real part of (a) incident wavefield, and the scattered wavefield by (b) a cylindrical cavity, and (c) a complex shape boundary.

curves for the geometry approximation, while it was necessary to use six order elements for the variable field.

4 CONCLUSIONS

This paper has proposed a novel formulation of the BEM to consider exact CAD geometry. The boundary geometry is described in the Bézier–Bernstein space which is commonly used in computer-aided design. Boundary elements are defined by the subdivision of Bézier curves according to an h-refinement strategy. The element geometry is evaluated and subdivided using the de Casteljau algorithm and the polar form of Bernstein polynomials. Geometric operations are easily performed without CAD interaction which entails a great flexibility for h-refinement.

The field variables are approximated independently of the geometry representation. Therefore, the proposed method is arbitrarily defined as subparametric, isoparametric or superparametric, depending on the problem properties and boundary geometry. The approximation of field variables has been generalised considering different shape functions families defined from node locations and element degree, using the Newton–Bernstein algorithm.

The boundary element integrals are solved according to distance from collocation points and the integration element. The asymptotic behaviour of the fundamental solution has given a critical radius below which the element integrals become singular. In this case, singular integrals have been numerically integrated with a polynomial transformation of the integration domain.

In particular, this work has been used to solve the Helmholtz equation in longitudinally invariant domains, but it is also possible to apply it in different two-dimensional or 2.5D physical problems, using an appropriate fundamental solution for the boundary element method.

ACKNOWLEDGEMENTS

The research work presented herein was supported by the Spanish Ministry of the Economy and Competitiveness (Ministerio de Economía y Competitividad) through research project BIA2016-75042-C2-1-R. It was also supported by project POCI-01-0247-FEDER-017759, funded by Portugal 2020 through the Operational Programme for Competitiveness Factors (COMPETE 2020). Financial support is gratefully acknowledged. The authors also wish to acknowledge the support provided by the Andalusian Scientific Computing Centre (CICA).

REFERENCES

[1] Atroshchenko, E. & Bordas, S.P.A., Geometry-independent field approximation for spline-based finite element methods. *11th World Congress on Computational Mechanics*, 2014.

[2] Farouki, R.T., The Bernstein polynomial basis: A centennial retrospective. *Computer Aided Geometric Design*, **29**(6), pp. 379–419, 2012.

[3] Farouki, R.T. & Rajan, V.T., Algorithms for polynomials in Bernstein form. *Computer Aided Geometric Design*, **5**(1), pp. 1–26, 1988.

[4] Farouki, R.T., Legendre–Bernstein basis transformations. *Journal of Computational and Applied Mathematics*, **119**(1–2), pp. 145–160, 2000.

[5] Bouchon, M. & Aki, K., Discrete wave-number representation of seismic source wavefields. *Bulletin of the Seismological Society of America*, **67**(2), pp. 259–277, 1977.

[6] Romero, A., Tadeu, A., Galvín, P. & António, J., 2.5D coupled BEM-FEM used to model fluid and solid scatterering wave. *International Journal for Numerical Methods in Engineering*, **101**(2), pp. 148–164, 2015.

[7] Ramshaw, L., Blossoming: A connect-the-dots approach to splines. Digital Equipment Corporation SRC Report No. 19.

[8] Seidel, H.-P., An Introduction to Polar Forms. *IEEE Compututer Graphics and Applications*, **13**(1), pp. 38–46, 1993.

[9] Dokken, T., Dæhlen, M., Lyche, T. & Mørken, K., Good approximation of circles by curvature-continuous Bézier curves. *Computer Aided Geometric Design*, **7**(1–4), pp. 33–41, 1990.

[10] Hur, S. & Kim, T.-W., The best G1 cubic and G2 quartic Bézier approximations of circular arcs. *Journal of Computational and Applied Mathematics*, **236**(6), pp. 1183–1192, 2011.

[11] Ainsworth, M. & Sánchez, M.A., Computing the Bézier control points of the lagrangian interpolant in arbitrary dimension. *SIAM Journal on Scientific Computing*, **38**(3), pp. A1682–A1700, 2016.

[12] Farouki, R.T., Goodman, T.N.T. & Sauer, T., Construction of orthogonal bases for polynomials in Bernstein form on triangular and simplex domains. *Computer Aided Geometric Design*, **20**(4), pp. 209–230, 2003.

[13] Romero, A., Galvín, P., António, J., Domínguez, J. & Tadeu, A., Modelling of acoustic and elastic wave propagation from underground structures using a 2.5D BEM-FEM approach. *Engineering Analysis with Boundary Elements*, **76**, pp. 26–39, 2017.

[14] Monegato, G. & Scuderi, L., Numerical integration of functions with boundary singularities. *Journal of Computational and Applied Mathematics*, **112**(1–2), pp. 201–214, 1999.

[15] Wu, T., *Boundary Element Acoustics Fundamentals and Computer Codes, Advances in Boundary Elements*, WIT Press, 2000.

BOUNDARY ELEMENT ANALYSIS OF THIN-WALLED STRUCTURES USING AN EXPANDING ELEMENT INTERPOLATION METHOD

JIANMING ZHANG, YUDONG ZHONG, LE YANG, BAOTAO CHI & CHUANMING JU
State Key Laboratory of Advanced Design and Manufacturing for Vehicle Body,
College of Mechanical and Vehicle Engineering, Hunan University, China

ABSTRACT

Boundary element analysis of thin-walled structures using a new expanding element interpolation method is presented. The element is obtained by adding virtual nodes based on a traditional discontinuous element. Coupled with the virtual nodes, the shape functions become higher order shape functions and are named fine shape functions. The original shape function of the discontinuous element is named raw shape function. With the expanding element, the interpolation accuracy can be increased by at least two orders compared with the original discontinuous element. The elements are able to naturally and accurately interpolate both continuous and discontinuous fields. The virtual nodes can be obtained by those of adjacent source points using the raw shape functions. Furthermore, the boundary integral equations are built up only at the inner nodes. Thus the size of the equations has not changed. Numerical examples are presented to verify our methods. Results demonstrate the accuracy and efficiency of the proposed method.

Keywords: boundary element method, expanding element, boundary integral equations.

1 INTRODUCTION

Accurate and efficient analysis of super thin structures is of crucial importance in engineering applications [1], [2]. The boundary element method (BEM) [3]–[9] has become recently an attractive alternative for problems in computational mechanics, and has been widely used to solve the problems of computational mechanics. Compared with the FEM, the BEM is a boundary method. BEM has evident advantages for solving problems such as linear elasticity [10], crack propagation [11], acoustics and contact problem [12], [13], and so on. In modern daily industrial settings, mesh preparation is the most labour intensive in numerical modelling, particularly for the FEM. The BEM needs not to deal with the interior mesh, so it is more effective in mesh preparation. In addition, the FEM requires the trial functions C^0-continuous, but the BEM does not require that. In addition, the BEM has the advantages of dimensionality reduction and high accuracy of calculation. In view of the above observations, we can know that the BEM is a more suitable method to analyse thin-super thin structures since the BEM just requires one to discretize the boundary and the stress can be accurately calculated. Then combined with the advantages of BEM, the super thin structure can be accurately approximated.

At present, there are mainly two kinds of implementation for the BEM [14], one is to use discontinuous element interpolation, and the other is to use continuous element interpolation. In the specific numerical realization process, the discontinuous element configures the node inside the element, which provides convenience for the realization process, which simplifies the system equation and reduces the difficulty of grid generation, but increases the calculation of the whole system equation. The continuous element directly arranges the points on the geometric vertices of the element. The number of nodes is less than the number of non-continuous nodes under the same order, and the calculation amount is small. However, in some cases, there are still some difficulties in processing the corner points [15].

WIT Transactions on Engineering Sciences, Vol 126, © 2019 WIT Press
www.witpress.com, ISSN 1743-3533 (on-line)
doi:10.2495/BE420031

In this paper, a new expanding element is proposed to synthesize the discontinuous and continuous element. The proposed element has all the advantages of previous element and overcoming their disadvantages, improving the interpolation accuracy at least by two without changing the degrees of freedom of the equation. Finally, a universal method is proposed to solve the near-singular integral of the thin structure, that is, the integral element is continuously subdivided according to the position of the source point, and each time subdivided into two identical sub-elements. Through this method, the near-singular integrals in thin structures can be accurately calculated.

The structure of the paper is as follows. Section 2 is to introduce the proposed element interpolation method. In Section 3, the new numerical implementation of BIE is described. Section 4 gives some numerical examples. Section 5 is the conclusions.

2 THE EXPANDING ELEMENT INTERPOLATION METHOD

2.1 The expanding elements

The expanding element is obtained by adding virtual nodes on the traditional discontinuous element. These elements have two kinds of shape functions: raw and fine shape function. As shown in Fig. 1, the two shape functions of the constant expanding element are as follows:

$$N^r = 1, \tag{1}$$

$$\begin{cases} N_1^f = -0.5\xi(1-\xi) \\ N_2^f = (1-\xi)(1+\xi) \\ N_3^f = 0.5\xi(1+\xi), \end{cases} \tag{2}$$

where N^r, N_1^f, N_2^f and N_3^f denote the raw and fine shape function, respectively.

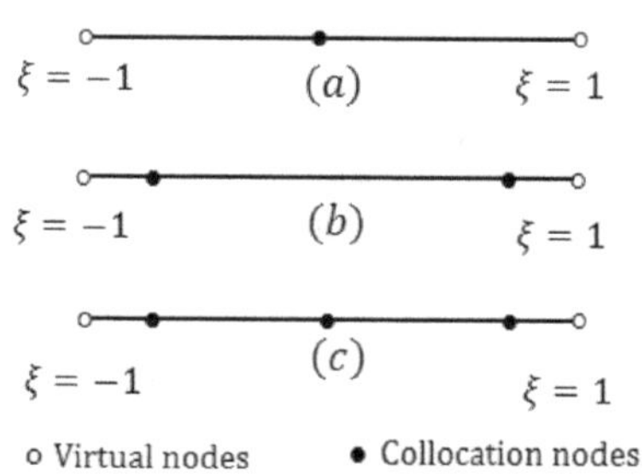

Figure 1: Constant, linear and quadratic expanding element.

2.2 Introduction of the new interpolation method

Firstly, we will dentally introduce the method of the calculation of the values of the virtual nodes (considering the example in Fig. 2).

When the expanding element is used to interpolate the known boundary conditions, the value of the virtual node is directly equal to the boundary conditions. When the expanding element is used to interpolate unknown boundary conditions, the node values of the virtual nodes can be obtained by the extrapolation of the raw shape functions.

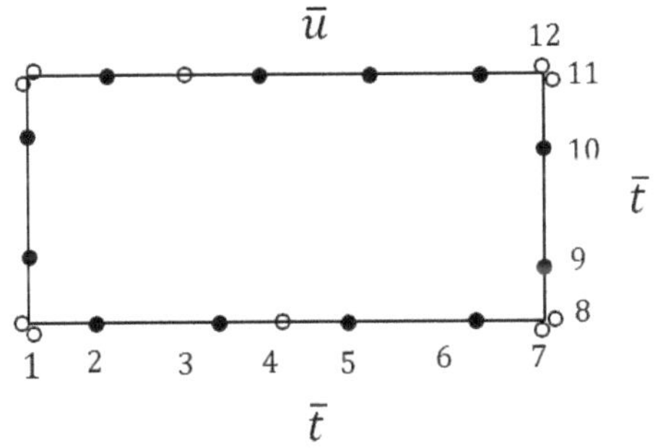

Figure 2: The geometric model for different boundary variables.

When the proposed method is used to interpolate the discontinuous boundary variables, we collocate two virtual nodes at a vertex shared by two adjacent elements. The nodal values of virtual nodes are calculated by eqn (3)

$$\begin{aligned} \mathbf{t}_7 &= \bar{\mathbf{t}}_7,\ \mathbf{t}_8 = \bar{\mathbf{t}}_8 \\ \mathbf{t}_{11} &= \bar{\mathbf{t}}_{11},\ \mathbf{t}_{12} = \bar{\mathbf{t}}_{12}. \end{aligned} \tag{3}$$

In this way, the discontinuous boundary conditions can be kept accurate. And for continuous fields (displacement boundary conditions), eqns (4) and (5) can be used

$$\mathbf{u}_{11} = \mathbf{u}_{12} = \bar{\mathbf{u}}_{11}, \tag{4}$$

$$\mathbf{u}_7 = \mathbf{u}_8 = \frac{1}{2}\left(N_5^r(1)\mathbf{u}_5 + N_6^r(1)\mathbf{u}_6 + N_9^r(-1)\mathbf{u}_9 + N_{10}^r(-1)\mathbf{u}_{10}\right). \tag{5}$$

After getting the values of virtual node, the physical variables in the BIE can be obtained by interpolating by the fine shape functions. So the **u** and **t** on the right edge can be obtained by

$$\begin{aligned} \mathbf{u} &= N_8^f\mathbf{u}_8 + N_9^f\mathbf{u}_9 + N_{10}^f\mathbf{u}_{10} + N_{11}^f\mathbf{u}_{11} \\ \mathbf{t} &= N_8^f\mathbf{t}_8 + N_9^f\mathbf{t}_9 + N_{10}^f\mathbf{t}_{10} + N_{11}^f\mathbf{t}_{11}. \end{aligned} \tag{6}$$

Eqn (6) can be further written as

$$\begin{aligned} \mathbf{u} &= \frac{1}{2}N_8^f N_5^r\mathbf{u}_5 + \frac{1}{2}N_8^f N_6^r\mathbf{u}_6 + N_{11}^f\bar{\mathbf{u}}_{11} \\ &\quad + \left(N_9^f + \frac{1}{2}N_8^f N_9^r\right)\mathbf{u}_9 + \left(N_{10}^f + \frac{1}{2}N_8^f N_{10}^r\right)\mathbf{u}_{10} \\ \mathbf{t} &= N_8^f\bar{\mathbf{t}}_8 + N_9^f\bar{\mathbf{t}}_9 + N_{10}^f\bar{\mathbf{t}}_{10} + N_{11}^f\bar{\mathbf{t}}_{11}. \end{aligned} \tag{7}$$

From above introduction, when the boundary condition is known, the value of the virtual node is directly equal to the boundary condition or directly obtained by extrapolation of the raw shape functions. As the degrees of freedom of the virtual nodes need to be eliminated by the source points, thus and the size of the matrix equation does not change. Moreover, no matter discontinuous and continuity boundary variables, the expanding elements can also approximate accurately.

3 NUMERICAL IMPLEMENTATION OF BIE

3.1 Nearly singular integration

Accurate and efficient calculation of nearly singular integrals is the key to numerical calculations. In this part, we proposed an adaptive element subdivision method to solve nearly singular integrals. First we need to calculate the side length l of the integral element, and then calculate the distance d from the source point to the center of the element. In the global coordinate system, when l is less than d, the integral of the element is a regular integral. Otherwise, the element is divided into two equal sub-elements (see Fig. 3). By using this method, the nearly singular integral can be accurately calculated, and no integral transformation is needed, and that is easily to implement.

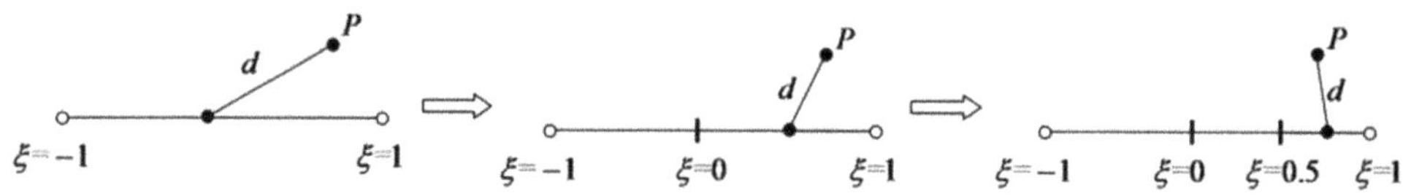

Figure 3: The method of element subdivision.

3.2 Numerical implementation

The well-known BIE for the solution of 2D linear elasticity problem in a bounded by Γ is

$$c(P)u(P)+\int_{\Gamma}q^{*}(P,Q)u(P)d\Gamma(Q)=\int_{\Gamma}u^{*}(P,Q)q(P)d\Gamma(Q), \tag{8}$$

where P and Q are the source and the field point, respectively. The Kelvin fundamental solutions $u_{ij}^{*}(P,Q)$ and $t_{ij}^{*}(P,Q)$ are given by:

$$u_{ij}^{*}(P,Q)=\frac{1}{8\pi\mu(1-v)}\left[(3-4v)\delta_{ij}\ln\frac{1}{r}+r_{,i}r_{,j}\right], \tag{9}$$

$$t_{ij}^{*}(P,Q)=-\frac{1}{4\pi(1-v)r}\left\{\frac{\partial r}{\partial n}\left[(1-2v)\delta_{ij}+2r_{,i}r_{,j}\right]-(1-2v)(r_{,i}n_{j}-r_{,j}n_{i})\right\}, \tag{10}$$

where μ and v are the shear modulus and the Poisson's ratio, respectively. r is the distance between the source and field point.

Eqn (8) is discretized with N expanding elements. The discretized equations for BIE can be obtained

$$c(P)u^{j}(P)+\sum_{i=1}^{N}\left(\int_{\Gamma_i}q^{*}(P,Q)\mathbf{N}^{f}d\Gamma(Q)\right)\mathbf{u}^{i}=\sum_{i=1}^{N}\left(\int_{\Gamma_i}u^{*}(P,Q)\mathbf{N}^{f}d\Gamma(Q)\right)\mathbf{t}^{i}. \tag{11}$$

Eqn (11) can be expressed in a matrix form as

$$\mathbf{Hu}=\mathbf{Gt}, \tag{12}$$

where vectors **u** and **t** consist of all nodal displacement and traction. **H** and **G** are shown as follows:

$$\mathbf{H}_{ij} = \int_{\Gamma_e} \mathbf{t}^*(P^i, Q) N_j^f(Q) d\Gamma(Q), \tag{13}$$

$$\mathbf{G}_{ij} = \int_{\Gamma} \mathbf{u}^*(P^i, Q) N_j^f(Q) d\Gamma(Q). \tag{14}$$

Distinguishing the different boundary conditions, eqn (12) can be further written as

$$\begin{bmatrix} \bar{\mathbf{H}}^s & \mathbf{H}^s & \bar{\mathbf{H}}^v & \mathbf{H}^v \end{bmatrix} \begin{bmatrix} \bar{\mathbf{u}}^s \\ \mathbf{u}^s \\ \bar{\mathbf{u}}^v \\ \mathbf{u}^v \end{bmatrix} = \begin{bmatrix} \mathbf{G}^s & \bar{\mathbf{G}}^s & \mathbf{G}^v & \bar{\mathbf{G}}^v \end{bmatrix} \begin{bmatrix} \mathbf{t}^s \\ \bar{\mathbf{t}}^s \\ \mathbf{t}^v \\ \bar{\mathbf{t}}^v \end{bmatrix}, \tag{15}$$

where $\bar{\mathbf{u}}^s$, $\bar{\mathbf{t}}^s$, $\bar{\mathbf{u}}^v$, $\bar{\mathbf{t}}^v$ and $\mathbf{u}^s$, $\mathbf{t}^s$, $\mathbf{u}^v$, $\mathbf{t}^v$ are the known and unknown boundary variables of the source points and virtual points, respectively.

From 2.2, we can see that the relationship of source points and virtual points are as follows:

$$\begin{aligned} \mathbf{u}^v &= \mathbf{N}_1^r \mathbf{u}^s \\ \mathbf{q}^v &= \mathbf{N}_2^r \mathbf{q}^s. \end{aligned} \tag{16}$$

Using eqn (16), eqn (15) can be written as

$$\mathbf{A}\mathbf{x} = \mathbf{f}, \tag{17}$$

where

$$\mathbf{A} = \begin{bmatrix} \mathbf{H}^s + \mathbf{H}^v \mathbf{N}_{r1} & -(\mathbf{G}^s + \mathbf{G}^v \mathbf{N}_{r2}) \end{bmatrix}, \tag{18}$$

$$\mathbf{x} = \begin{bmatrix} \mathbf{u}^s \\ \mathbf{q}^s \end{bmatrix}, \tag{19}$$

$$\mathbf{f} = \begin{bmatrix} \bar{\mathbf{G}}^s \bar{\mathbf{q}}^s + \bar{\mathbf{G}}^v \bar{\mathbf{q}}^v - \bar{\mathbf{H}}^s \bar{\mathbf{u}}^s - \bar{\mathbf{H}}^v \bar{\mathbf{u}}^v \end{bmatrix}. \tag{20}$$

Matrix **A** is a square matrix of order *n*; **x** is unknown at the source points; **f** is the known vector on the right-hand side.

From eqns (17)–(20), one can see that the degrees of freedom of the virtual nodes are eliminated by the source points, so the size of the system equation does not change. This point is the main advantage of the proposed method.

4 NUMERICAL EXAMPLES

To verify the advantages of the proposed method for thin-walled structures, a few illustrative numerical results are given in this section. In the following examples, e represents relative errors and n stands for the number of the nodes

$$e = \frac{1}{|u|_{\max}} \sqrt{\frac{1}{N} \sum_{i=1}^{N} (u_i^{\text{num}} - u_i^{\text{exact}})}. \tag{21}$$

In all figures, ExpdConst, ExpdLinear, ExpdQuad, TradConst, TradLinear, and TradQuad represent the proposed expanding elements and traditional continuous elements (constant, linear, quadratic elements) interpolation methods with different orders, respectively.

4.1 Example 1: Concentric circles subject to a displacement field problem

In this case, the proposed method is used to analyze a thin coating on a shaft, and the geometry model is as shown in Fig. 4 (these radii are r_1 and r_2, respectively). The thickness of coating is h (varies in the range of $10^{-1}r_1 - 10^{-6}r_1$). It is subject to the boundary conditions of eqn (22). Plane strain case with $E = 1$, $v = 0.25$ (in consistent units). 60 nodes are used on each boundary. The relative errors and convergence rates of t_y are shown in Figs 5 and 6.

$$u_x = y^3 - 3yx^2, \quad u_y = -x^3 + 3xy^2 . \tag{22}$$

It can be seen from Figs 5 and 6, when the thickness of the coating becomes smaller, the results by using the traditional element interpolation method are not correct, and that is difficult to improve the accuracy with too many nodes. Overall, when the coating is gradually thinned, the accuracy of using the proposed method is significantly higher and the convergence is better than the conventional discontinuous element interpolation method.

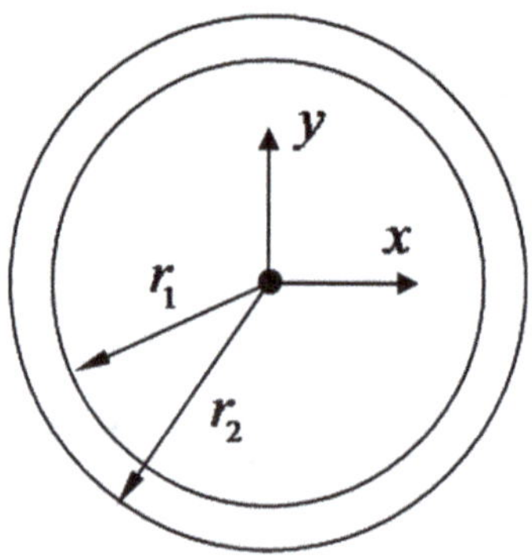

Figure 4: The simplified geometry model.

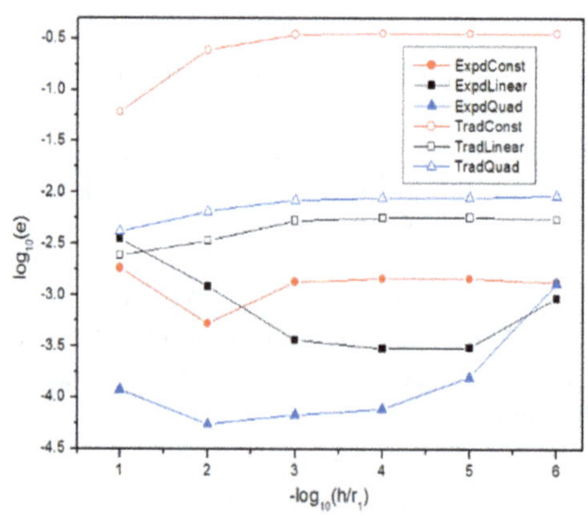

Figure 5: The relative errors of t_y on the boundary of the inner circle.

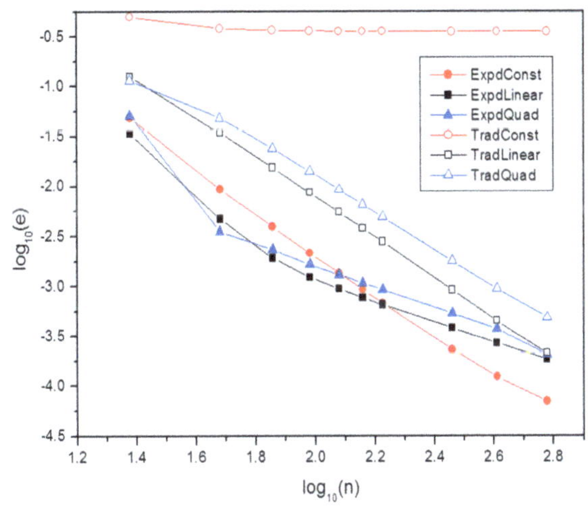

Figure 6: The convergence of t_y on the boundary of the inner circle for h/r_1 = 10-6.

4.2 Example 2: cantilever beam

The expanding element interpolation method is also studied in the cantilever beam problem. Consider a beam of length $L = 10$ and 100, respectively, height $h = 1$, $E = 1.1\times10^5$, $v = 0.25$, subjected to a parabolic traction at the free end with $P(y) = 1.0$ (as shown in Fig. 7). The analytical solution of the displacement is given by

$$u_x = -\frac{P}{6\bar{E}I}(y-\frac{h}{2})\left[(6L-3x)x+(2+\bar{v})(y^2-2hy)\right]$$
$$u_y = \frac{P}{6\bar{E}I}\left[3\bar{v}(y^2-2hy+\frac{1}{2}h^2)(L-x)+\frac{1}{4}(4+5\bar{v})h^2x+(L-\frac{1}{3}x)3x^2\right], \tag{23}$$

where

$$I=\frac{h^3}{12}, \qquad \bar{E}=\frac{E}{1-v^2}, \qquad \bar{v}=\frac{v}{1-v^2}, \tag{24}$$

and 50 elements are used on edge ab and cd, respectively. On edge ad and bc, we collocate only one element. The results of u_y on the edge ab, together with the analytical solution are shown in Fig. 8. To study the accuracy and convergence rate, a series number of nodes are used to research that and the results are shown in Fig. 9.

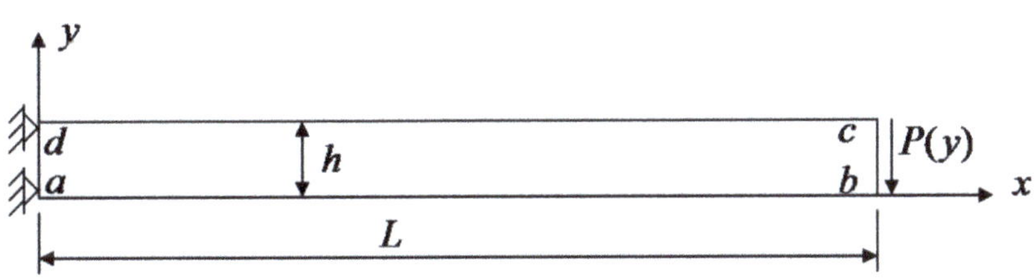

Figure 7: Cantilever beam problem.

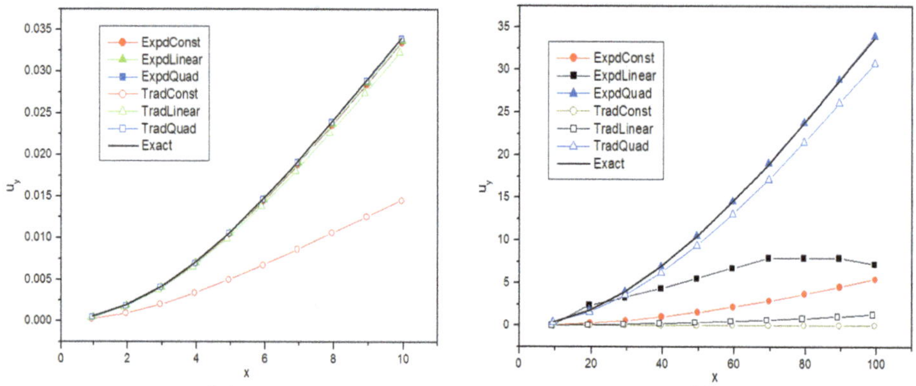

(a) (b)

Figure 8: The u_y along the edge *ab*. (a) L = 10; (b) L = 100.

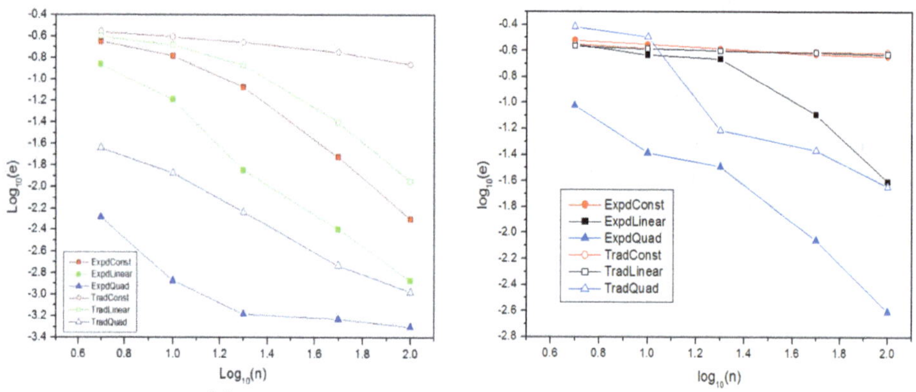

(a) (b)

Figure 9: Convergence of u_y on the edge *ab*. (a) L = 10; (b) L = 100.

From Figs 8 and 9, one can see that the accuracy and astringency rate by proposed method are higher than those by traditional elements interpolation method. The advantages of the proposed method are demonstrated again.

4.3 Example 3: structures with negative Poisson's ratio

To verify the universality of the proposed method, in this example, the proposed method is used to analysis a negative Poisson's ratio structure. Plane strain case with P = 1 Mpa, E = 1,000 MPa, Poisson's ratio v = 0.25. The corresponding geometric parameters are given as follows: structure length l = 90, height h = 60, t = 2 as shown in Fig. 10.

"ExpdQuad684" and "FEM1650" in Fig. 11 represent to use 684 and 1,650 nodes with different methods, respectively. The result of u_y on the edge a are as shown in Fig. 11. The result of 3,232,854 nodes by the FEM is taken as the reference solution. The von Mises stress by the proposed method and the FEM are as shown in Figs 12 and 13, respectively.

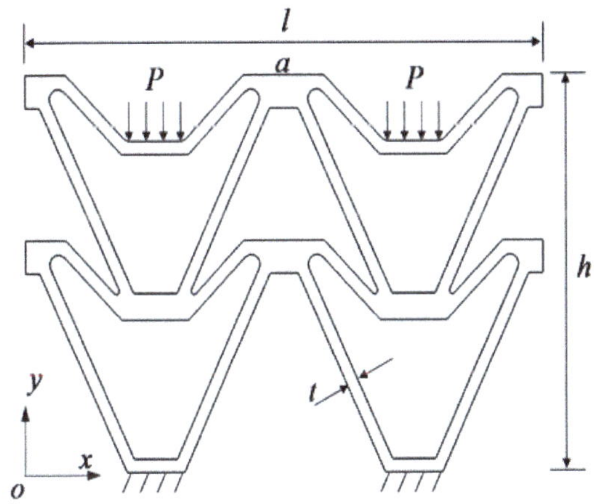

Figure 10: The geometric model of the structures.

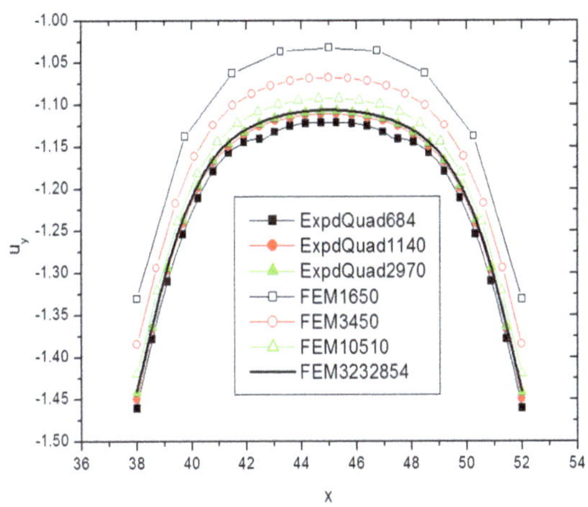

Figure 11: u_y on the edge a.

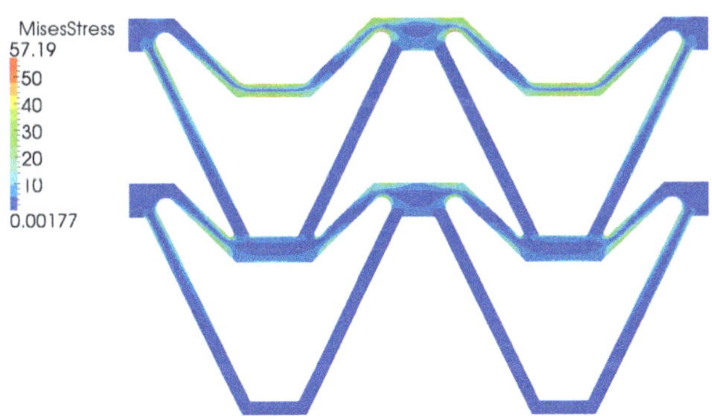

Figure 12: The result by using the proposed method with 1,140 nodes.

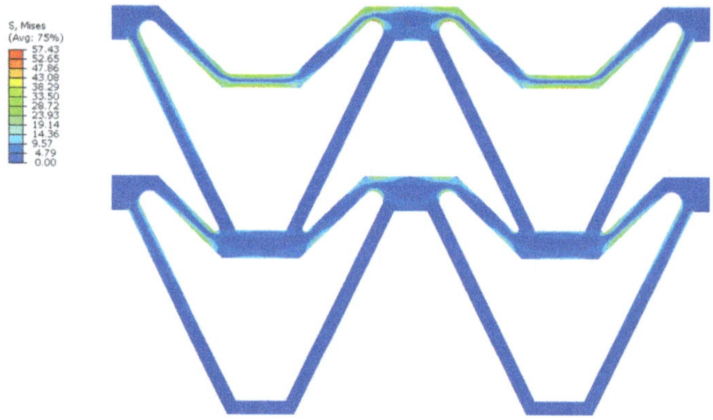

Figure 13: The result by using the FEM with 3,232,854 nodes.

5 CONCLUSIONS

This paper has presented a boundary element analysis of super thin structures using the expanding element interpolation method. Compared with FEM, there is no need for any assumptions in the BEM, and high accuracy can be obtained without using large numbers of elements. The proposed method not only has all the merits of the discontinuous element, but also guarantees the interelement continuity of field variables. Both continuous and discontinuous fields can be accurately interpolated by the proposed method. Moreover, as the degrees of freedom of the virtual nodes need to be eliminated, the size of the matrix equations does not increase. Results demonstrate that the proposed method possesses satisfactory accuracy and convergence rate compared with the traditional method. Extending our method to solve three-dimensional problems is ongoing.

ACKNOWLEDGEMENT

The research was supported by National Science Foundation of China (Grant No. 11772125).

REFERENCES

[1] Wempner, G., Mechanics and finite elements of shells. *Applied Mechanics Reviews*, **42**(5), pp. 129–142, 1989.

[2] Yao, Z.H., A new type of high-accuracy BEM and local stress analysis of real beam, plate and shell structures. *Engineering Analysis with Boundary Elements*, **65**, pp. 1–17, 2016.

[3] Zhang, J.M., Han, L., Lin, W.C., Dong, Y.Q. & Ju, C.M., A new implementation of BEM by an expanding element interpolation method. *Engineering Analysis with Boundary Elements*, **78**, pp. 1–7, 2017.

[4] Tanaka, M., Zhang, J.M. & Matsumoto, T., Boundary-type meshless solution of potential problems: Comparison between singular and regular formulations in hybrid BNM Transactions of JASCOME. *Journal of Boundary Element Methods*, **20**, pp. 21–26, 2003.

[5] Zhou, F.L., Zhang, J.M., Sheng, X.M. & Li, G.Y., A dual reciprocity boundary face method for 3D non-homogeneous elasticity problems. *Engineering Analysis with Boundary Elements*, **36**, pp. 1301–1310, 2012.

[6] Zhang, J.M., Qin, X.Y., Han, X. & Li, G.Y., A boundary face method for potential problems in three dimensions. *International Journal for Numerical Methods in Engineering*, **80**(3), pp. 320–337, 2009.

[7] Xie, G.Z., Zhang, J.M., Dong, Y.Q., Huang, C. & Li, G.Y., An improved exponential transformation for nearly singular boundary element integrals in elasticity problems. *International Journal of Solids and Structures*, **51**(6), pp. 1322–1329, 2013.

[8] Luo, J.F., Liu, Y.J. & Berger, E.J., Analysis of two-dimensional thin structures (from micro- to nano-scales) using the boundary element method. *Computational Mechanics*, **22**(5), pp. 404–412, 1998.

[9] Liu, Y.J., Analysis of shell-like structures by the boundary element method based on 3-D elasticity: Formulation and verification. *International Journal for Numerical Methods in Engineering*, **41**(3), pp. 541–558, 1998.

[10] Manolis, G.D. & Banerjee, P.K., Conforming versus non-conforming boundary elements in three-dimensional elastostatics. *International Journal for Numerical Methods in Engineering*, **23**(10), pp. 1885–1904, 1986.

[11] Xie, G.Z., Zhang, J.M., Huang, C., Lu, C.J. & Li, G.Y., A direct traction boundary integral equation method for three-dimension crack problems in infinite and finite domains. *Computational Mechanics*, **53**, pp. 575–586, 2014.

[12] Soares, Jr. D. & Mansur, W.J., An efficient stabilized boundary element formulation for 2D time-domain acoustics and elastodynamics. *Computational Mechanics*, **40**, pp. 355–365, 2007.

[13] Blázquez, A., París, F. & Mantič, V., BEM solution of two-dimensional contact problems by weak application of contact conditions with non-conforming discretizations. *International Journal of Solids and Structures*, **35**, pp. 3259–3278, 1998.

[14] Parreira, P., On the accuracy of continuous and discontinuous boundary elements. *Engineering Analysis*, **5**(4), pp. 205–211, 1988.

[15] Mukhopadhyay, S. & Majumdar, N., A study of three-dimensional edge and corner problems using the neBEM solver. *Engineering Analysis with Boundary Elements*, **33**(2), pp. 105–119, 2009.

ACOUSTIC SHAPE OPTIMIZATION BASED ON ISOGEOMETRIC BEM WITH ADJOINT VARIABLE METHOD

HAIBO CHEN[1], JIE WANG[1], CHANGJUN ZHEN[2] & LEILEI CHEN[3]
[1]CAS Key Laboratory of Mechanical Behavior and Design of Materials, University of Science and Technology of China, China
[2]Institute of Sound and Vibration Research, Hefei University of Technology, China
[3]College of Civil Engineering, Xinyang Normal University, China

ABSTRACT

The isogeometric analysis (IGA) has been applied to the boundary element method (BEM), forming the IGA BEM. In this work, we introduce the IGA BEM to 2D acoustic shape optimization. The key treatment is the acoustic sensitivity analysis using the adjoint variable method (AVM). Compared with the direct differentiation method (DDM), the AVM is more suitable for problems with a large number of design variables. The gradient-based optimization solver is applied in order to update the design variables during the optimization iteration process and the Burton-Miller method is adopted to conquer the fictitious eigen-frequency problem in solving exterior acoustic problems. An example of scattering by an infinite rigid cylinder is presented to demonstrate the improved accuracy of IGA BEM. Then we verify the efficiency of the developed sensitivity algorithms through the example and demonstrate their potential in solving large-scale engineering problems. Finally, an optimization example is provided to validate the proposed optimization procedure. Numerical tests show that the optimal results are strongly frequency dependent.

Keywords: isogeometric analysis, boundary element method, acoustics sensitivity analysis, shape optimization.

1 INTRODUCTION

Reducing sound emission of structures is of great concern in engineering problems, and shape optimization has been proven to be an effective approach [1]. Compared with the conventional BEM, the IGA BEM offers a higher accuracy in solving acoustic problems [2], as the geometry is accurately described by using the control points. Since the geometry can be easily changed by changing the control points, it is very convenient to carry out the shape optimization by the IGA BEM. However, gradient-based optimization solvers usually require gradient information, which leads to an additional step, i.e. the sensitivity analysis. This yields extra computational efforts and thus makes the efficient analysis techniques indispensable.

In recent years, several works about sensitivity analysis and optimization have been made for acoustic problems. Zheng et al. [3]–[5] has made many efforts in sensitivity analysis methods by BEM in 3D acoustics. Chen et al. [6]–[9] also investigated sensitivity analysis with single design parameter by FEM-BEM both in 2D and 3D acoustics. Liu et al. [2] has carried out the shape optimization of noise barrier by IGA BEM with the differentiation method (DDM) in sensitivity analysis. However, these aforementioned works mainly use the DDM with conventional BEM for the sensitivity analysis. But in many engineering problems, multi design variables should be adopted to describe the shape accurately, which yields more computational efforts in the sensitivity analysis. To this end, this paper presents a shape optimization procedure based on the IGA BEM in 2D acoustics. The formulations of the design sensitivity analysis by using the adjoint variable method

WIT Transactions on Engineering Sciences, Vol 126, © 2019 WIT Press
www.witpress.com, ISSN 1743-3533 (on-line)
doi:10.2495/BE420041

(AVM) are derived. The gradient-based optimization solver is applied to update the design variables for minimizing the sound pressure at observation points.

2 IGA BEM IN 2D ACOUSTICS

For the IGA BEM in 2D acoustics, a series of efforts have been made by Chen et al. [1] and Liu et al [2]. To conquer the fictitious eigen-frequency problem, we use the conventional boundary integral equation (CBIE) and the hypersingular BIE (HBIE) as follows:

$$CBIE: c(x)p(x) + \int F(x,y)p(y)dS(y) = \int G(x,y)q(y)dS(y), \tag{1}$$

$$HBIE: c(x)q(x) + \int F^1(x,y)p(y)dS(y) = \int G^1(x,y)q(y)dS(y), \tag{2}$$

where $G(x,y)$、$F(x,y)$、$G^1(x,y)$、$F^1(x,y)$ are the kernel functions. According to the definition of IGA, we have the NURBS interpolation [2] as follows:

$$\begin{aligned} x(\xi) &= \textstyle\sum_{i=1}^{n} R_{i,p_g}^{g}(\xi)X_i, \\ p(\xi) &= \textstyle\sum_{i=1}^{n_f} R_{i,p_f}^{f}(\xi)p_i, \\ q(\xi) &= \textstyle\sum_{i=1}^{n_f} R_{i,p_f}^{f}(\xi)q_i. \end{aligned} \tag{3}$$

Substituting eqn (3) into the CBIE and HBIE, the discretized isogeometric BIEs can be obtained as:

$$c\left(\boldsymbol{x}(\hat{\xi}_i)\right)\sum_{j=1}^{n_f} R_{j,p_f}^{f}(\hat{\xi}_i)p_j + \sum_{e=1}^{N_e}\sum_{j=1}^{n_f}\left[\int_{\xi_e}^{\xi_{e+1}} F\left(\boldsymbol{x}(\hat{\xi}_i),\boldsymbol{y}(\xi)\right) R_{j,p_f}^{f}(\xi)J(\xi)d\xi\right]p_j$$
$$= \textstyle\sum_{e=1}^{N_e}\sum_{j=1}^{n_f}\left[\int_{\xi_e}^{\xi_{e+1}} G\left(\boldsymbol{x}(\hat{\xi}_i),\boldsymbol{y}(\xi)\right) R_{j,p_f}^{f}(\xi)J(\xi)d\xi\right]q_j \quad i = 1,2,\cdots,n_f, \tag{4}$$

$$c\left(\boldsymbol{x}(\hat{\xi}_i)\right)\sum_{j=1}^{n_f} R_{j,p_f}^{f}(\hat{\xi}_i)q_j + \sum_{e=1}^{N_e}\sum_{j=1}^{n_f}\left[\int_{\xi_e}^{\xi_{e+1}} F^1\left(\boldsymbol{x}(\hat{\xi}_i),\boldsymbol{y}(\xi)\right) R_{j,p_f}^{f}(\xi)J(\xi)d\xi\right]p_j$$
$$= \textstyle\sum_{e=1}^{N_e}\sum_{j=1}^{n_f}\left[\int_{\xi_e}^{\xi_{e+1}} G^1\left(\boldsymbol{x}(\hat{\xi}_i),\boldsymbol{y}(\xi)\right) R_{j,p_f}^{f}(\xi)J(\xi)d\xi\right]q_j \quad i = 1,2,\cdots,n_f, \tag{5}$$

where N_e is the number of NURBS elements, and $[\xi_e, \xi_{e+1}]$ is a NURBS element.

According to the Burton-Miller method, giving coefficient α as follows:

$$\begin{aligned} k < 1, \alpha &= i, \\ k \geq 1, \alpha &= \frac{i}{k}, \end{aligned} \tag{6}$$

where k is the wave number. Then we have the following equation:

$$CBIE + \alpha HBIE = 0. \tag{7}$$

Performing linear combination to eqn (7), the following linear algebraic equations in matrix form are obtained:

$$\boldsymbol{H}p = \boldsymbol{G}q. \tag{8}$$

Moving all the unknowns of eqn (8) to the left-hand side and all the known values to the right-hand side, the linear equations are given by:

$$\boldsymbol{A}X = b. \tag{9}$$

Finally, we obtain the unknowns by solving eqn (9).

3 ACOUSTIC SENSITIVITY ANALYSIS WITH ADJOINT VARIABLE METHOD

Sensitivity analysis plays a great role in the shape optimization. Here the coordinates of control points in IGA can be chosen as the design variables. Some works about sensitivity analysis using the DDM [1]–[3], [5]–[7] have been developed. According to the DDM, eqns (1) and (2) can be differentiated with respect to the design variables as follows:

$$c(x)\dot{p}(x) + \int F(x,y)\dot{p}(y)dS(y) + \int \dot{F}(x,y)p(y)dS(y) + \int F(x,y)p(y)d\dot{S}(y)$$

$$= \int G(x,y)\dot{q}(y)dS(y) + \int \dot{G}(x,y)q(y)dS(y) + \int G(x,y)q(y)d\dot{S}(y), \tag{10}$$

$$c(x)\dot{q}(x) + \int F^1(x,y)\dot{p}(y)dS(y) + \int \dot{F}^1(x,y)p(y)dS(y) + \int F^1(x,y)p(y)d\dot{S}(y)$$

$$= \int G^1(x,y)\dot{q}(y)dS(y) + \int \dot{G}^1(x,y)q(y)dS(y) + \int G^1(x,y)q(y)d\dot{S}(y), \tag{11}$$

where the upper dot $(\dot{\ })$ denotes the differentiation with respect to the design variables.

Substituting eqn (3) into eqns (10) and (11), and still applying the Burton-Miller method, we get the linear algebraic equations in matrix form as:

$$\dot{\boldsymbol{H}}p + \boldsymbol{H}\dot{p} = \dot{\boldsymbol{G}}q + \boldsymbol{G}\dot{q}, \tag{12}$$

where the matrix $\boldsymbol{H}$、$\boldsymbol{G}$ have been calculated in the IGA BEM. By swapping the unknowns, we can obtain the unknown sensitivity values by the following equation:

$$\boldsymbol{A}\dot{\psi} = \boldsymbol{B}\dot{\phi} - \dot{\boldsymbol{H}}p + \dot{\boldsymbol{G}}q, \tag{13}$$

where $\dot{\psi}$ is the unknown sensitivity values. If the sound pressures at the field points are included in the objective function, the sensitivity values are given by:

$$\dot{p}_{inner}(x) = \{u\}^T\dot{\psi} + \{k\}^T\ \dot{\phi}\ + \{\dot{e}\}^T q - \{\dot{b}\}^T p. \tag{14}$$

However, if the objective function in optimization is only related to the field points, AVM could be used. According to eqn (13), we have:

$$\dot{\psi} = \boldsymbol{A}^{-1}\{\boldsymbol{B}\dot{\phi} - \dot{\boldsymbol{H}}p + \dot{\boldsymbol{G}}q\}. \tag{15}$$

Introducing the adjoint equations as follows:

$$\boldsymbol{A}^T\eta = u, \\ \{\eta\}^T = \{u\}^T\boldsymbol{A}^{-1}. \tag{16}$$

In addition, by substituting eqns (15) and (16) into eqn (14), the new formulation of field points' sensitivities can be obtained as

$$\dot{p}_{inner}(x) = \{\eta\}^T\{\boldsymbol{B}\dot{\phi} - \dot{\boldsymbol{H}}p + \dot{\boldsymbol{G}}q\} + \{k\}^T\,\dot{\phi}\, + \{\dot{e}\}^T q - \{\dot{b}\}^T p. \tag{17}$$

As we can see, eqn (16) is independent with all of the design variables. Hence, we only need to compute it once even if there exist a larger number of design variables. This will significantly improve the efficiency of sensitivity analysis, which will play a great role in the following optimization process.

4 SHAPE OPTIMIZATION PROCESS

After the sensitivity analysis by the AVM, we choose the method of moving asymptotes (MMA) as the optimization solver. The minimization of sound pressure on a specified reference plane is set as the design objective. The optimization model can be formulated as follows:

$$\begin{gathered} min\colon \Pi(\boldsymbol{x}) = \bar{p}_s(\boldsymbol{x})p_s(\boldsymbol{x}) \\ s.t.\colon A(\boldsymbol{x}) \le A_0 \\ bound\colon x_{mini} \le x_i \le x_{maxi}, i = 1,2,\cdots,n_d \\ iteration\colon \left|\frac{\Pi^{m+1}-\Pi^m}{\Pi^m}\right| < \tau, \end{gathered} \tag{18}$$

where $\boldsymbol{x}$ is a vector of control points' coordinates $x_i(i = 1,2,\cdots,n_d)$ that are set as design variables; $\Pi(\boldsymbol{x})$ is the objective function; $p_s(\boldsymbol{x})$ denotes the sound pressure of field points which located on the reference plane, and $\bar{p}_s(\boldsymbol{x})$ denotes the conjugate transpose of $p_s(\boldsymbol{x})$. A_0 is the initial area of structures, which is called constraint. x_{mini} and x_{maxi} are the lower and upper bounds for every design variable x_i; τ is the given value which decide the iteration termination.

5 NUMERICAL EXAMPLES

To validate the proposed approach, some numerical examples are carried out. We consider an infinite rigid cylinder excited by plane wave (wave speed c = 340 m/s) as shown in Fig. 1.

Firstly we compare the IGA BEM with the conventional BEM with different Lagrange basis functions. As Fig. 2 shows, IGA BEM achieves a higher accuracy than the constant, discontinuous linear and discontinuous quadratic elements (DBE21, DBE22 and DBE33, respectively). Then Fig. 3 shows the correctness of the sensitivity analysis. We compare the two analytical methods, i.e., the DDM and AVM, with finite difference method (FDM), and all of them give the same results. Finally we carry out the shape optimization of a rectangle structure shown in Fig. 4. A unit monopole is located at the point (0,1). The thickness of the structure is set as 0.2 m. Seventy-seven observation points are evenly distributed on the reference plane. The average sound pressure at these points is used as the objective function, as shown in eqn (18).

To demonstrate the advantage of AVM, we choose 19 design variables (from control point P_2 to P_{20}) as shown in Fig. 5. The lower and upper bounds of these control points' horizontal coordinates are 4.9 and 5.1, respectively. The parameter τ is set as 10^{-4}. And the constraint is that the structure should not go beyond the initial area.

For excitation frequencies 100 Hz, 400 Hz, 700 Hz and 1000 Hz are considered, and the optimized shapes are illustrated in Fig. 6. From the figure, it can be seen that different frequency produce different shape which indicates the optimized shape is strongly frequency dependent. Further, the optimized shape tends to be more complex as frequency increases, which will bring more peaks and valleys.

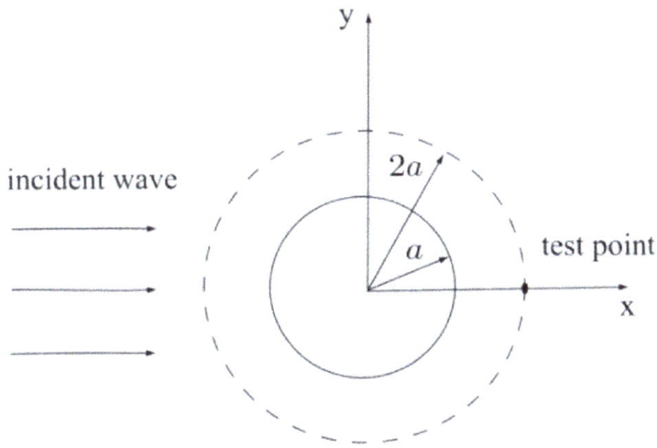

Figure 1: Scattering from an infinite rigid cylinder.

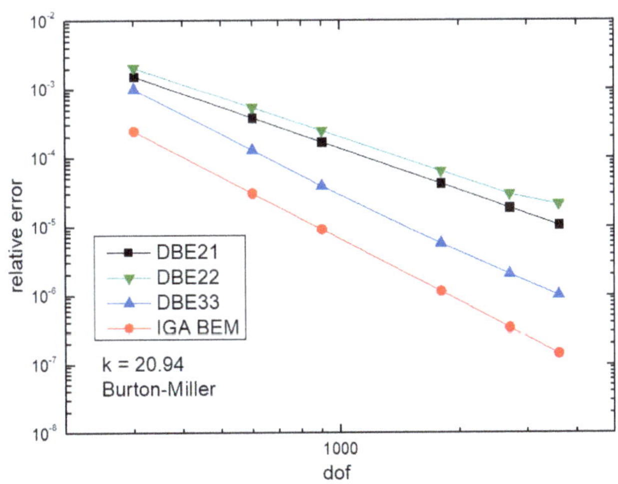

Figure 2: Comparison of the relative error of discontinued BEM and IGA BEM.

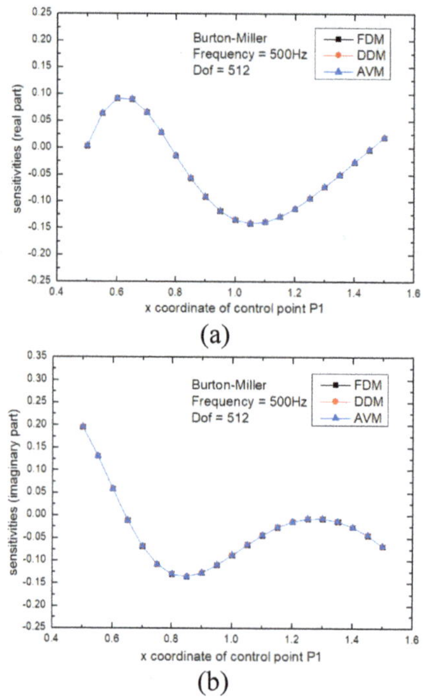

Figure 3: Sound pressure sensitivities at the field point (2,0) with respect to the horizontal coordinate of control point P_1 (1,0). (a) Real part; and (b) Imaginary part.

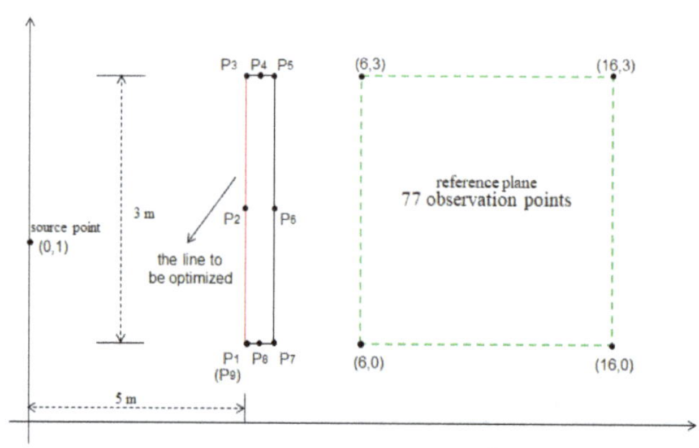

Figure 4: Initial rectangle structure.

The reduction of sound pressure level (SPL) is presented in Fig. 7. From the figure, we can see that the SPL values at 100Hz reduce less than those at higher frequencies, which means that the optimization achieves better SPL reduction at higher frequencies. This can be explained that the higher the frequency, the shorter the wavelength, the less the diffraction by structures.

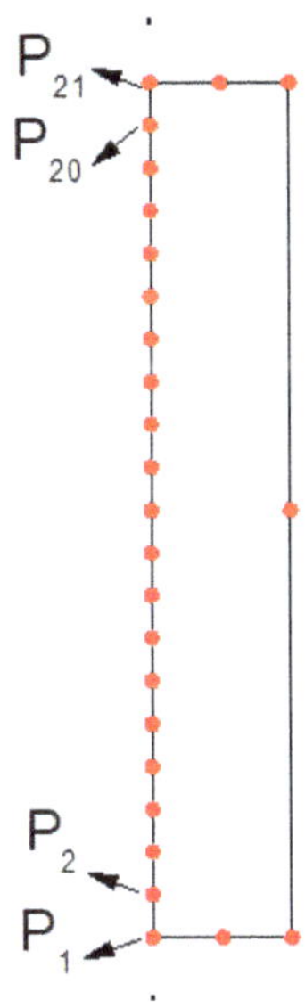

Figure 5: Design variables.

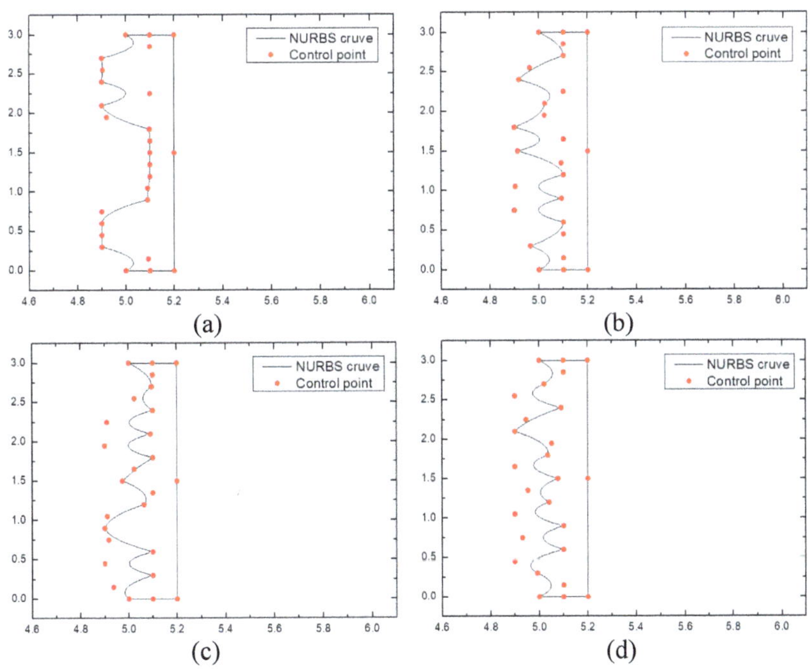

Figure 6: Shape optimization results with 19 design variables at different frequencies. (a) Frequency = 100Hz; (b) Frequency = 400Hz; (c) Frequency = 700Hz; and (d) Frequency = 1000Hz.

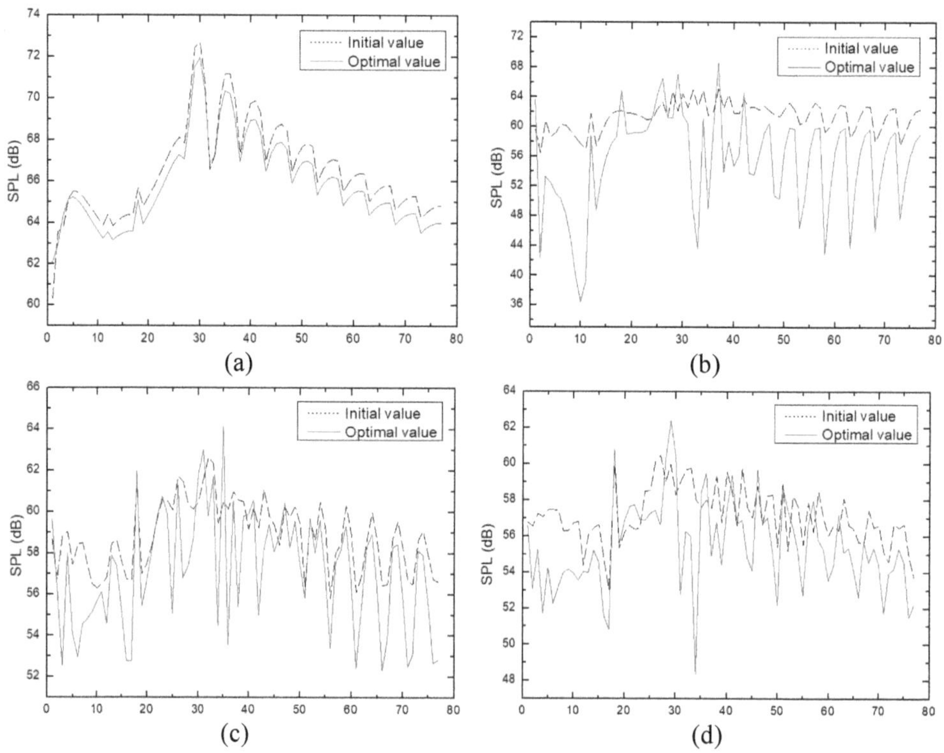

Figure 7: Reduction of SPL with 19 design variables at different frequencies. (a) Frequency = 100 Hz; (b) Frequency = 400 Hz; (c) Frequency = 700 Hz; and (d) Frequency = 1000 Hz.

6 CONCLUSIONS

This work presents a shape optimization approach based on the IGA BEM in 2D acoustics. The key treatment is the use of adjoint variable method for the design sensitivity analysis. Numerical tests show the validity of the developed approach. Furthermore, numerical simulation shows that IGA BEM is able to solve acoustic problems with high accuracy, and the proposed sensitivity analysis method is able to calculate the sensitivity values efficiently. The shape optimization example shows that the optimal results are strongly frequency dependent. The proposed optimization approach has the potential in application for engineering problems.

ACKNOWLEDGEMENTS

This work is financially supported by the National Natural Science Foundation of China (NSFC) under Grant Nos. 11772322 and 11872168, and the Strategic Priority Research Program of the Chinese Academy of Sciences under Grant No.XDB22040502.

REFERENCES

[1] Chen, L.L., Liu, C., Zhao, W.C. & Liu, L.C., An isogeometric approach of two dimensional acoustic design sensitivity analysis and topology optimization analysis for absorbing material distribution. *Comput. Methods Appl. Mech. Engrg.*, **336**(3), pp. 507–532, 2018.

[2] Liu, C., Chen, L.L., Zhao, W.C. & Chen, H.B., Shape optimization of sound barrier using an isogeometric fast multipole boundary element method in two dimensions. *Engineering Analysis with Boundary Elements*, **85**(9), pp. 142–157, 2017.

[3] Zheng, C.J., Matsumoto, T., Takahashi, T. & Chen, H.B., Explicit evaluation of hypersingular boundary integral equations for acoustic sensitivity analysis based on direct differentiation method. *Engineering Analysis with Boundary Elements*, **35**(11), pp. 1225–1235, 2011.

[4] Zheng, C.J., Chen, H.B., Matsumoto, T. & Takahashi, T., 3D Acoustic shape sensitivity analysis using fast multipole boundary element method. *International Journal of Computational Methods*, **9**(1), pp. 1240004, 2012.

[5] Zheng, C.J., Matsumoto, T., Takahashi, T. & Chen, H.B., A wideband fast multipole boundary element method for three dimensional acoustic shape sensitivity analysis based on direct differentiation method. *Engineering Analysis with Boundary Elements*, **36**(3), pp. 361–371, 2012.

[6] Chen, L.L., Zheng, C.J. & Chen, H.B., A wideband FMBEM for 2D acoustic design sensitivity analysis based on direct differentiation method. *Computation Mechanics*, **52**(3), pp. 631–648, 2013.

[7] Chen, L.L., Chen, H.B. & Marburg, S., Structural-acoustic design sensitivity analysis based on direct differentiation method with different element types. *Computer Modeling in Engineering & Science,* 1**07**(3), pp. 249–276, 2015.

[8] Chen, L., Chen, H., Zheng, C. & Marburg, S., Structural-acoustic sensitivity analysis of radiated sound power using a finite element / discontinuous fast multipole boundary element scheme. *International Journal for Numerical Methods in Fluids*, **82**(12), pp. 858–878, 2016.

[9] Chen, L.L., Liu, L.C., Zhao, W.C. & Chen, H.B., 2D Acoustic design sensitivity analysis based on adjoint variable method using different types of boundary elements. *Acoustics Australia*, **44**(2), pp. 343–357, 2016.

[10] Zheng, C.J., Bi, C.X., Zhang, C.Z., Zhang, Y.B. & Chen, H.B., Fictitious eigenfrequencies in the BEM for interior acoustic problems. *Engineering Analysis with Boundary Elements*, **104**(3), pp. 170–182, 2019.

NUMERICAL COMPUTATION OF DOUBLE SURFACE INTEGRALS OVER TRIANGULAR CELLS FOR VORTEX SHEET INTENSITY RECONSTRUCTION ON BODY SURFACE IN 3D VORTEX METHODS

ILIA MARCHEVSKY[1,2] & GEORGY SHCHEGLOV[1,2]
[1]Bauman Moscow State Technical University, Russia
[2]Ivannikov Institute for System Programming of the RAS, Russia

ABSTRACT

In this paper, a new approach is developed for the computation of vortex sheet intensity in vortex methods for 3D flow simulation. The problem is reduced to a boundary integral equation of the second kind on the body surface with respect to an unknown vector variable. The proposed technique makes it possible to improve the accuracy significantly in comparison to the technique traditionally implemented in vortex methods. A Galerkin-type approach is used with piecewise-constant basis functions. The coefficients of the resulting algebraic system are expressed through double surface integrals, calculated over the mesh cells. A semi-analytic technique is developed for the integrals calculation; integration over one cell is carried out analytically, while at the integration over the other cell (having common edge or vertex with the first one), the integrand is singular. Analytic expressions are obtained for singular parts of integrands and for the results of their integration. The regular parts of the integrands are integrated numerically. The developed approach provides less than 0.1% error. The regularization technique is developed for a divergence-free vortex sheet reconstruction on the body surface. The developed approach works well on coarse and non-uniform surface meshes for complex-shaped bodies, which is important for engineering applications.

Keywords: vortex methods, double layer potential, vortex sheet, boundary integral equation, singularity exclusion, numerical integration.

1 INTRODUCTION

The problem of 3D incompressible flow simulation around rigid and immovable body is considered. The governing equations for the outer flow around the body are the Navier–Stokes equations with the no-slip boundary conditions on the body surface K and perturbation decay conditions on infinity [1].

It is well-known that in order to take into account the presence of the body in the flow, it is possible to replace it with the vortex sheet of unknown intensity $\boldsymbol{\gamma}(\mathbf{r})$, placed on the body surface, $\mathbf{r} \in K$, which generates the velocity field $\mathbf{V}_\gamma(\mathbf{r})$. Than the summary velocity field is the superposition of the incident flow velocity $\mathbf{V}_\infty$, velocity field, generated by vorticity in the flow domain $\mathbf{V}_\Omega(\mathbf{r})$, and the introduced field $\mathbf{V}_\gamma(\mathbf{r})$:

$$\mathbf{V}(\mathbf{r}) = \mathbf{V}_\infty + \mathbf{V}_\Omega(\mathbf{r}) + \mathbf{V}_\gamma(\mathbf{r}), \qquad \mathbf{r} \in S, \tag{1}$$

where S is fluid domain, velocities $\mathbf{V}_\Omega(\mathbf{r})$ and $\mathbf{V}_\gamma(\mathbf{r})$ can be reconstructed through vorticity distribution according to the Biot–Savart law:

$$\mathbf{V}_\Omega = \int_S \frac{\boldsymbol{\Omega}(\boldsymbol{\xi}) \times (\mathbf{r} - \boldsymbol{\xi})}{4\pi |\mathbf{r} - \boldsymbol{\xi}|^3} dS_\xi, \qquad \mathbf{V}_\gamma = \oint_K \frac{\boldsymbol{\gamma}(\boldsymbol{\xi}) \times (\mathbf{r} - \boldsymbol{\xi})}{4\pi |\mathbf{r} - \boldsymbol{\xi}|^3} dK_\xi.$$

doi:10.2495/BE420051

From a mathematical point of view, the potential of the velocity field $\mathbf{V}_\gamma$ can be expressed through unknown double layer potential density $g(\mathbf{r})$ [2]:

$$\mathbf{V}_\gamma(\mathbf{r}) = \nabla\Phi(\mathbf{r}) = \oint_K g(\boldsymbol{\xi}) \frac{\partial}{\partial \mathbf{n}(\mathbf{r})} \frac{\partial}{\partial \mathbf{n}(\boldsymbol{\xi})} \frac{1}{4\pi|\mathbf{r}-\boldsymbol{\xi}|} dS_\xi. \tag{2}$$

Note that the velocity field, which corresponds to this potential, can also be written down in the following form [2]:

$$\mathbf{V}_\gamma(\mathbf{r}) = \nabla\Phi(\mathbf{r}) = \oint_K \frac{\boldsymbol{\gamma}(\boldsymbol{\xi}) \times (\mathbf{r}-\boldsymbol{\xi})}{4\pi|\mathbf{r}-\boldsymbol{\xi}|^3} dS_\xi, \tag{3}$$

where vector $\boldsymbol{\gamma}(\mathbf{r}) = -\operatorname{Grad} g(\mathbf{r}) \times \mathbf{n}(\mathbf{r})$; "Grad" means surface gradient operator. One can notice, that the expression (3) coincides with the Biot–Savart law for incompressible flows. So the potential $g(\mathbf{r})$ is closely connected with vortex sheet intensity $\boldsymbol{\gamma}(\mathbf{r})$.

The velocity field $\mathbf{V}(\mathbf{r})$ given by (1), has discontinuity at the body surface; its limit value in the considered case of rigid immovable body is the following:

$$\mathbf{V}_-(\mathbf{r}) = \mathbf{V}(\mathbf{r}) - \frac{\operatorname{Grad} g(\mathbf{r})}{2} = \mathbf{V}(\mathbf{r}) - \frac{\boldsymbol{\gamma}(\mathbf{r}) \times \mathbf{n}(\mathbf{r})}{2}, \quad \mathbf{r} \in K.$$

Taking into account the no-slip boundary condition in the form $\mathbf{V}_- = \mathbf{0}$ at the body surface, we obtain from (2) and (3) two forms of the integral equation:

$$\oint_K g(\boldsymbol{\xi}) \frac{\partial}{\partial \mathbf{n}(\mathbf{r})} \frac{\partial}{\partial \mathbf{n}(\boldsymbol{\xi})} \frac{1}{4\pi|\mathbf{r}-\boldsymbol{\xi}|} dS_\xi - \frac{\operatorname{Grad} g(\mathbf{r})}{2} = -\left(\mathbf{V}_\infty + \mathbf{V}_\Omega(\mathbf{r})\right), \quad \mathbf{r} \in K, \tag{4}$$

or

$$\oint_K \frac{\boldsymbol{\gamma}(\boldsymbol{\xi}) \times (\mathbf{r}-\boldsymbol{\xi})}{4\pi|\mathbf{r}-\boldsymbol{\xi}|^3} dS_\xi - \frac{\boldsymbol{\gamma}(\mathbf{r}) \times \mathbf{n}(\mathbf{r})}{2} = -\left(\mathbf{V}_\infty + \mathbf{V}_\Omega(\mathbf{r})\right), \quad \mathbf{r} \in K. \tag{5}$$

It is proven in Kempka et al. [3], that in order to satisfy vector integral eqs (4) and (5), it is enough to satisfy the corresponding equations, being projected either onto surface normal unit vector or the tangential plane. According to the first approach, i.e., satisfying the boundary condition for normal velocity components, we obtain scalar boundary integral equation, while in the framework of projection onto the tangential plane the equation remains vectorial.

2 DIRECT RECONSTRUCTION OF THE DOUBLE LAYER POTENTIAL DENSITY

The most common approach, which is normally used in vortex methods, is the equation (4) projection onto normal unit vector, that leads to the hypersingular integral equation with respect to the double layer potential. Its approximate solution as a rule is considered to be piecewise-constant distribution of the double layer density on surface mesh cells. The efficient numerical algorithm for calculation of hypersingular integrals in sense of the Hadamard principal values is developed and proved by Lifanov et al. [2].

The idea of this algorithm is based on the fact, that the i-th polygonal panel with double layer potential density $g_i = \text{const}$ put exactly the same contribution $\mathbf{V}_\gamma^{(i)}$ to the velocity field given by (1) as a closed vortex filament with circulation $\Gamma_i = g_i$, placed on the panel circumfery. So it is possible to consider that the body surface is replaced by system of closed vortex lines, and the corresponding vorticity distribution is divergence-free, so it corresponds to the Helmholtz fundamental theorems [4]. A similar approach is used in Timofeev [5].

Numerical experiments show this approach works satisfactory for flows simulation around smooth bodies of rather simple shape, when the surface mesh is close to uniform. However, even in this case the directions of vortex line on the body surface is determined by the mesh topology, and can differ significantly from true vorticity surface distribution. This can lead to significant error in velocity field reconstruction in neighborhood of the body surface, that in turn leads to error in pressure distribution reconstruction. The mentioned problems can be partially overcome by representing of the surface vorticity distribution as a system of closed vortex filaments – vortex loops, which positions and circulations can be found according to [6]–[8] based on known double layer potential distribution.

Such approach works perfect, for example, in computer graphics applications [6], where it is enough to provide only qualitative results and high accuracy is not required. Its usage for flow simulation and hydrodynamic forces calculation is restricted, again, with rather simple body geometries and uniform meshes [7]. For essentially non-uniform meshes main problems are connected with the poor accuracy of the hypersingular boundary equation numerical solution and the large error, that can arise from re-interpolation procedure due to high cells area ratio.

3 VORTEX SHEET INTENSITY RECONSTRUCTION

The other way to the boundary condition satisfaction is developed in Kempka et al. [3] and it supposes the eqn (5) projection onto the tangential plane. Such approach leads to the vector boundary integral equation of the second kind,

$$\mathbf{n}(\mathbf{r})\times\left(\oint_K \frac{\boldsymbol{\gamma}(\boldsymbol{\xi})\times(\mathbf{r}-\boldsymbol{\xi})}{4\pi|\mathbf{r}-\boldsymbol{\xi}|^3}dS_\xi\times\mathbf{n}(\mathbf{r})\right)-\frac{\boldsymbol{\gamma}(\mathbf{r})\times\mathbf{n}(\mathbf{r})}{2}=\mathbf{f}(\mathbf{r}),\quad \mathbf{r}\in K, \tag{6}$$

where the right-hand side $\mathbf{f}(\mathbf{r})$ is known vector function, which in the case of immovable body surface depends on the vortex wake influence and the incident flow velocity:

$$\mathbf{f}(\mathbf{r})=-\mathbf{n}(\mathbf{r})\times\big((\mathbf{V}_\infty+\mathbf{V}_\Omega(\mathbf{r}))\times\mathbf{n}(\mathbf{r})\big).$$

Note, that the kernel of the equation (6) is unbounded when $|\mathbf{r}-\boldsymbol{\xi}|\to 0$.

In order to solve it numerically with high accuracy the following assumptions are introduced:

1. The body surface is discretized into N triangular panels K_i with areas A_i and unit normal vectors $\mathbf{n}_i$, $i=1,\dots,N$.
2. The unknown vortex sheet intensity on the i-th panel is assumed to be constant vector $\boldsymbol{\gamma}_i$, $i=1,\dots,N$, which lies in the plane of the i-th panel, i.e., $\boldsymbol{\gamma}_i\cdot\mathbf{n}_i=0$.
3. The integral equation (6) is satisfied on average over the panels, or, the same, in Galerkin sense: its residual is orthogonal to the basis functions, where the j-th basis function is equal to 1 on the j-th panel and equal to 0 on all other panels.

According to these assumptions, the discrete analogue of the eqn (6) can be derived:

$$\frac{1}{A_i}\sum_{j=1}^{N}\int_{K_i}\left(\int_{K_j}\left(\mathbf{n}_i\times\frac{\boldsymbol{\gamma}_j\times(\mathbf{r}-\boldsymbol{\xi})}{4\pi|\mathbf{r}-\boldsymbol{\xi}|^3}\right)dS_\xi\times\mathbf{n}_i\right)dS_r-\frac{\boldsymbol{\gamma}_i\times\mathbf{n}_i}{2}=\frac{1}{A_i}\int_{K_i}\mathbf{f}(\mathbf{r},t)dS_r,\quad i=1,\dots,N. \tag{7}$$

To write down the eqn (7) in the form of a linear algebraic system, we choose local orthonormal basis on every cell $(\boldsymbol{\tau}_i^{(1)},\boldsymbol{\tau}_i^{(2)},\mathbf{n}_i)$, where tangent vectors $\boldsymbol{\tau}_i^{(1)}$, $\boldsymbol{\tau}_i^{(2)}$ can be chosen arbitrary (in the plane of the cell) and $\boldsymbol{\tau}_i^{(1)}\times\boldsymbol{\tau}_i^{(2)}=\mathbf{n}_i$, so $\boldsymbol{\gamma}_i=\gamma_i^{(1)}\boldsymbol{\tau}_i^{(1)}+\gamma_i^{(2)}\boldsymbol{\tau}_i^{(2)}$, and we can project (7) for every i-th panel onto directions $\boldsymbol{\tau}_i^{(1)}$ and $\boldsymbol{\tau}_i^{(2)}$ [9]–[11].

Note, that the obtained algebraic system has infinite set of solutions; in order to select the unique solution, the additional condition for the total vorticity (the integral from the vortex sheet intensity over the body surface) should be satisfied:

$$\oint_K \boldsymbol{\gamma}(\mathbf{r},t)dS_r=\mathbf{0},$$

which can be easily written down in the discretized form.

The resulting algebraic system is overdetermined, so it should be regularized similarly to [2] by introducing the regularization vector $\mathbf{R}=(R_1,R_2,R_3)^T$:

$$\left.\begin{aligned}
&\frac{1}{A_i}\boldsymbol{\tau}_i^{(1)}\cdot\left(\sum_{j=1}^{N}\gamma_j^{(1)}\mathbf{v}_{ij}^{(1)}+\sum_{j=1}^{N}\gamma_j^{(2)}\mathbf{v}_{ij}^{(2)}\right)-\frac{\gamma_i^{(2)}}{2}+\mathbf{R}\cdot\boldsymbol{\tau}_i^{(2)}=\frac{b_i^{(1)}}{A_i},\\
&\frac{1}{A_i}\boldsymbol{\tau}_i^{(2)}\cdot\left(\sum_{j=1}^{N}\gamma_j^{(1)}\mathbf{v}_{ij}^{(1)}+\sum_{j=1}^{N}\gamma_j^{(2)}\mathbf{v}_{ij}^{(2)}\right)+\frac{\gamma_i^{(1)}}{2}+\mathbf{R}\cdot\boldsymbol{\tau}_i^{(1)}=\frac{b_i^{(2)}}{A_i},\quad i=1,\dots,N,\\
&\sum_{j=1}^{N}A_j\left(\gamma_j^{(1)}\boldsymbol{\tau}_j^{(1)}+\gamma_j^{(2)}\boldsymbol{\tau}_j^{(2)}\right)=0,
\end{aligned}\right\}\tag{8}$$

where the coefficients $\mathbf{v}_{ij}^{(k)}$ and b_i^k, $i,j=1,\dots,N$, $k=1,2$, are expressed as the following:

$$\mathbf{v}_{ij}^{(k)}=\int_{K_i}\left(\int_{K_j}\frac{\boldsymbol{\tau}_j^{(k)}\times(\mathbf{r}-\boldsymbol{\xi})}{4\pi|\mathbf{r}-\boldsymbol{\xi}|^3}dS_\xi\right)dS_r,\quad b_i^{(k)}=\int_{K_i}\boldsymbol{\tau}_i^{(k)}\cdot\mathbf{f}(\mathbf{r})dS_r.\tag{9}$$

Similar approach, based on the projection of the boundary integral equation onto the tangential direction, has been investigated in Kuzmina et al. [12] for 2D modifications of vortex methods, and it is shown, that it allows for significant improvement of the accuracy of numerical solution. It should be noted, that in 2D case the resulting linear algebraic system has $(N+1)\times(N+1)$ size (N is number of panels on the 2D airfoil surface line discretization; the regularization variable R is scalar), but in the discussed 3D case matrix size is twice as number of panels: $(2N+3)\times(2N+3)$.

In contrast to 2D case, when the exact analytic expressions can be derived [13] for all the coefficients of the linear system, which is an analogue of the system (8), for 3D case it is impossible to calculate the integrals in expressions (9), but it is possible to develop efficient semi-analytical algorithm [9]–[11], which provides rather high accuracy.

4 SEMI-ANALYTICAL TECHNIQUE FOR THE COEFFICIENTS CALCULATION

The coefficients $\mathbf{v}_{ij}^{(k)}$ of the linear system (8) should be calculated for all the pairs of the panels on the surface mesh; however note, that $\mathbf{v}_{ii}^{(k)}=\mathbf{0}$ due to the flatness of the panels, while their self-influence is taken into account through the second (non-integral) term the eqn (7), which corresponds to the terms $\pm\gamma_i^{(k)}/2$ in (8). In the case of $i\neq j$ let us denote:

$$\mathbf{v}_{ij}^{(k)} = \boldsymbol{\tau}_j^{(k)} \times \int_{K_i} \left(\int_{K_j} \frac{\mathbf{r}-\boldsymbol{\xi}}{4\pi\,|\mathbf{r}-\boldsymbol{\xi}|^3} dS_\xi \right) dS_r = \mathbf{I}_{ij} \times \boldsymbol{\tau}_j^{(k)}, \quad k=1,2, \quad i,j=1,\ldots,N. \tag{10}$$

The integral $\mathbf{I}_{ij}$ in (10) is calculated over the triangular panels K_i and K_j; the i-th panel we call hereinafter "control panel", and the k -th one – "influence panel".

The problem, that is very similar to calculation of the $\mathbf{I}_{ij}$ coefficients, arises in the boundary element method (BEM, [14]). There are some known approaches to approximate calculation of such integrals, for example the Taylor – Duffy method, but the corresponding numerical procedures have high computational cost [15].

The inner integral in (10), calculated over the influence panel K_j,

$$\mathbf{J}_j(\mathbf{r}) = -\int_{K_j} \frac{\mathbf{r}-\boldsymbol{\xi}}{4\pi\,|\mathbf{r}-\boldsymbol{\xi}|^3} dS_\xi\,, \tag{11}$$

can be calculated analytically using computer algebra system, such as *Wolfram Mathematica*, or some classic handbooks [16]. In order to write down the resulting formula for it, note that it is expressed through the vectors $\mathbf{s}_k = \mathbf{r}_k^{(j)} - \mathbf{r}$, $k=1,2,3$, where $\mathbf{r}$ is the position vector of the observation point, at which the integral (11) is calculated; $\mathbf{r}_k^{(j)}$ are position vectors of the vertices of the influence panel K_j. Introducing unit vectors (hereinafter we denote the unit vector, corresponding to arbitrary vector $\mathbf{r}$ as $\operatorname{ort}(\mathbf{r}) = \mathbf{r}/|\mathbf{r}|$),

$$\mathbf{e}_k^{(j)} = \operatorname{ort}(\mathbf{s}_{k+1} - \mathbf{s}_k) = \operatorname{ort}(\mathbf{r}_{k+1}^{(j)} - \mathbf{r}_k^{(j)}), \quad \boldsymbol{\sigma}_k = \operatorname{ort}(\mathbf{s}_k), \quad k=1,2,3,$$

and having in mind that all indices are calculated in modulo 3, the expression for (11) now is:

$$\mathbf{J}_j(\mathbf{r}) = \frac{1}{4\pi}(\Theta_j \mathbf{n}_j + \boldsymbol{\Psi}_j \times \mathbf{n}_j), \quad j=1,\ldots,N, \tag{12}$$

where Θ_j means the signed solid angle at which the panel K_j is observed, equal to [17]:

$$\Theta_j = 2\operatorname{atan}\left(\mathbf{s}_1\mathbf{s}_2\mathbf{s}_3,\ |\mathbf{s}_1|\cdot|\mathbf{s}_2|\cdot|\mathbf{s}_3| + (\mathbf{s}_1\cdot\mathbf{s}_2)|\mathbf{s}_3| + (\mathbf{s}_2\cdot\mathbf{s}_3)|\mathbf{s}_1| + (\mathbf{s}_3\cdot\mathbf{s}_1)|\mathbf{s}_2|\right),$$

where $\mathbf{s}_1\mathbf{s}_2\mathbf{s}_3$ means scalar triple product of the vectors. The function $\phi = \operatorname{atan}(y,x)$ hereinafter corresponds to usually used in programming languages `atan2(y, x)` function, which means in turn the argument of a complex number $z = x + iy$, or, the same, the angle from the interval $\phi \in (-\pi,\pi]$, for which $\sin\phi = y/\sqrt{x^2+y^2}$, $\cos\phi = x/\sqrt{x^2+y^2}$.

The vector-function $\boldsymbol{\Psi}_j$ has the form $\boldsymbol{\Psi}_j = \sum_{k=1}^{3} \ln \psi_k^{(j)} \mathbf{e}_k$, where

$$\psi_k^{(j)} = \frac{|\mathbf{s}_k|}{|\mathbf{s}_{k+1}|}\left(\frac{1+\mathbf{e}_k^{(j)}\cdot\boldsymbol{\sigma}_k}{1+\mathbf{e}_k^{(j)}\cdot\boldsymbol{\sigma}_{k+1}}\right)$$

except the case $\mathbf{e}_k^{(j)}\cdot\boldsymbol{\sigma}_k = \mathbf{e}_k^{(j)}\cdot\boldsymbol{\sigma}_{k+1} = -1$, for which the limit value is $\psi_k^{(j)} = \dfrac{|\mathbf{s}_{k+1}|}{|\mathbf{s}_k|}$.

4.1 The outer integral calculation in the general case

The outer integral in (10) over the control mesh cell,

$$\mathbf{I}_{ij} = \int_{K_i} \mathbf{J}_j(\mathbf{r}) dS_r, \tag{13}$$

cannot be analytically expressed through elementary functions in general case, so for the approximate computation of the vectors $\mathbf{I}_{ij}$ Gaussian quadrature formulae can be used:

$$\mathbf{I}_{ij} = \int_{K_i} \mathbf{J}_j(\mathbf{r}) dS_r \approx A_i \sum_{p=1}^{N_{GP}} \omega_p \mathbf{J}_j(\boldsymbol{\eta}_p), \tag{14}$$

where N_{GP} is number of Gaussian points; ω_p are weights; $\boldsymbol{\eta}_p$ are positions of the Gaussian points for triangular mesh cell [18].

The described approach for integrals $\mathbf{I}_{ij}$ computation is applicable, however, only for those pairs of the panels, which are non-neighboring, i.e., if they do not have common edges and vertices, because in the case of neighboring panels the integrand in (13) becomes unbounded in proximity to the common edge/vertex, and the outer integral itself become improper (but convergent). Its calculation with high precision straightforwardly by using quadrature formulae is difficult, so it is suggested to exclude the singularity from the integrand. We write down the integrand $\mathbf{J}_j(\mathbf{r})$ as a sum of two terms

$$\mathbf{J}_j(\mathbf{r}) = \mathbf{J}_j^{\text{reg}}(\mathbf{r}) + \mathbf{J}_j^{\text{sing}}(\mathbf{r}),$$

where the regular part

$$\mathbf{J}_j^{\text{reg}}(\mathbf{r}) = \frac{1}{4\pi}\Big(\big(\Theta_j(\mathbf{r}) - \Theta_j^{\text{sing}}(\mathbf{r})\big)\mathbf{n}_j + \big(\boldsymbol{\Psi}_j(\mathbf{r}) - \boldsymbol{\Psi}_j^{\text{sing}}(\mathbf{r})\big)\times \mathbf{n}_j\Big)$$

is a smooth function, which can be easily integrated numerically by using above mentioned Gaussian quadratures. For the singular part it is possible to integrate analytically the following terms:

$$\int_{K_i} \mathbf{J}_j^{\text{sing}}(\mathbf{r}) dS_r = \frac{1}{4\pi}\left(\left(\int_{K_i} \Theta_j^{\text{sing}}(\mathbf{r}) dS_r\right)\mathbf{n}_j + \left(\int_{K_i} \boldsymbol{\Psi}_j^{\text{sing}}(\mathbf{r}) dS_r\right)\times \mathbf{n}_j\right).$$

4.2 The outer integral calculation in the case of common edge

If the panels K_i and K_j have a common edge with unit direction vector $\mathbf{e}_3$ (Fig. 1), the singular terms have the form (the upper index (j) for the unit vectors $\mathbf{e}_k^{(j)}$ is omitted):

$$\Theta_j^{\text{sing}}(\mathbf{r}) = 2\sum_{k=1}^{2} \operatorname{atan}\big(\mathbf{a}_k \mathbf{e}_k \mathbf{e}_3,\ (\mathbf{e}_3 - \mathbf{e}_k)\cdot(\mathbf{a}_k - \mathbf{e}_k)\big),$$

$$\boldsymbol{\Psi}_j^{\text{sing}}(\mathbf{r}) = \mathbf{e}_3 \ln\left(\frac{m_1}{m_2}\frac{\mathbf{e}_3\cdot(\mathbf{e}_3 + \mathbf{a}_1)}{\mathbf{e}_3\cdot(\mathbf{e}_3 - \mathbf{a}_2)}\right) + \sum_{k=1}^{2} \mathbf{e}_k \ln\left(\frac{m_k}{m_3}\mathbf{e}_k\cdot(\mathbf{e}_k - \mathbf{a}_k)\right),$$

where $m_1 = |\mathbf{r} - \mathbf{r}_1^{(j)}|$, $m_2 = |\mathbf{r} - \mathbf{r}_3^{(j)}|$, $m_3 = |\mathbf{r}_1^{(j)} - \mathbf{r}_3^{(j)}|$, $\mathbf{a}_1 = \operatorname{ort}(\mathbf{r} - \mathbf{r}_1^{(j)})$, $\mathbf{a}_2 = -\operatorname{ort}(\mathbf{r} - \mathbf{r}_3^{(j)})$.

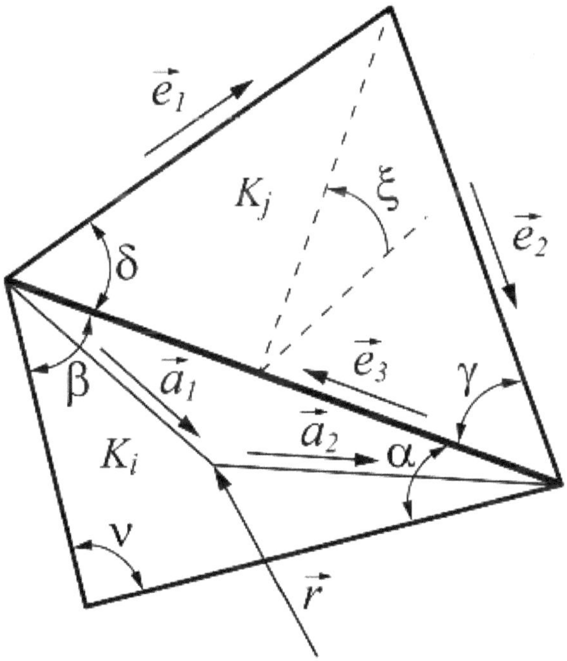

Figure 1: Mesh cells with common edge.

The expression for Θ_j^{sing}, as well as for scalar multipliers in $\mathbf{\Psi}_j^{\text{sing}}$, can be integrated analytically over the control panel K_i:

$$\int_{K_i} \Theta_j^{\text{sing}}(\mathbf{r})dS_r = -2A_i(q_0(\xi,\alpha,\beta,\mu,\gamma,\lambda) + q_0(\xi,\beta,\alpha,\sigma,\delta,\theta)),$$

$$\int_{K_i} \mathbf{\Psi}_j^{\text{sing}}(\mathbf{r})dS_r = A_i(q_{12}(\xi,\alpha,\beta,\mu,\gamma,\lambda)\mathbf{e}_1 + q_{12}(\xi,\beta,\alpha,\sigma,\delta,\theta)\mathbf{e}_2 + q_3(\alpha,\beta)\mathbf{e}_3).$$

The expressions for the auxiliary functions q_0, q_{12} and q_3 have the following form:

$$q_0(\xi,\alpha,\beta,\mu,\gamma,\lambda) = \phi^* + \frac{\sin\gamma\sin\nu}{\sin^2\mu\sin\alpha}\Bigg((\cos\beta\sin\gamma - \sin\beta\cos\gamma\cos\xi)\phi^{**} +$$

$$+\sin\xi\sin\beta\left(\cos^2\frac{\mu}{2}\ln\frac{\cos\beta/2}{\sin\nu/2} + \sin^2\frac{\mu}{2}\ln\frac{\sin\beta/2}{\cos\nu/2} + \ln\frac{\cos\lambda/2}{\sin\gamma/2}\right)\Bigg),$$

$$q_{12}(\xi,\alpha,\beta,\mu,\gamma,\lambda) = -\frac{3}{2} + \frac{1}{\sin\alpha\sin^2\mu}\Bigg(\sin\beta(\cos\nu + \cos\mu\cos\lambda)\ln(1+\cos\lambda) +$$

$$+\sin\nu(\cos\beta + \cos\gamma\cos\mu)\ln(1-\cos\gamma) + \sin\beta(1-\cos\mu)(\cos\nu - \cos\lambda)\ln\frac{\sin\beta}{\sin\nu} -$$

$$-\sin\nu\sin\beta\left(-2\phi^{**}\sin\xi\sin\gamma+(\sin\gamma\cos\beta\cos\xi-\sin\beta\cos\gamma)\ln\frac{1-\cos\nu}{1+\cos\beta}\right)\Bigg),$$

$$q_3(\alpha,\beta)=\frac{\sin\nu}{\sin\beta}\ln\left(\tan\frac{\alpha}{2}\tan\frac{\nu}{2}\right)+\frac{\sin\nu}{\sin\alpha}\ln\left(\tan\frac{\beta}{2}\tan\frac{\nu}{2}\right)+\ln\left(\tan\frac{\alpha}{2}\tan\frac{\beta}{2}\right).$$

Here α and β are angles of the control triangular cell K_i, which adjoin the common edge; $\nu=\pi-\alpha-\beta$; γ and δ are angles of the influence triangular cell K_j, which adjoin the common edge; ξ is the angle between the planes of the cells K_i and K_j (Fig. 1); the other angles are given by the expressions:

$$\sigma=\pi-\arccos(\cos\alpha\cos\delta+\cos\xi\sin\alpha\sin\delta),\qquad \mu=\pi-\arccos(\cos\beta\cos\gamma+\cos\xi\sin\beta\sin\gamma),$$

$$\lambda=\pi-\arccos(\cos\alpha\cos\gamma-\cos\xi\sin\alpha\sin\gamma),\qquad \theta=\pi-\arccos(\cos\beta\cos\delta-\cos\xi\sin\beta\sin\delta),$$

$$\phi^{*}=\mathrm{atan}(\sin\xi\sin\alpha\sin\gamma,\,1-\cos\alpha+\cos\gamma+\cos\lambda),\quad \phi^{**}=\mathrm{atan}(\sin\xi\sin\alpha\sin\gamma,\,1+\cos\alpha-\cos\gamma+\cos\lambda).$$

The given formulae are valid for arbitrary mutual configuration of the panels K_i and K_j, except the particular case $\xi=\beta-\gamma=0$, for which they lead to uncertainty of $0/0$ type. However, carrying out the passage to the limit, it is possible to obtain for this case rather simple expressions for q_0 and q_{12} functions:

$$q_0(\xi,\alpha,\beta,\mu,\gamma,\lambda)=\phi^{*},$$

$$q_{12}(\xi,\alpha,\beta,\mu,\gamma,\lambda)=-\frac{3}{2}+\frac{1}{2\sin\alpha}(\cos\nu\sin\beta\ln(1+\cos\nu)+\sin\nu\cos\beta\ln(1-\cos\beta)).$$

4.3 The outer integral calculation in the case of common vertex

The scheme, which illustrates the configuration of the panels having common vertex with introduced designations is shown in Fig. 2.

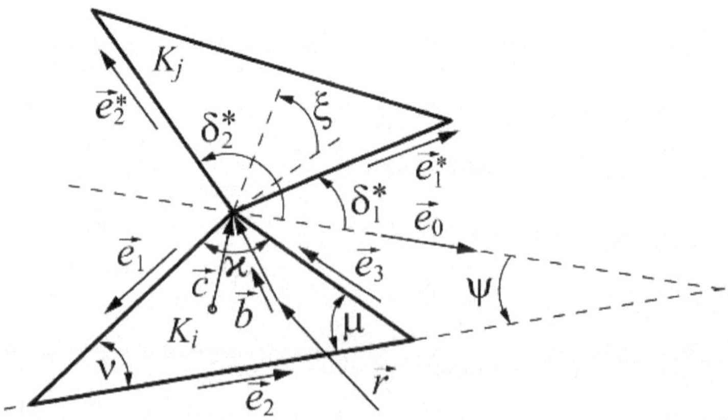

Figure 2: Mesh cells with common edge.

The triangular panels K_i and K_j lie in planes, which intersection line has direction unit vector $\mathbf{e}_0$, collinear to the vector $\mathbf{n}_i \times \mathbf{n}_j$. The direction of the vector $\mathbf{e}_0$ is chosen in such a way, that scalar triple product $\mathbf{e}_0\mathbf{c}\mathbf{n}_i > 0$, where vector $\mathbf{c}$ connects the center of the panel K_i with the common vertex. If the panels K_i and K_j are complanar, i.e., $\mathbf{n}_i \times \mathbf{n}_j = \mathbf{0}$, the direction of the unit vector $\mathbf{e}_0$ can be chosen arbitrary; to be more specific, we choose it as $\mathbf{e}_0 = -\mathbf{e}_1$ if $\mathbf{e}_2 \cdot \mathbf{e}_3 \le 0$ and $\mathbf{e}_0 = \mathbf{e}_3$ if $\mathbf{e}_2 \cdot \mathbf{e}_3 > 0$, where $\mathbf{e}_k$ are unit vectors, which are now directed along the edges of the control panel K_i under the assumption, that the common vertex has index $k = 1$ (i.e., its position vector is $\mathbf{r}_1^{(i)}$):

$$\mathbf{e}_k = \mathrm{ort}(\mathbf{r}_{k+1}^{(i)} - \mathbf{r}_k^{(i)}), \; k = 1,2,3 .$$

The orientation of the influence panel K_j is defined by the unit vectors

$$\mathbf{e}_1^* = \mathrm{ort}(\mathbf{r}_2^{(j)} - \mathbf{r}_1^{(j)}), \; \mathbf{e}_2^* = \mathrm{ort}(\mathbf{r}_3^{(j)} - \mathbf{r}_1^{(j)}),$$

and the angles:

$$\delta_k^* = \angle(\mathbf{e}_0, \mathbf{e}_k^*) = \mathrm{atan}(\mathbf{e}_0\mathbf{e}_k^*\mathbf{n}_j, \; \mathbf{e}_0 \cdot \mathbf{e}_k^*), \; k = 1,2 .$$

In particular cases when $\delta_1^* = \pi$ or $\delta_2^* = \pi$ the direction of the vector $\mathbf{e}_0$ should be reversed, and the angles δ_k^* should be recalculated for its new orientation. The same operations should be performed in the case, when $\delta_1^*\delta_2^* < 0$ and $|\delta_1^* - \delta_2^*| > \pi$ simultaneously.

Since the vector $\mathbf{e}_0$ is defined, the signed angle between the planes K_i and K_j is:

$$\xi = \mathrm{atan}(\mathbf{n}_i\mathbf{n}_j\mathbf{e}_0, \; \mathbf{n}_i \cdot \mathbf{n}_j),$$

its sign defines the mutual orientation of the triangular cells.

Considering the singular part $\mathbf{J}_j^{\mathrm{sing}}(\mathbf{r})$ of

$$\mathbf{J}_j(\mathbf{r}) = \frac{1}{4\pi}\left(\Theta_j \mathbf{n}_j + \boldsymbol{\Psi}_j \times \mathbf{n}_j\right),$$

we write down the singular parts of its terms. For the first term

$$\Theta_j^{\mathrm{sing}}(\mathbf{r}) = -2(\Theta_2^* - \Theta_1^* + \tilde{\Theta}),$$

where Θ_1^* and Θ_2^* are expressed through vectors $\mathbf{u}_k^* = \mathbf{e}_0 + \mathbf{e}_k^*$, $\mathbf{v}_k^* = (\mathbf{e}_0 + \mathbf{e}_k^*) \times \mathbf{e}_0$ and the unit vector $\mathbf{b} = \mathrm{ort}(\mathbf{r}_1^{(i)} - \mathbf{r})$:

$$\Theta_k^* = \mathrm{atan}(\mathbf{p}_k \cdot \mathbf{v}_k^*, \; \mathbf{p}_k \cdot \mathbf{u}_k^*),$$

where

$$\mathbf{p}_k = ((\mathbf{b} + \mathbf{e}_0) \cdot \mathbf{u}_k^*)\mathbf{u}_k^* + ((\mathbf{b} + \mathbf{e}_0) \cdot \mathbf{v}_k^*)\mathbf{v}_k^*, \; k = 1,2 .$$

For the particular case at $|\mathbf{p}_k|=0$ the limit value is $\Theta_k^* = \mathrm{atan}(\mathbf{b}\cdot\mathbf{n}_j,\ \mathbf{bn}_j\mathbf{e}_0)$. The term $\tilde{\Theta}=0$ at $|\mathbf{p}_k|\neq 0$ and $\tilde{\Theta}=\pi\mathrm{sign}\xi$ at $|\mathbf{p}_k|=0$.

For the singular part of the second term in $\mathbf{J}_j(\mathbf{r})$ we obtain

$$\mathbf{\Psi}_j^{\mathrm{sing}}(\mathbf{r}) = -(\Psi_2^* - \Psi_1^*),$$

where

$$\Psi_k^* = \mathbf{e}_k^* \ln\left(\frac{m_b}{\sqrt{A_j}}\mathbf{e}_k^*\cdot(\mathbf{b}+\mathbf{e}_k^*)\right),\quad m_b = |\mathbf{r}_1^{(i)} - \mathbf{r}|.$$

As for the panels with common edge, the singular parts $\Theta_j^{\mathrm{sing}}(\mathbf{r})$ and $\mathbf{\Psi}_j^{\mathrm{sing}}(\mathbf{r})$ can be integrated analytically over the control mesh cell K_i:

$$\int_{K_i}\Theta_j^{\mathrm{sing}}(\mathbf{r})dS_r = -2A_i\left(q_4(\delta_2^*) - q_4(\delta_1^*) + \tilde{q}q_4(0)\right),$$

$$\int_{K_i}\mathbf{\Psi}_j^{\mathrm{sing}}(\mathbf{r})dS_r = -A_i\left(q_5(\delta_2^*)\mathbf{e}_2^* - q_5(\delta_1^*)\mathbf{e}_1^*\right),$$

where the coefficient $\tilde{q}=0$ at $\delta_1^*\delta_2^*\geq 0$ and $\tilde{q}=-2\mathrm{sign}\xi$ at $\delta_1^*\delta_2^*<0$, and the auxiliary functions q_4 and q_5 are the following:

$$q_4(\delta) = \frac{q_{41}(\delta) - q_{42}(\delta)}{\sin\psi\sin\kappa},\quad q_5(\delta) = q_{51}(\delta) - q_{52}(\delta).$$

The expressions for their terms are rather complicated, but they are expressed through elementary functions:

$$q_{41}(\delta) = \sin\mu\sin(\nu+\psi)\mathrm{atan}\left(\sin\xi\sin\frac{\delta}{2},\ \cos\xi\sin\frac{\delta}{2} + \cos\frac{\delta}{2}\cot\frac{\nu+\psi}{2}\right) -$$

$$-\sin\nu\sin(\mu-\psi)\mathrm{atan}\left(\sin\xi\sin\frac{\delta}{2},\ \cos\xi\sin\frac{\delta}{2} + \cos\frac{\delta}{2}\tan\frac{\mu-\psi}{2}\right),$$

$$q_{42}(\delta) = \frac{\sin\mu\sin\nu\sin\delta}{D}\left(\omega\cos\eta + \frac{\sin\psi\sin\xi}{2}\left(\ln\left(\frac{1+\cos\lambda}{1+\cos\theta}\frac{\sin\nu}{\sin\mu}\right) - \cos\sigma\ln\left(\tan\frac{\nu}{2}\tan\frac{\mu}{2}\right)\right)\right),$$

$$q_{51}(\delta) = -\frac{3-\ln 2}{2} + \frac{\sin\mu\sin(\nu+\psi)\ln(1+\cos\theta) - \sin\nu\sin(\mu-\psi)\ln(1+\cos\lambda)}{\sin\kappa\sin\psi} +$$

$$+\frac{1}{2}\ln\left(\frac{\sin\nu\sin\mu}{\sin\kappa}\right) - \frac{\cos\nu\sin\mu}{\sin\kappa}\ln(\sin\nu) - \frac{\cos\mu\sin\nu}{\sin\kappa}\ln(\sin\mu),$$

$$q_{52}(\delta) = \frac{\sin\nu\sin\mu}{D\sin\kappa}\left(\frac{\sin\delta\cos\eta}{\sin\psi}\ln\frac{1+\cos\theta}{1+\cos\lambda} - \cos\chi\ln\left(\tan\frac{\nu}{2}\tan\frac{\mu}{2}\right) + \right.$$

$$\left. +2\omega\sin\delta\sin\xi + \frac{1}{2}G\ln\frac{\sin\nu}{\sin\mu}\right).$$

Here the following designations are introduced:

$$D = \sin^2(\delta - \psi) + 4\sin\delta\sin\psi\left(\sin\delta\sin\psi\cos^2\frac{\xi}{2} + \cos\delta\cos\psi\right)\sin^2\frac{\xi}{2},$$

$$G = \sin 2(\delta - \psi) - 4\sin\delta\left(\sin\delta\sin 2\psi\cos^2\frac{\xi}{2} + \cos\delta\cos 2\psi\right)\sin^2\frac{\xi}{2},$$

$$\omega = \operatorname{atan}\left(\sin\xi\sin\delta\sin\frac{\kappa}{2},\ \cos\xi\sin\delta\sin\left(\nu + \psi + \frac{\kappa}{2}\right) + \cos\frac{\kappa}{2} - \cos\delta\cos\left(\mu - \psi + \frac{\kappa}{2}\right)\right),$$

value κ means the angle of the triangular cell K_i which adjoins to the common vertex of the panels K_i and K_j; μ and ν the other angles of the cell K_i; ξ as earlier is the signed angle between the planes of the triangles K_i and K_j; ψ is the angle between the vector $\mathbf{e}_0$ and the side of the triangle K_i, which is opposite to the common vertex; δ_1^* and δ_2^* are the angles between the vector $\mathbf{e}_0$ and vectors $\mathbf{e}_1^*$ and $\mathbf{e}_2^*$, respectively (Fig. 2); for $\cos\sigma$, $\cos\lambda$, $\cos\theta$, $\cos\eta$ and $\cos\chi$ the following expressions are valid:

$$\cos\sigma = \sin\delta\sin\psi\cos\xi + \cos\delta\cos\psi, \qquad \cos\theta = \sin\delta\sin(\nu + \psi)\cos\xi + \cos\delta\cos(\nu + \psi),$$
$$\cos\chi = \sin\delta\cos\psi\cos\xi - \cos\delta\sin\psi, \qquad \cos\lambda = \sin\delta\sin(\mu - \psi)\cos\xi - \cos\delta\cos(\mu - \psi),$$
$$\cos\eta = \cos\delta\sin\psi\cos\xi - \sin\delta\cos\psi.$$

These formulae for the integrals from the singular part $\mathbf{\Psi}_j^{\text{sing}}(\mathbf{r})$ remain correct for arbitrary mutual configuration of the triangular cells with common vertex except the case of $\xi = \sin(\delta - \psi) = 0$. For this particular case:

$$q_{42}(\delta) = 0, \quad q_{52}(\delta) = \frac{1}{2}\left(\sin\psi\left(\frac{1 + \cos\nu}{\sin\nu} - \frac{\sin\mu}{1 + \cos\mu}\right) - \cos\psi\ln\frac{1 - \cos\nu}{1 + \cos\mu}\right).$$

5 DOUBLE LAYER POTENTIAL RECONSTRUCTION

After solution of the linear system (8) with respect to cell values of the vortex sheet intensity, it is important to provide divergence-free reconstruction of the solution. That means, that the numerical solution of (8) in general case doesn't satisfy the following condition, which is analogue of the condition of zero total vorticity flux $\sum_{q=1}^{3}(\boldsymbol{\gamma}_i - \boldsymbol{\gamma}_i^{\text{neib},q})l_q$, where $\boldsymbol{\gamma}_i^{\text{neib},q}$ is the vortex sheet intensity on the panel, which is neighbor to the i-th cell against the q-th edge; l_q is the length of the q-th edge.

In order to satisfy this condition and provide divergence-free numerical solution, it is suggested firstly to reconstruct the double layer density distribution on the body surface, and secondly calculate its surface gradient. Moreover, if the double layer density distribution is known, it is easy to represent vorticity distribution on the body surface as a set vortex loops, that have been mentioned in Section 2.

The solution of the linear system (8) corresponds to piecewise-constant vortex sheet intensity distribution over the body surface. From the other side, vortex sheet intensity is surface gradient of the double layer potential density, multiplied by normal unit vector. It means, that the most convenient way to double layer potential density reconstruction is its approximation by a continuous function, which is piecewise-linear with respect to all coordinates on every panel. To do it, we consider the nodal values of the potential g_j, $j=1,\dots,M$ to be unknown; M is number of vertices of the surface mesh. Then the potential density can be represented as FEM-type interpolation using the 1-st order shape functions. Let us note positions of all the vertices of the surface mesh as $\boldsymbol{\rho}_j$, $j=1,\dots,M$, and consider that the vertices of the i-th triangular panel have indices p_i^k, $k=1,2,3$. So the shape functions, defined on the i-th panel, coincide with barycentric coordinates on the mesh cell:

$$\phi_i^{(k)}(\boldsymbol{\rho}) = \frac{|(\boldsymbol{\rho}_{p_i^l} - \boldsymbol{\rho}) \times (\boldsymbol{\rho}_{p_i^m} - \boldsymbol{\rho})|}{|(\boldsymbol{\rho}_{p_i^l} - \boldsymbol{\rho}_{p_i^k}) \times (\boldsymbol{\rho}_{p_i^l} - \boldsymbol{\rho}_{p_i^k})|}, \quad \boldsymbol{\rho} \in K_i,$$

where $(k,l,m) = (1,2,3)$, or $(2,3,1)$ or $(3,1,2)$.

Then the double layer density on the *i*-th panel is linear function with respect to $\boldsymbol{\rho}$ and has the following form:

$$g(\boldsymbol{\rho}) = \sum_{k=1}^{3} g_{p_i^k} \phi_i^{(k)}(\boldsymbol{\rho}), \quad \boldsymbol{\rho} \in K_i.$$

The surface gradient of this distribution of the double layer on the every i-th panel, multiplied by unit normal vector $\mathbf{n}_i$ gives thus the constant vector, which physical sense is approximate value of divergence-free vortex sheet intensity on the corresponding panel:

$$\boldsymbol{\gamma}_i^* = -\sum_{k=1}^{3} g_{p_i^k} (\mathrm{Grad}\,\phi_i^{(k)} \times \mathbf{n}_i), \quad \boldsymbol{\rho} \in K_i,$$

where the surface gradients of the shape functions $\mathrm{Grad}\,\phi_i^{(k)}$ are constant vectors.

Now the unknown values g_j can be found from the least-squares procedure:

$$\Psi = \sum_{i=1}^{N} |\boldsymbol{\gamma}_i - \boldsymbol{\gamma}_i^*|^2 \to \min.$$

Taking partial derivatives of Ψ with respect to g_j, $j=1,\dots,M$, and making them equal to zero, we obtain linear system of $M \times M$ size with symmetric, but not positively-defined matrix. Moreover, it is clear, that this system is singular (in practice, due to the truncation errors, it can be non-singular, but ill-conditioned), that follows from the fact, that the potential density value can be chosen arbitrary at some specified point. To be more specific, we assume $g_M = 0$, that means that the last row and last column in the least-squares matrix should be nullified, the diagonal coefficient can be chosen arbitrary (non-zero) and the last coefficient in the right-hand side also should be nullified. The resulting regularized matrix is now symmetric and positively defined.

6 NUMERICAL RESULTS

In order to estimate the accuracy of the developed technique of singularity exclusion and its analytical integration, some model problems were considered for the bodies, shown in Fig. 3. For all the bodies (sphere, wing of the finite span, complex-shaped body) unstructured essentially non-uniform triangular mesh was generated by using preprocessor SALOME and NETGEN algorithm. Some parameters of the surface meshes are given in the Table 1.

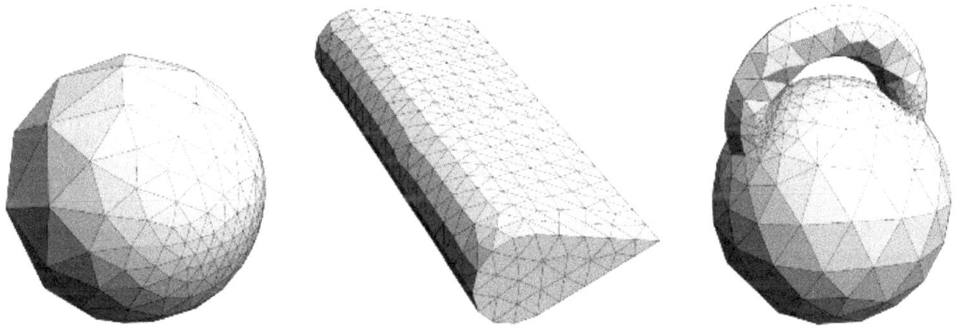

Figure 3: Sphere, wing of the finite span, complex-shaped body and the surface triangular meshes on their surfaces.

Table 1: Mesh parameters and relative errors for the coefficients $\mathbf{I}_{ij}$ calculation.

	Sphere	Wing	Complex-shaped body
Number of cells	342	742	636
Max cells area ratio	33.9	2.9	1007.9
Max cells area ratio for neighbouring cells	3.4	2.9	9.3
Max cells sides ratio	2.0	1.7	2.4
Min/max cell angle	30.2°/98.7°	35.6°/91.9°	22.4°/122.6°
δ, method 1	0.40	2.28	0.65
δ, method 2	0.054	0.053	0.052
δ, method 3	0.00092	0.0018	0.00098

The exact values of the $\mathbf{I}_{ij}$ coefficients are unknown, so instead of exact ones we considered approximate numerical values of the corresponding integrals (13), calculated by using computer algebra system Wolfram Mathematica to be reference values.

In Table 1 the values of the maximal relative error

$$\delta = \max_i \left(\frac{\max_j |\mathbf{I}_{ij} - \mathbf{I}_{ij}^*|}{\max_j |\mathbf{I}_{ij}^*|} \right), \quad i,j = 1,\ldots,N,$$

are given for three different methods of the integrals (13) calculation:

1. By using formula (13), where for $\mathbf{J}_j(\mathbf{r})$ calculation in (11) also the Gaussian quadrature formulae are used;
2. According to formulae (13), where the inner integral $\mathbf{J}_j(\mathbf{r})$ is calculated analytically (exactly) using formula (12);
3. Semi-analytically according to the developed technique of singularity exclusion and its analytical integrating for the mesh cells having common edge or common vertex.

It is seen, that the developed technique allows for significant improvement of the accuracy of the boundary integral equation approximation.

7 CONCLUSION

A procedure of the vortex sheet intensity reconstruction is proposed, which consists of two steps. Firstly, the boundary integral equation of the second order is solved by using Galerkin approach, which expresses the equality between the tangential components of flow velocity limit value and the body surface velocity. Secondly, the least-squares procedure is implemented, which permits one to find nodal values of the double layer potential and then reconstruct divergence-free vortex sheet intensity distribution.

For the coeffcients of the linear system, which approximates the boundary integral equation, a semi-analytical approach is suggested, which allows for their calculation with high accuracy. It is shown that the developed algorithm permits one to improve significantly the accuracy of solution for the complex-shaped bodies with low-quality surface meshes.

This procedure be used in vortex methods implementations where it is necessary to simulate vorticity generation process on the body surface at every time step.

ACKNOWLEDGEMENT

The research was supported by the Russian Science Foundation (grant No. 17-79-20445).

REFERENCES

[1] Cottet, G.-H. & Koumoutsakos, P., *Vortex Methods: Theory and Practice*, Cambridge: Cambridge University Press, 2000.

[2] Lifanov, I.K., Poltavskii, L.N. & Vainikko, G.M., *Hypersingular Integral Equations and their Applications*, Boca Raton: CRC Press, 2003.

[3] Kempka, S.N., Glass, M.W., Peery, J.S., Strickland, J.H. & Ingber, M.S., Accuracy consideration for implementing velocity boundary conditions in vorticity formulations. *SANDIA REPORT,* SAND96-0583 UC-700, 52 p., 1996.

[4] Saffman, P.G., *Vortex Dynamics.* Cambridge: Cambridge University Press, 1992.

[5] Timofeev, V.N., Mathematical simulation of the subsonic flow around the lengthening bodies with the flow separation in the region of ground shear with the use of an equivalent body. *Journal of Physics: Conference Series*, **1141**, art. 012095, 2018.

[6] Weissmann, S. & Pinkall, U., Filament-based smoke with vortex shedding and variational reconnection. *ACM Trans. Graph.* **29**(4), pp. 1, 2010.

[7] Shcheglov, G.A. & Dergachev, S.A. Hydrodynamic loads simulation for 3D bluff bodies by using the vortex loops based modification of the vortex particle method. *Proc. of V Intern. Conf. on Particle-based Methods PARTICLES 2017*, pp. 725–731, 2017.

[8] Shcheglov, G.A. & Dergachev, S.A., Vortex loops based method for subsonic aerodynamic loads calculation. *3rd International Conference on Design and Manufacturing Engineering*, **221**, art. 05004, 2018.

[9] Marchevsky, I.K. & Shcheglov, G.A., Efficient semi-analytical integration of vortex sheet influence in 3d vortex method. *Proceedings of the V International Conference on Particle-based Methods PARTICLES 2017,* pp. 703–714, 2017.

[10] Marchevsky, I.K. & Shcheglov, G.A., Semi-analytical influence computation for vortex sheet with piecewise constant intensity distribution in 3D vortex methods. *Proceedings of the 6th European Conference on Computational Mechanics (ECCM 6), 7th European Conference on Computational Fluid Dynamics (ECFD 7)*, 2018.

www.eccm-ecfd2018.org/admin/files/filePaper/p2059.pdf. Accessed on: 10 Mar. 2019.

[11] Marchevsky, I., Dergachev, S.A. & Shcheglov, G.A., New modification of 3D meshless Lagrangian vortex method with improved boundary condition satisfaction and divergence-free vorticity representation. *Proceedings Topical Problems of Fluid Mechanics 2019*, pp. 151–160, 2019.

[12] Kuzmina, K.S., Marchevskii, I.K. & Moreva, V.S., Vortex sheet intensity computation in incompressible flow simulation around an airfoil by using vortex methods,. *Mathematical Models and Computer Simulations*, **10**(3), pp. 276–287, 2018.

[13] Kuzmina, K.S., Marchevsky, I.K. & Ryatina, E.P., Exact analytical formulae for linearly distributed vortex and source sheets influence computation in 2D vortex methods. *Journal of Physics: Conference Series*, **918**(1), art. 012013, 2017.

[14] Brebbia, C.A., Telles, J.C.F. & Wrobel, L.C., (eds), *Boundary Element Techniques*, Berlin and New York: Springer-Verlag, 1984.

[15] Reid, M.T.H, White, J.K. & Johnson, S.G., Generalized Taylor-Duffy method for efficient evaluation of galerkin integrals in boundary-element method computations. *IEEE Transactions on Antennas and Propagation*, **63**(1), pp. 195–209, 2015.

[16] Gradshteyn, I.S., Ryzhik, I.M., Geronimus, Y.V., Tseytlin, M.Y. & Jeffrey, A., *Table of Integrals, Series, and Products*. Academic Press, 2015.

[17] Van Oosterom, A. & Strackee, J. The solid angle of a plane triangle. *IEEE Transactions on Biomedical Engineering*, **BME-30**, pp. 125–126, 1983.

[18] Zienkiewicz, O.C. & Taylor, L.R. *The Finite Element Method. Volume 1: The Basis*. Butterworth-Heinemann, 2000.

NEW BOUNDARY ELEMENT FORMULATION FOR THE SOLUTION OF LAPLACE'S EQUATION

JACQUES LOBRY
Department of General Physics, Faculty of Engineering, University of Mons, Belgium

ABSTRACT

The boundary element method (BEM) is typically used for solving potential problems and has several advantages over the traditional finite element method (FEM). However, the normal derivative of the potential appears explicitly as an unknown, with some inherent ambiguity at the corner nodes. Moreover, the BEM requires time consuming numerical integration (Gaussian quadrature) along the boundary of the domain. In this paper, we propose a new approach of BEM by introducing the weak nodal "cap" flux approach defined in the context of the finite element method. The domain integrals are eliminated at the discrete level by introducing the FEM approximation of the fundamental solutions at every node of the related mesh as basic functions in the Galerkin formulation of the BVP under study. The implementation of this new technique appears to be simpler as no numerical integration on the boundary of the domain is required so that the method leads to a substantially reduced computational burden. Our method is compared to the classical BEM for the numerical solution of the two-dimensional Laplace equation. It is observed that the normal flux presents a better behaviour at corners. A loss of accuracy may occur but it is compensated by a smaller execution time, allowing a finer mesh.
Keywords: finite element method, boundary element method, green functions, electrostatics.

1 INTRODUCTION

The numerical solution of potential boundary value problems (BVP) is traditionally obtained from the finite element method (FEM) or the boundary element method (BEM) that are well documented in the literature as e.g. [1] and [2]. BEM has several advantages over the FEM such as a discretization of the surface rather than the volume, consequently reducing the number of unknown values and the ability of taking into account domains extending to infinity. This method has proved to give very accurate results in particular in case of stress concentration or with re-entrant corner configurations. Conversely to the FEM, the normal derivative of the potential appears explicitly as an unknown value. It is discretized in the same way, e.g. linearly, as the potential along the boundary but appears ill-defined at the corner nodes of the boundary mesh. In order to avoid this problem, the nodes of the two linear elements meeting at the corner can be shifted inside the two elements. Yet it appears as a trick to solve the ambiguity of the flux definition.

In this paper, we propose a new BEM formulation based on the FEM solution of fundamental solutions (Green's functions) [3], [4] and on a rational definition of nodal flux, as presented in the paper of Bossavit [5]. The domain integrals are eliminated by combining the Galerkin formulation of the BVP under study and the one associated to the fundamental solution for Dirac delta loading every node of the related mesh. This is in contrast with the standard BEM that exploit Green's second identity. In order to avoid a huge time consuming FEM computation of nodal Green's functions, which would lead to an inapplicability of the method for open boundary problem anyway, their exact value is used instead as they are known analytically. The main difference with the classical BEM is that finite domain elements connected to the boundary are needed in order to well define the nodal fluxes. This approach leads to a linear system of equations with a fully populated matrix similarly to BEM. The complexity of the method remains as $O(n^2)$ for the assembly of the system but the implementation is simpler as no numerical integration (Gaussian quadrature) on the

WIT Transactions on Engineering Sciences, Vol 126, © 2019 WIT Press
www.witpress.com, ISSN 1743-3533 (on-line)
doi:10.2495/BE420061

boundary of the domain is required. Our method is compared to both FEM and BEM for the numerical solution of the two-dimensional Laplace equation.

2 MATHEMATICAL FORMULATION

2.1 The classical BEM

The boundary element method has been used for many years for the solution of 2-D and 3-D engineering problems with some advantages over the finite element method such as a reduced mesh task or a natural open problem treatment capability. In BEM, only the boundary of the domain is discretized and the degrees of freedom are typically potential and flux nodal values that are interpolated over the boundary elements. A system of linear equations is derived for the unknown functions at the boundary points and the potential values at internal points in the domain are post-processed. Let us recall the classical boundary formulation for the Laplace equation defined on a 2-D domain Ω ($\partial\Omega = \Gamma_u \cup \Gamma_n$) as illustrated in Fig. 1(a).

$$\nabla^2 u = 0 \text{ in } \Omega, \text{ with } u = u_0 \text{ on } \Gamma_u \text{ and } \frac{\partial u}{\partial n} = 0 \text{ on } \Gamma_n. \tag{1}$$

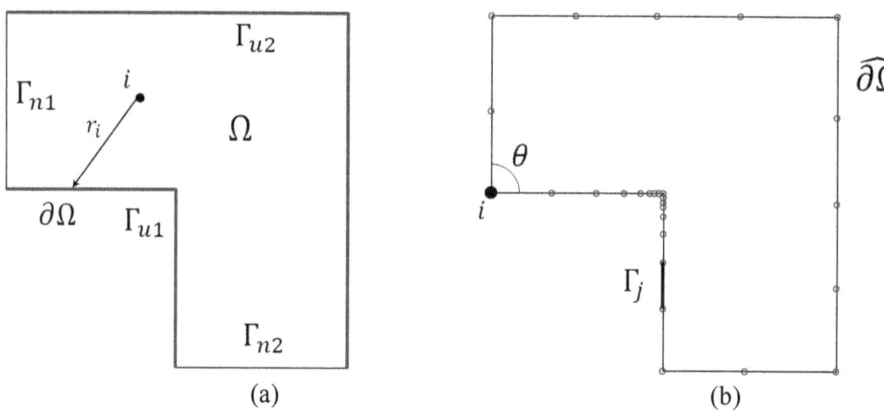

Figure 1: (a) Typical 2-D domain Ω for the Laplace problem; and (b) Associated boundary element discretization.

In order to establish the formulation, the fundamental solution $\mathcal{G}_i$ regarding the governing equation is employed. It is defined from the Dirichlet problem:

$$\nabla^2 \mathcal{G}_i = -\delta_i \text{ in } \Omega, \quad \text{with } \mathcal{G}_i = \frac{1}{2\pi} \ln \frac{1}{r_i} \text{ on } \partial\Omega, \tag{2}$$

where δ_i is the Dirac delta function at any point i, assumed internal to begin with. Then the domain integrals are eliminated using Green's second identity so that there is obtained:

$$u_i = \oint_{\partial\Omega} \left(u \frac{\partial \mathcal{G}_i}{\partial n} - \mathcal{G}_i \frac{\partial u}{\partial n} \right) d\Gamma. \tag{3}$$

The discretization $\widehat{\partial\Omega}$ of the boundary $\partial\Omega$ into elements Γ_j (Fig. 1(b)) leads to:

$$\hat{u}_i = \sum_{j \in \widehat{\partial\Omega}} \int_{\Gamma_j} \frac{\partial \mathcal{G}_i}{\partial n} \hat{u} \, d\Gamma - \sum_{j \in \widehat{\partial\Omega}} \int_{\Gamma_j} \mathcal{G}_i \hat{q} \, d\Gamma, \tag{4}$$

where $\hat{q}$ denotes the approximate normal derivative. In order to assembly a consistent system of equations, eqn (4) must be written by choosing every node i of the boundary $\widehat{\partial\Omega}$ as the Dirac point. In this situation, a geometric factor c_i is affected to the left-hand side of (4) so that:

$$c_i\, \hat{u}_i = \sum_{j\in\widehat{\partial\Omega}} \int_{\Gamma_j} \frac{\partial \mathcal{G}_i}{\partial n}\, \hat{u}\, d\Gamma - \sum_{j\in\widehat{\partial\Omega}} \int_{\Gamma_j} \mathcal{G}_i\, \hat{q}\, d\Gamma. \tag{5}$$

It is equal to the relative part of the Dirac density embedded in the domain Ω at each node i, i.e. the ratio $\theta/2\pi$ (θ is the internal angle at node i – see Fig. 1(b)). This factor is e.g. equal to ½ in case of a smooth boundary and 1 at internal nodes. The boundary potential $\hat{u}$ and flux $\hat{q}$ are typically interpolated linearly on every element Γ_j. The various integrals are either solved analytically or numerically using a standard Gaussian quadrature, depending on whether the node i belongs or not to the element Γ_j. After prescribing the boundary conditions, the procedure leads to a fully populated linear system of equations of which the solution consists of the unknown nodal values $\hat{u}_i$ and $\hat{q}_i$.

The computational effort to construct the matrix is an important part in BEM by contrast with FEM where it is relatively inexpensive to assemble. It scales as $\mathcal{O}(n^2)$ where n is the size of the boundary mesh.

2.2 A new BEM formulation

The finite element analysis of the Laplace problem (1) is based on the Galerkin problem:

$$\int_{\hat{\Omega}} \nabla\hat{u}\, \nabla N_j = 0 \quad \forall\, j \in \hat{\Omega} \setminus \hat{\Gamma}_u, \quad \hat{u} = u_0 \text{ on } \Gamma_u, \tag{6}$$

where the N_j's are the classical nodal basis "hat" functions defined on a mesh $\hat{\Omega}$ of the domain Ω (Fig. 2). The solution u is approached by:

$$\hat{u} = \sum_{k\in\hat{\Omega}} N_k\, \hat{u}_k\, . \tag{7}$$

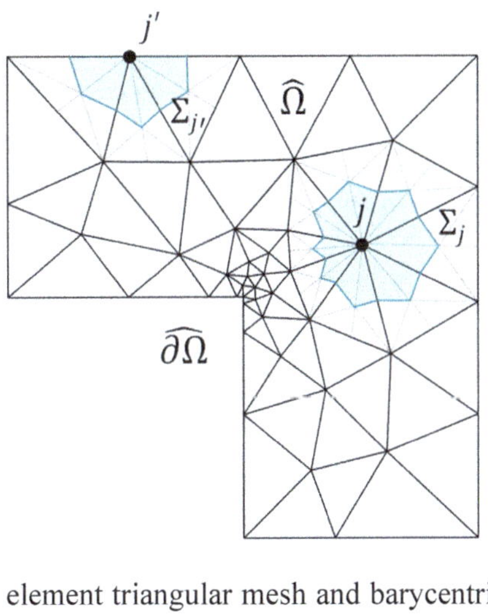

Figure 2: Typical finite element triangular mesh and barycentric boxes Σ_j ("cap" $\Sigma_{j'}$) for an internal (boundary) node j (j').

It leads to a sparse system of linear equations that involves the computation of the integrals of the form:

$$s_{jk} = \int_{\widehat{\Omega}} \nabla N_j \, \nabla N_k \, d\Omega. \tag{8}$$

On the other hand, in the paper of Bossavit [5], the term:

$$\int_{\widehat{\Omega}} \nabla \hat{u} \, \nabla N_j \, d\Omega = \sum_{k \in \widehat{\Omega}} s_{jk} \, \hat{u}_k = \Phi_{\hat{u},\Sigma_j} \quad \forall \, j \in \widehat{\Omega} \tag{9}$$

is shown to be equal to the inward flux of $\nabla \hat{u}$ across the "box" Σ_j associated to node j in the dual mesh obtained from the barycentric subdivision of the primal mesh as shown in Fig. 2. Those weak "box" fluxes vanish for nodes $j \in \partial\Omega \setminus \Gamma_u$ due to eqn (6) and are interpreted as a part of the flux of $\nabla \hat{u}$ through "caps" associated to Dirichlet boundary nodes like $j' \in \Gamma_u$. This gives rise to the best ("variationally" correct) estimate of surface charges in electrostatics for e.g. the direct computation of capacitance. In this case, the result is shown to be exactly the same as the approximate energy value $\hat{\mathbf{u}}^t \mathbf{S} \, \hat{\mathbf{u}}$, where $\mathbf{S} = (s_{jk})$ is the stiffness matrix of the problem. In the following, the parameters Φ_{Σ_j} are used in place of the approximation of the normal derivative $\hat{q}$ given by the standard BEM.

In order to eliminate the domain integral, the 2-D fundamental solution $\mathcal{G}_i$ of Laplace's equation is employed, as in the standard BEM. In spite of the singularity of the solution, i.e. $\mathcal{G}_i$ lies outside Sobolev space H^1, a FEM solution exists [3], [4] and is derived from the Galerkin problem associated to eqn (2):

$$\int_{\widehat{\Omega}} \nabla \hat{\mathcal{G}}_i \, \nabla N_j \, d\Omega = \int_{\widehat{\Omega}} \delta_i \, N_j \, d\Omega \quad \forall \, j \in \widehat{\Omega} \setminus \widehat{\partial\Omega}, \;\; \hat{\mathcal{G}}_i = \frac{1}{2\pi} \ln \frac{1}{r_i} \text{ on } \widehat{\partial\Omega}. \tag{10}$$

Consider point i be any internal node of the mesh $\widehat{\Omega}$ then, in the same way as for $\hat{u}$ we can write the nodal approximation:

$$\int_{\widehat{\Omega}} \nabla \hat{\mathcal{G}}_i \, \nabla N_j \, d\Omega = \Phi_{\hat{\mathcal{G}}_i,\Sigma_j} \quad \forall \, j \in \widehat{\Omega}, \tag{11}$$

that is the equivalent of (9). Note that we have $\Phi_{\hat{\mathcal{G}}_i,\Sigma_i} = 1$. The FEM solution $\hat{\mathcal{G}}_i$ can be expressed as:

$$\hat{\mathcal{G}}_i = \sum_{j \in \widehat{\Omega}} N_j \, \hat{\mathcal{G}}_{i,j} = \sum_{j \in \widehat{\Omega} \setminus \hat{\Gamma}_u} N_j \, \hat{\mathcal{G}}_{i,j} + \sum_{j \in \hat{\Gamma}_u} N_j \, \hat{\mathcal{G}}_{i,j}. \tag{12}$$

It is permitted to write, using successive eqns (7), (12), (8), (6) and (9):

$$\int_{\widehat{\Omega}} \nabla \hat{u} \, \nabla \hat{\mathcal{G}}_i \, d\Omega = \sum_{j \in \widehat{\Omega} \setminus \hat{\Gamma}_u} \Big(\sum_{k \in \widehat{\Omega}} s_{jk} \, \hat{u}_k \Big) \hat{\mathcal{G}}_{i,j} + \sum_{j \in \hat{\Gamma}_u} \Big(\sum_{k \in \widehat{\Omega}} s_{jk} \, \hat{u}_k \Big) \hat{\mathcal{G}}_{i,j} = \sum_{j \in \hat{\Gamma}_u} \Phi_{\hat{u},\Sigma_j} \, \hat{\mathcal{G}}_{i,j}. \tag{13}$$

Conversely, using eqns (7), adapted to $\hat{\mathcal{G}}_i$, (11) instead of (9) and the boundary conditions for u, we can write:

$$\begin{aligned} \int_{\Omega} \nabla \hat{\mathcal{G}}_i \, \nabla \hat{u} \, d\Omega &= \sum_{j \notin \widehat{\partial\Omega}} \Bigg(\sum_{k \in \widehat{\Omega}} s_{jk} \, \hat{\mathcal{G}}_{i,k} \Bigg) \hat{u}_j + \sum_{j \in \widehat{\partial\Omega}} \Bigg(\sum_{k \in \widehat{\Omega}} s_{jk} \, \hat{\mathcal{G}}_{i,k} \Bigg) \hat{u}_j \\ &= \hat{u}_i + \sum_{j \in \hat{\Gamma}_n} \Phi_{\hat{\mathcal{G}}_i,\Sigma_j} \, \hat{u}_j + \sum_{j \in \hat{\Gamma}_u} \Phi_{\hat{\mathcal{G}}_i,\Sigma_j} \, \hat{u}_0 \, . \end{aligned} \tag{14}$$

Equating (13) and (14), we obtain the following expression where the domain integral is eliminated as expected:

$$\hat{u}_i + \sum_{j \in \hat{\Gamma}_n} \Phi_{\hat{\mathcal{G}}_i,\Sigma_j} \, \hat{u}_j + \sum_{j \in \hat{\Gamma}_u} \Phi_{\hat{\mathcal{G}}_i,\Sigma_j} \, \hat{u}_0 = \sum_{j \in \hat{\Gamma}_u} \hat{\mathcal{G}}_{i,j} \, \Phi_{\hat{u},\Sigma_j}. \tag{15}$$

A consistent linear system of equations is derived by writing eqn (15) for all the nodes i belonging to the boundary $\widehat{\partial\Omega}$. In this situation, the boundary condition of the problem (2) must be changed as:

$$\mathcal{G}_i = \frac{1}{2\pi} \ln \frac{1}{r_i} \text{ on } \partial\Omega \backslash \{i\} \quad \text{and} \quad \frac{\partial \mathcal{G}_i}{\partial n} = 0 \text{ at } \{i\}, \tag{16}$$

so that the Galerkin formulation becomes:

$$\int_\Omega \nabla \hat{\mathcal{G}}_i \, \nabla N_j \, d\Omega = \delta_{ij} \, c_i \quad \forall \, j \in (\hat{\Omega} \backslash \widehat{\partial\Omega}) \cup \{i\}, \quad \hat{\mathcal{G}}_i = \frac{1}{2\pi} \ln \frac{1}{r_i} \text{ on } \widehat{\partial\Omega} \backslash \{i\}, \tag{17}$$

where δ_{ij} is the common Kronecker symbol and c_i is the same geometric factor as used in classical BEM. In the FEM context it can be easily shown that:

$$c_i = \Phi_{\hat{\mathcal{G}}_i, \Sigma_i} \quad \text{and} \quad c_i + \textstyle\sum_{j \in \partial\hat{\Omega} \backslash \{i\}} \Phi_{\hat{\mathcal{G}}_i, \Sigma_j} = 0, \tag{18}$$

i.e. the Gauss law at the discrete level.

Finally the boundary element scheme amounts to solve simultaneously:

$$c_i \, \hat{u}_i + \sum_{\substack{j \in \hat{\Gamma}_n \\ \neq i}} \Phi_{\hat{\mathcal{G}}_i, \Sigma_j} \, \hat{u}_j + \sum_{\substack{j \in \hat{\Gamma}_u \\ \neq i}} \Phi_{\hat{\mathcal{G}}_i, \Sigma_j} \, \hat{u}_0 = \sum_{j \in \hat{\Gamma}_u} \hat{\mathcal{G}}_{i,j} \, \Phi_{\hat{u}, \Sigma_j} \qquad \forall \, i \in \widehat{\partial\Omega}, \tag{19}$$

or, in matrix form:

$$\mathbf{H} . \hat{\mathbf{u}} = \mathbf{G} . \boldsymbol{\Phi}_{\hat{u}}. \tag{20}$$

Rearranging terms, it yields a linear system that is to be solved for unknown potential values $\hat{u}_j$ on $\hat{\Gamma}_n$ and flux values $\Phi_{\hat{u}, \Sigma_j}$ on $\hat{\Gamma}_u$.

At this point, the exact FEM solution of the BVP (1) is obtained since we have derived the eqns (19) by choosing test functions as a linear combination of the hat functions N_j in (13) and (14). It is clear that there is no interest in doing so. Moreover, this requires the computation of the FEM approximations $\hat{\mathcal{G}}_i$ that is highly time consuming. The trick is then to use the exact fundamental solutions $\mathcal{G}_i$ instead of $\hat{\mathcal{G}}_i$, i.e. we use the interpolant:

$$\mathcal{G}_{i,I} = \textstyle\sum_{j \in \hat{\Omega}} N_j \, \mathcal{G}_i(j) = \sum_{j \in \hat{\Omega}} N_j \frac{1}{2\pi} \ln \frac{1}{r_{ij}}, \tag{21}$$

where r_{ij} is the distance between nodes i and j. Due to the singularity, the infinite value $\mathcal{G}_i(i)$ is replaced by a value $\mathcal{G}^*_{i,i}$ from the pair of eqns (18), i.e.:

$$\hat{c}_i = \textstyle\sum_{k \neq i} s_{ik} \, \mathcal{G}_i(k) + s_{ii} \, \mathcal{G}^*_{i,i} \tag{22a}$$

and

$$\hat{c}_i + \textstyle\sum_{j \in \widehat{\partial\Omega} \backslash \{i\}} \left(\sum_{k \neq i} s_{jk} \, \mathcal{G}_i(k) + s_{ji} \, \mathcal{G}^*_{i,i} \right) = 0. \tag{22b}$$

Indeed, (22a) express the "cap" flux at any boundary node i and (22b) stands for the weak flux inside the domain Ω along the boundary $\widehat{\partial\Omega}$, due to the Dirac delta δ_i. Solving simultaneously these equations leads to the desired value $\mathcal{G}^*_{i,i}$, along with an approximate factor $\hat{c}_i$. The modified geometric factor $\hat{c}_i$ is not equal to the exact one obtained previously. Although we could force $\hat{c}_i \triangleq c_i$, the numerical simulations indicate a better accuracy by using the above statements. This is probably due to the consistency brought by the discrete Gauss laws given by (22a) and (22b).

It is important to notice that it is not necessary to mesh the whole domain since the internal nodes disappear in the boundary formulation (19). A single layer of finite elements along the boundary is sufficient. This is even mandatory in the case of an open boundary problem. In this case, the right-hand side of eqn (22b) is equal to unity due to the integral at infinity.

Finally, notice that the computational effort to construct the matrix scales as $O(n^2)$ as in classical BEM. However it does not require any integration so that a reduction by a factor of about N_G is expected when using a N_G-point Gaussian quadrature.

3 NUMERICAL RESULTS

The method has been applied to several simple electrostatic problems in order to estimate the basic performance of our numerical scheme. We have implemented both our method and the classical BEM with MATLAB® software for the ease of analysis and exploitation of the results. The boundary integrals are calculated with a Gaussian quadrature method with seven points in our BEM implementation. COMSOL Multiphysics® has been used for the generation of the various meshes required by the simulations.

3.1 A closed geometry example: the square coaxial capacitor

First of all, we consider the square coaxial air capacitor arrangement depicted in Fig. 3. The governing equation is given by eqn (1), up to the absolute permittivity ε_0. The prescribed potential values on the Dirichlet boundaries Γ_{u1} and Γ_{u2} are $u_1 = 1$ V and $u_2 = 0$ V, respectively. The problem can be analysed either on the whole geometry or on a reduced geometry considering some symmetry. In the following, both approaches are investigated in order to assess the various corner effects on the normal flux value such as the re-entrant corner and the Dirichlet to Neumann change in boundary condition.

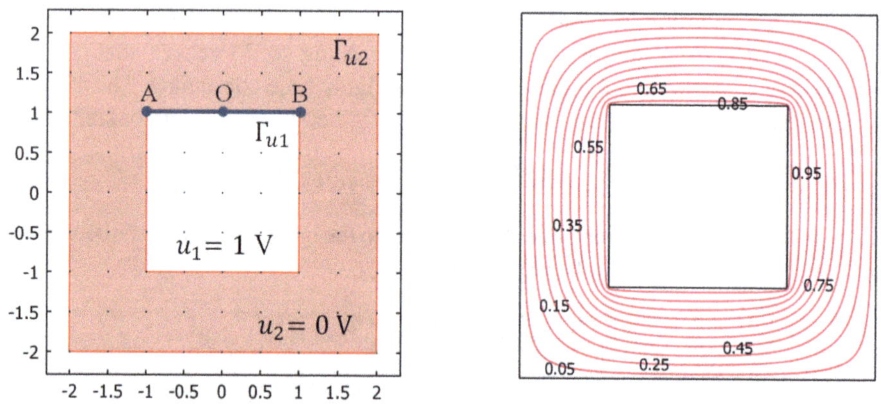

Figure 3: A square coaxial configuration. (a) Geometry; and (b) Solution.

3.1.1 Full geometry configuration

We consider the complete geometry of the problem of which a finite element mesh is shown in Fig. 4(a). Uniform meshing is applied in order to get a reference convergence behaviour. A comparison has been carried out between our method and the classical BEM. First, we have computed the distribution of the surface charge density $\sigma = \varepsilon_0 . \partial u / \partial n$ along the inner

Dirichlet boundary. Classical BEM gives directly this result since it relates to the $\hat{q}_i$ parameters at every node of the boundary mesh. In our method, the nodal normal derivatives can be reasonably deduced from the nodal "cap" flux value $\Phi_{\hat{u},\Sigma_i}$ by using the following ansatz:

$$\hat{q}_i = \frac{\Phi_{\hat{u},\Sigma_i}}{s_i}, \tag{23}$$

where the cross section s_i is chosen as shown in Fig. 4(b), though it is not strictly correct as explained in [5].

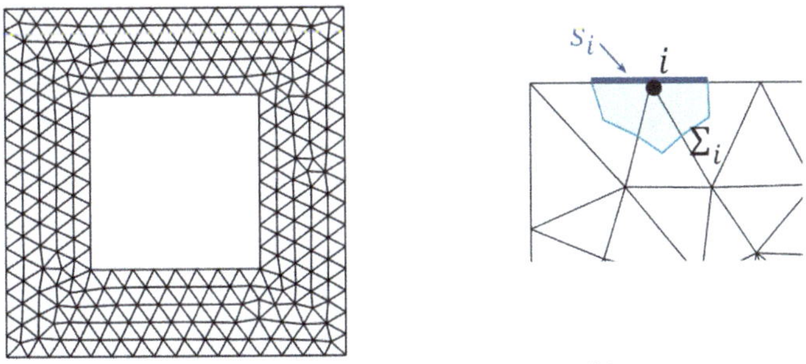

(a) (b)

Figure 4: (a) FE-mesh; and (b) Computation of the normal derivative from nodal "cap" flux.

The results are plotted in Fig. 5 along the path AB depicted in Fig. 3(a) for a boundary mesh of size $n = 768$, superimposed to the "exact" one, obtained from a FEM simulation with a very fine mesh. It is clear that the normal derivative at re-entrant corners A and B presents a bad accuracy with standard BEM though it is better with the new technique. The computation of the surface charge value Q distributed on the Dirichlet boundary Γ_{u1} has also been performed. It is obtained by summing the "cap" flux values along the boundary in the proposed method, and a boundary integration of the normal flux $\hat{q}$ for BEM. The reference value $Q = 10.23409$ C is obtained analytically following the method described in [6]. Fig. 6(a) presents the convergence results where it can be observed a loss in accuracy despite the better behaviour of the normal derivative reported in Fig. 5. However, as explained in Section 2, since the computational effort is reduced in a factor of N_G chosen equal to 7 here, our technique remains competitive in this respect as shown in Fig. 6(b) where the results of BEM have been shifted to the right accordingly that is, taking into account the complexity of $O(n^2)$, by $\log(\sqrt{N_G})$.

The charge has also been computed by the finite element method. It can be derived from either the electrostatic energy W, i.e. $Q = 2W/U$, or by summing the "cap" flux values along the boundary. Both method leads exactly to the same value for the charge Q as suggested in [5]. The convergence of FEM appears to be very close to our method as observed in Fig. 6(a).

3.1.2 L-shaped geometry configuration

A quarter of the geometry of the coaxial configuration is now considered as presented in Fig. 7(a). From symmetry argument, the Neumann condition is applied on the parts Γ_{n1} and Γ_{n2}. Again, a uniform triangle mesh is applied. The distribution of the normal derivative on the path OB (Fig. 7(b)) is again computed. Here we point out a discrepancy at point O due to

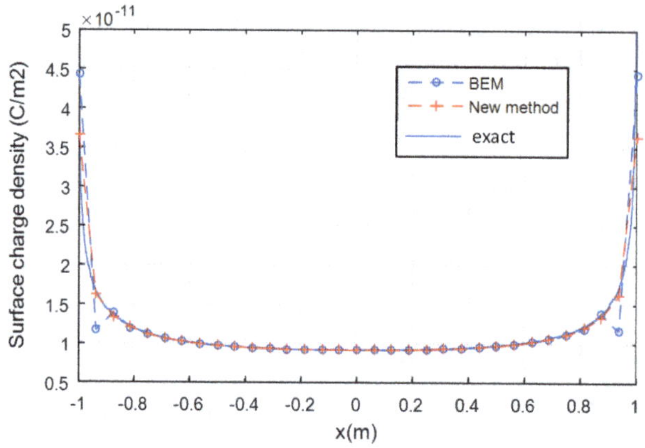

Figure 5: Square coaxial configuration: surface charge density distribution along path AB (n = 768). Comparison of numerical (symbols) and exact (full line) results.

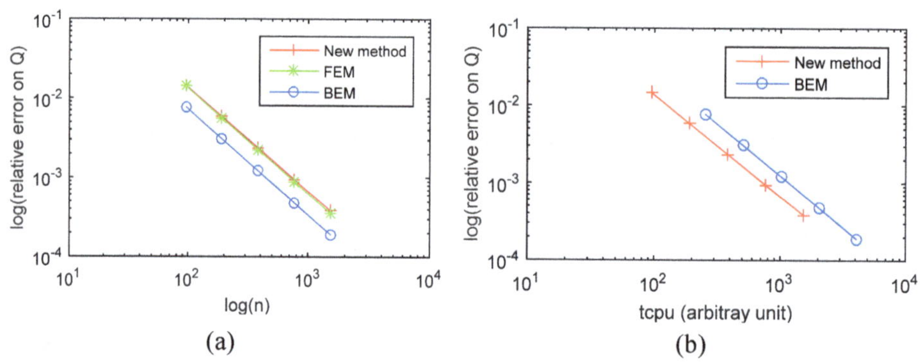

Figure 6: Square coaxial configuration: relative error on the surface charge Q with respect to (a) The size n; and (b) The computational burden t_{cpu}.

the change in boundary condition as well as at re-entrant corner B as in Section 3.1.1. The computation of the surface charge value Q distributed on the Dirichlet boundary Γ_{u1} gives the error results shown in Fig. 8. Now it is clear that our method outperforms the classical methods for the computation of this global quantity. The loss in accuracy of standard BEM is due to the underestimation of the normal flux at corner O. The results obtained from a FEM analysis are also plotted and lie between the two BEM techniques.

The same comparison study has been conducted for the potential value at internal points 1, 2 and 3 that are chosen as indicated in Fig. 7(a). The numerical results are compared to reference values that are deduced from a FEM analysis with a dense triangulation: $u_1 =$ 0.8787 V, $u_2 = 0.2865$ V, $u_3 = 0.7916$ V. We now restrict the comparison to the new method and the BEM. Fig. 9 shows the results for equal computational burden. They are not better for all points with our method. Clearly, accuracy at point 3 is far better due to a bad accuracy at the nearby Dirichlet/Neumann corner with BEM but is not so good for point 1 although re-entrant corner presents problem with BEM. From an execution time point of view, the comparison indicates that our method remains competitive.

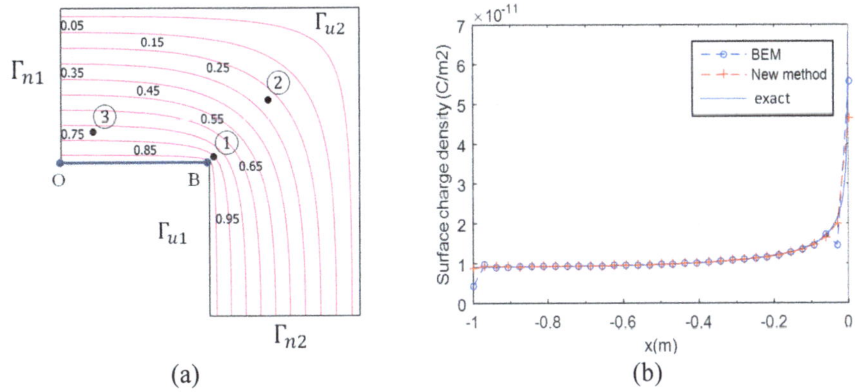

Figure 7: L-shaped configuration. (a) Solution; and (b) Surface charge density distribution along path OB (n = 248). Comparison of numerical (symbols) and exact (full line) results.

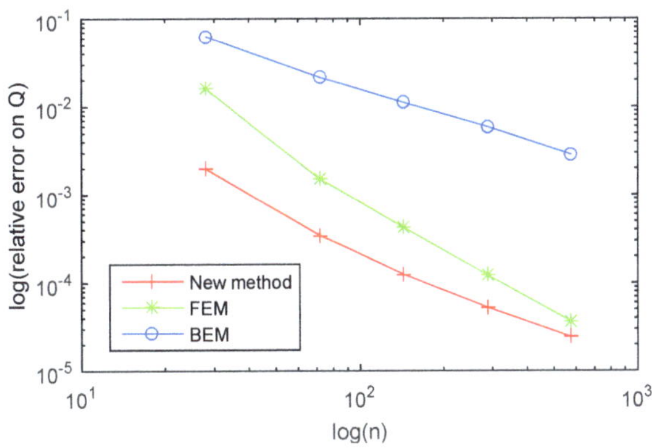

Figure 8: L-shaped configuration: relative error on the surface charge.

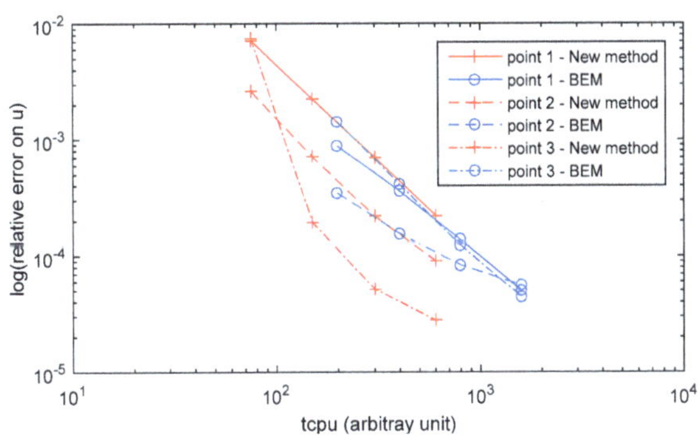

Figure 9: L-shaped configuration: relative error on the potential value at internal points 1, 2 and 3.

3.2 An open boundary problem

As an advantage of BEM lies in the ability of treating open boundary problems, we consider the situation of a symmetric parallel wire capacitor as shown in Fig. 10. The prescribed potential values on the Dirichlet boundaries Γ_{u1} and Γ_{u2} are $u_1 = -1/2$ V and $u_2 = 1/2$ V, respectively. Similar numerical analysis has been conducted concerning the surface charge density along the boundary Γ_{u1} and the related global charge Q. Classical electromagnetism gives analytical results for these quantities so that an error analysis can be performed. The results are shown in Fig. 11.

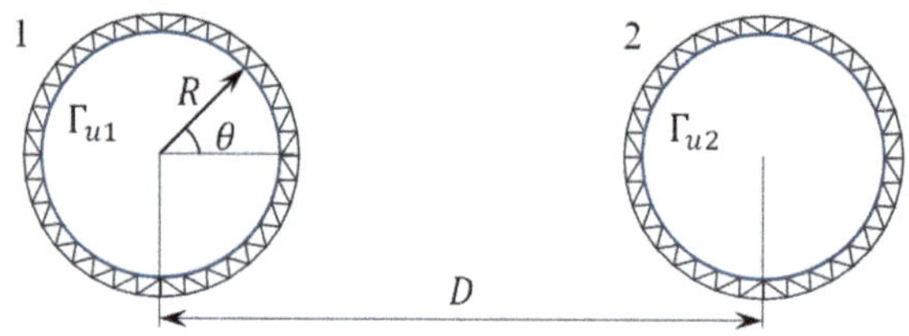

Figure 10: Parallel wire capacitor (R = 0.2 m, D = 1 m).

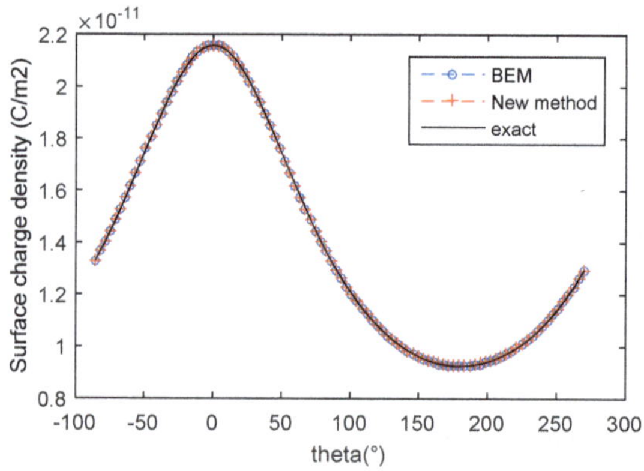

(a)

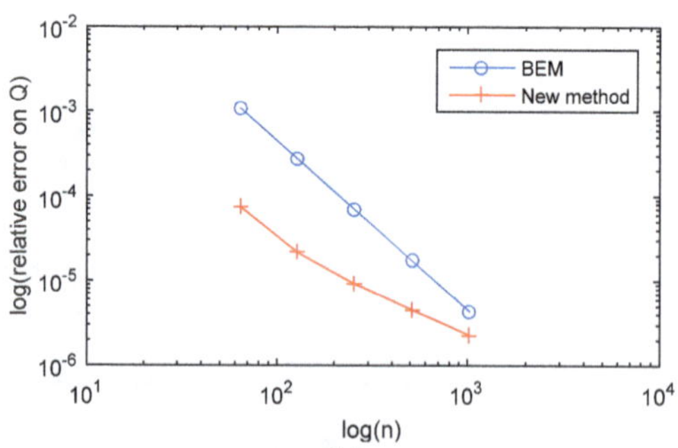

(b)

Figure 11: Parallel wire capacitor. (a) Surface charge density (n = 192); and (b) Error on the charge Q on conductor 1.

Fig. 11(a) shows that surface charge density is very well reproduced by both methods. From Fig. 11(b), comparison of errors on the total charge value is shown to be lower with our method pointing out a better accuracy though convergence rate appears to be higher with standard BEM for more refined mesh. This last point should be further investigated.

4 DISCUSSION AND CONCLUSION

In this paper, we have proposed a new BEM formulation based on the FEM solution of the fundamental solution and on a "variationally" correct definition of nodal flux. The main difference with the classical BEM is that at least a layer of finite domain elements connected to the boundary is needed. The implementation is simpler as no numerical integration on the boundary of the domain is required. Numerical examples have been conducted in order to compare both our method and the classical BEM. The new scheme exhibits the advantage of an unambiguous definition of the normal flux when corners are present in the geometry. However, the computation of global quantities such as the charge or potential at internal nodes are sometimes less accurate. Further investigation is needed in order to analyse the influence of the quality of the finite element mesh that is necessary along the boundary. Indeed, numerical experiments show that it affects the accuracy. This point requires some improvements and clarifications. Yet the method is competitive since the computational burden is significantly reduced. Extension of the method to the Poisson equation and multi-domain configurations are straightforward and have been tested but not presented in the paper. Moreover, the proposed technique can be easily interfaced with finite elements in a FEM/BEM coupling since the "cap" flux is intrinsically related to the FEM. This will be the subject of a future paper.

REFERENCES

[1] Silvester, P.P. & Ferrari, R.L., *Finite Elements for Electrical Engineers*, Cambridge University Press, 1996.

[2] Brebbia, C.A. & Dominguez, J., *Boundary Elements: An Introductory Course*, WIT Press: Southampton and Boston, 1992.

[3] Hartmann, F., *Green's Functions and Finite Elements*, Springer, 2013.

[4] Araya, R., Behrens, E. & Rodriguez, R., A posteriori error estimates for elliptic problems with Dirac delta source terms. *Numerische Mathematik*, **105**(2), pp. 193–216, 2006.

[5] Bossavit, A., How weak is the "weak solution" in finite element methods? *IEEE Transactions on Magnetics*, **34**(5), pp. 2429–2432, 1998.

[6] Riblet, J.H., Expansion for the capacitance of a square in a square with a comparison. *IEEE Transactions on Microwave Theory and Techniques*, **44**(2), pp. 338–340, 1996.

PRECISE AND ROBUST MAGNETIC FIELD COMPUTATIONS FOR HIGH-END SMART SENSOR APPLICATIONS

ANDRÉ BUCHAU
Institute of Smart Sensors, University of Stuttgart, Germany

ABSTRACT

Highly precise magnetic field computations are essential for the design of magnets and coils of high-end smart sensors in the context of nuclear magnetic resonance (NMR) and electron spin resonance (ESR) applications. Here, requirements for numerical magnetic field computations are that the methods are precise and mainly robust in the sense of reliable results. Non-linear magnetic materials, induced eddy currents, and the infinite surrounding air domain must be accurately considered in the numerical model. To this end, a smart combination of the finite element method (FEM) with the boundary element method (BEM) including a powerful post-processing is proposed with a focus on the region of interest (ROI), which contains the studied sample. Separate BEM domains are defined for the infinite surrounding air domain and the ROI. Then, the computational costs are reduced due to the relatively small number of relevant boundary elements of the surface of the ROI and a precise and efficient evaluation of magnetic fields based on a meshfree BEM post-processing in the ROI is possible. As an example, the numerical formulation, the application of the proposed approach, and numerical results are shown for the homogeneous field of a permanent magnet.

Keywords: boundary element method, finite element method, smart sensors, magnetic fields, permanent magnets, nuclear magnetic resonance spectroscopy.

1 INTRODUCTION

The concept of a smart sensor is to combine a sensor, which measures a physical value, with a low-noise analogue interface circuit and the digital signal processing in a chip [1]. An implementation of such a smart sensor within an application specific integrated circuit (ASIC) is very advantageous, since it enables in-situ measurements with a high signal-to-noise ratio. An example for a high-end smart sensor application is to measure free radicals in blood samples using a mobile electron spin resonance (ESR) spectrometer [2]. A similar design can be applied for nuclear magnetic resonance (NMR) microscopy [3].

Both for ESR and NMR applications, homogeneous magnetic fields in the region of interest (ROI) are required, which directly influence the quality of the measured signal. Hence, the design of permanent magnets with highly homogenous magnetic fields is of great importance for these applications. To this end, a shimming based on accurate numerical field computations including optimization techniques and precise measurements is necessary [4]. Then, manufacturing tolerances can be compensated, too. Concerning numerical field computations, accuracy of the numerical methods is more important than the computational costs. However, a significant criterion for the selection of an appropriate numerical method is that the numerical model is simple and robust. That means, possible inaccuracies caused by the discretized model of the magnet should be minimized.

Typically, the magnet consists of permanent magnets, which are embedded into a yoke with defined geometrical shape of the pole shoes as a result of the shimming process. The yoke is made of magnetic material with a non-linear relation between the magnetic field strength $\boldsymbol{H}$ and the magnetic flux density $\boldsymbol{B}$. Hence, the finite element method (FEM) is well suited for accurate numerical computations of the magnetic field inside the magnet. A three-dimensional FEM model of the magnet takes all relevant effects including stray fields into

WIT Transactions on Engineering Sciences, Vol 126, © 2019 WIT Press
www.witpress.com, ISSN 1743-3533 (on-line)
doi:10.2495/BE420071

account. Furthermore, arbitrary complex numerical models of the magnetic material properties can be included in the design process of the magnet. However, the truncated surrounding air domain affects the magnetic field in the ROI significantly.

A very elegant approach to tackle the problem of the infinite air domain is to couple the FEM with the boundary element method (BEM). Then, the magnetic fields in the air domain are computed using the BEM. Furthermore, a discretization of the ROI is not necessary and the magnetic fields are not interpolated but directly computed from the related boundary integrals with a high accuracy.

In total, a coupled problem must be solved. Modern numerical techniques like iterative solvers for the linear system of equations (LSE), enhanced integration techniques for singular or nearly singular integrals of the BEM, and matrix compression techniques for the dense BEM matrix blocks enable an efficient solution of the problem.

A very important step in numerical computations of magnetic fields for smart sensor applications is the post-processing. The magnetic fields in the ROI must be computed with high accuracy for an evaluation of the homogeneity of the magnetic field. To this end, a meshfree post-processing of the BEM in the ROI is very advantageous, since no discretization and approximation of the fields is necessary. A modular software concept enables a combination of the computed fields for a visualization of the fields both inside the yoke, which are computed using the FEM, and inside the BEM controlled ROI simultaneously.

Here, an accurate and robust approach for the computation of magnetic fields for smart sensor applications is presented. First, a scalar potential formulation is discussed, which is well suited for this kind of field problems for a solution based on the FEM or the BEM. Starting from this overview, a strategy for a coupled application of both methods is shown. The focus is on an optimization of the BEM domains related to efficiency and accuracy. There, compressed matrix blocks of the corresponding linear equation system as well as boundary integrals for a meshfree post-processing are considered. In total, a computational method is presented, which is easy to use for a smart sensor designer and which delivers reliable and accurate numerical results.

2 FORMULATION

A mathematical and numerical formulation for the computation of static magnetic fields is required, which enables the determination of magnetic fields in the ROI reliably with a high accuracy. There, non-linear material properties of the yoke, small geometrical details of the pole shoes, and the infinite air domain must be considered. Furthermore, the formulation should be well suited both for the FEM and the BEM.

Here, a magnetic scalar potential formulation was chosen, since only configurations with permanent magnets as field sources are studied. Then, a scalar potential formulation reduces the computational costs in comparison to a more general applicable vector potential formulation significantly without a loss of accuracy.

First, the scalar potential formulation is revisited with respect to the computation of magnetic fields for NMR and ESR applications. Then, numerical models for the non-linear magnetic material properties are discussed. Finally, the basic equations for the applied FEM and BEM are shown to lay the foundation for the proposed robust field computation approach.

2.1 Scalar magnetic potential

Since only permanent magnets are used as sources for the magnetic field, a scalar magnetic potential ψ suffices to compute the magnetic field strength $\boldsymbol{H}$

$$\boldsymbol{H} = -\nabla\psi. \tag{1}$$

An important advantage of the scalar magnetic potential is that the computational costs are significantly smaller than for a more general applicable magnetic vector potential $\boldsymbol{A}$. Hence, a much finer discretization can be applied, especially near corners, edges, or at geometrical small structures, which is important in terms of accuracy.

In a domain with magnetic material properties, the magnetic flux density $\boldsymbol{B}$ is obtained directly from $\boldsymbol{H}$ and the permanent magnet's magnetization $\boldsymbol{M}$

$$\boldsymbol{B} = \mu_r\mu_0\boldsymbol{H} + \boldsymbol{M}, \tag{2}$$

with relative permeability μ_r and vacuum permeability μ_0. Starting from

$$\nabla\boldsymbol{B} = 0, \tag{3}$$

a partial differential equation for ψ must be solved in each computational domain

$$\nabla\mu_r\nabla\psi = \frac{1}{\mu_0}\nabla\boldsymbol{M}. \tag{4}$$

The magnetic potential is continuous at domain boundaries

$$\psi_1 = \psi_2. \tag{5}$$

Tangential component of $\boldsymbol{H}$ is continuous

$$\boldsymbol{n}_{12} \times (\boldsymbol{H}_2 - \boldsymbol{H}_1) = \boldsymbol{0} \tag{6}$$

and its normal component is discontinuous

$$\mu_0\mu_{r1}H_{n1} + M_{n1} = \mu_0\mu_{r2}H_{n2} + M_{n2}. \tag{7}$$

From eqn (7) it can be seen that the magnetic field strength in high permeable materials is much smaller than in the adjacent air domain. This yields to significant numerical problems, if ψ and with it $\boldsymbol{H}$ is not computed accurate enough. In that cases a difference field concept is useful to increase the accuracy of the solution [5].

2.2 Non-linear material properties

A crucial point for accurate magnetic field computations is the model of the non-linear magnetic material properties of the yoke and of the pole shoes. In general, a vector hysteresis model is necessary to model the magnetization effects within the magnetic domains accurate enough. Here, it is sufficient to apply a scalar model and to approximate the relation between $|\boldsymbol{B}|$ and $|\boldsymbol{H}|$ with a non-linear function (Fig. 1).

Between the measured values, which are drawn in black in Fig. 1, the non-linear function is interpolated by piecewise linear functions. For large values of $|\boldsymbol{H}|$ the non-linear function is linearly extrapolated.

2.3 Finite element method

Second order Lagrange elements are used to discretize the computational domain for a solution of the magnetic field problem with the FEM. Applying the Galerkin method yields to

$$[K(\psi)]\{\psi\} = \{b\}. \tag{8}$$

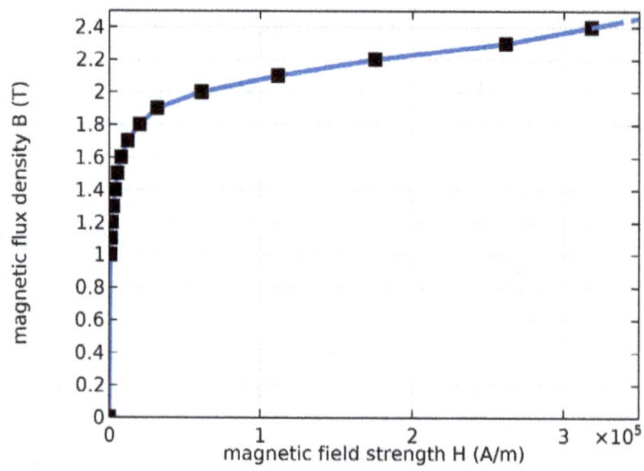

Figure 1: Scalar non-linear model of magnetic material properties.

$[K(\psi)]\{\psi\} = \{b\}$The stiffness matrix $[K(\psi)]$ depends on the magnetic material properties and therefore on ψ. The right-hand side $\{b\}$ contains indirectly the defined magnetization of the permanent magnets. During the derivation of (8) it is assumed that the magnetic flux through the infinite boundary vanishes and that is $\frac{\partial\psi}{\partial n} = 0$ respectively.

The non-linear system of equations in (8) is solved using a Newton–Raphson method with re-computation of the tangential matrix within each iteration step. An adaptive control of the damping factor ensures fast convergence.

Within each non-linear iteration step, conjugate gradient (CG) method with algebraic multigrid as preconditioner is used to solve the linear equation system.

After the computation of ψ, $\boldsymbol{B}$ is evaluated in the ROI based on the nodal values of ψ of the finite elements in the ROI. To this end, (1) and (2) are applied with analytical derivatives of the Ansatz functions. However, due to the second order Lagrange polynomials for ψ only first order polynomials for $\boldsymbol{B}$ are obtained. Furthermore, $\boldsymbol{B}$ is discontinuous at element interfaces. There exist several techniques like the patch recovery method to improve the accuracy of $\boldsymbol{B}$ [6]. Another often applied approach is to refine the mesh in the ROI.

2.4 Boundary element method

In the case of linear magnetic materials, the BEM is very well suited to solve the Laplace equation for the magnetic scalar potential

$$\Delta\psi = 0 \tag{9}$$

efficiently with high accuracy. Then, only the domain boundaries are discretized with second order Lagrange elements. Applying Galerkin method and Green's theorem

$$\psi(\boldsymbol{r}) = \int_{\partial\Omega} G(\boldsymbol{r},\boldsymbol{r}')\frac{\partial\psi(\boldsymbol{r}')}{\partial n'}\mathrm{d}A' - \int_{\partial\Omega}\frac{\partial G(\boldsymbol{r},\boldsymbol{r}')}{\partial n'}\psi(\boldsymbol{r}')\mathrm{d}A' \tag{10}$$

leads to a linear system of equations

$$[G]\left\{\frac{\partial\psi}{\partial n}\right\} = [H]\{\psi\} \tag{11}$$

with dense matrix blocks. The infinite domain boundary is implicitly considered by Green's function of three-dimensional Laplace equation

$$G(\boldsymbol{r},\boldsymbol{r}') = \frac{1}{4\pi}\frac{1}{|\boldsymbol{r}-\boldsymbol{r}'|}. \quad (12)$$

Eqn (10) is valid for each computational domain, i.e. for each material domain in practice. Further domains are created by additional cuts of a single domain, which is advantageous for instance in the ROI. The computational domains are coupled applying the boundary conditions (5) and (7). Furthermore, it is not necessary to compute all domains with the BEM, which is the starting point for the proposed coupling with the FEM in the next section.

Considering accuracy of the BEM, it is very important to compute singular and strong singular integrals accurately. Introducing polar coordinates on the boundary elements around the evaluation point, here the integration point of the Galerkin integral, is very advantageous to increase accuracy with acceptable computational costs [7]. These integration techniques are applicable to nearly singular and nearly strong singular integrals, too, which is relevant for accurate field evaluations near domain boundaries, e.g. near the pole shoes in the ROI.

To overcome the large memory requirements and computational costs of the dense matrix blocks in (10), established matrix compression techniques like the adaptive cross approximation (ACA) [8] or the fast multipole method (FMM) [9] are used. The ACA is well suited to solve (10) efficiently, especially in the context of a coupling of the BEM with the FEM [10]. The FMM can be used both for the solution of the equation system and for the post-processing [11]. Furthermore, a modified FMM is the foundation for an efficient meshfree post-processing [12], which is applied here to improve the accuracy in the ROI.

3 ROBUST AND PRECISE MAGENTIC FIELD COMPUTATION

A robust and precise method for the computation of magnetic fields is presented here. Besides the solution of the problem, a focus is set on the evaluation of the magnetic fields in the ROI. Starting from the fundamentals of the FEM and the BEM summarized in previous section, an approach is developed, which combines both methods very advantageous for the studied class of field problems. Finally, latest advances in both methods enable a new coupled post-processing and evaluation of magnetic fields in the ROI.

First, some limits in practical applications of the FEM are shown for precise magnetic field computations. Then, an efficient approach for the solution of the equation system is shown, which prepares an enhanced post-processing, too. This smart coupling of computational methods results then in a powerful and accurate post-processing.

3.1 Limits of a pure finite element method

As shown in the previous section, the FEM is in principal very well suited for the type of examined magnetic field problems. All material properties can be considered with desired accuracy and complexity. A refinement of the volume mesh along with a sufficiently large surrounding air domain ensures reliability of the computed fields. However, robustness in the sense of simple to use numerical models is missing. For precise computations, the size of the air domain and the discretization of the ROI must be chosen very carefully.

Here, a three-dimensional model of a magnetic circuit is considered (Fig. 2). Its yoke is made of iron with a non-linear magnetization curve as depicted in Fig. 1. Above and below the air gap for the ROI, permanent magnets with a remanence flux density of 0.6 T are embedded into the yoke.

Four different numerical cases have been investigated. The radius of the air domain was set to the same value as the size of the yoke and in the second parameter study it was set five times the size of the yoke. Furthermore, two different meshes in the ROI have been created. The magnetic flux density along a line in the ROI for all four cases is shown in Fig. 3. There, differences between the four studied cases are visible. That means, the quality of the results strongly depends on the size of the finite elements in the ROI and of the size of the air domain.

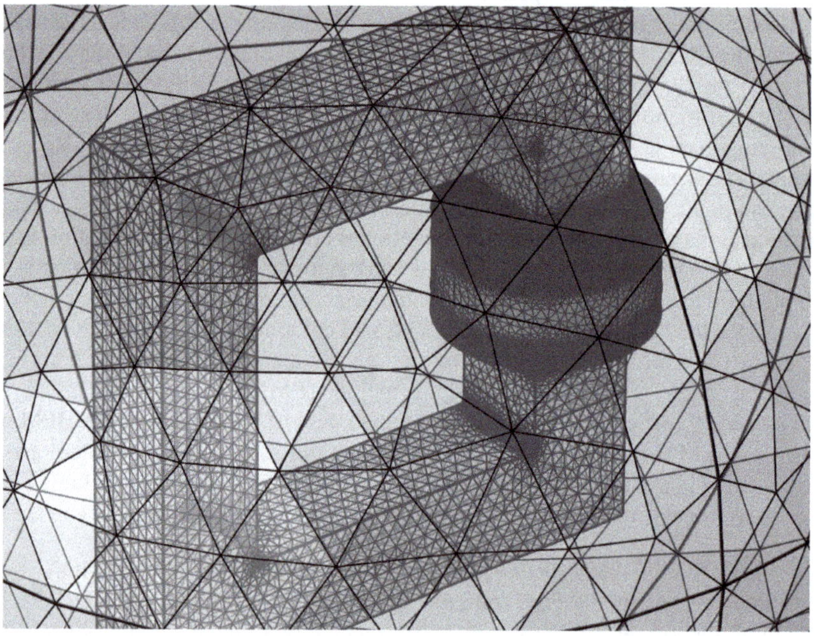

Figure 2: Three-dimensional discretized model of a magnetic circuit.

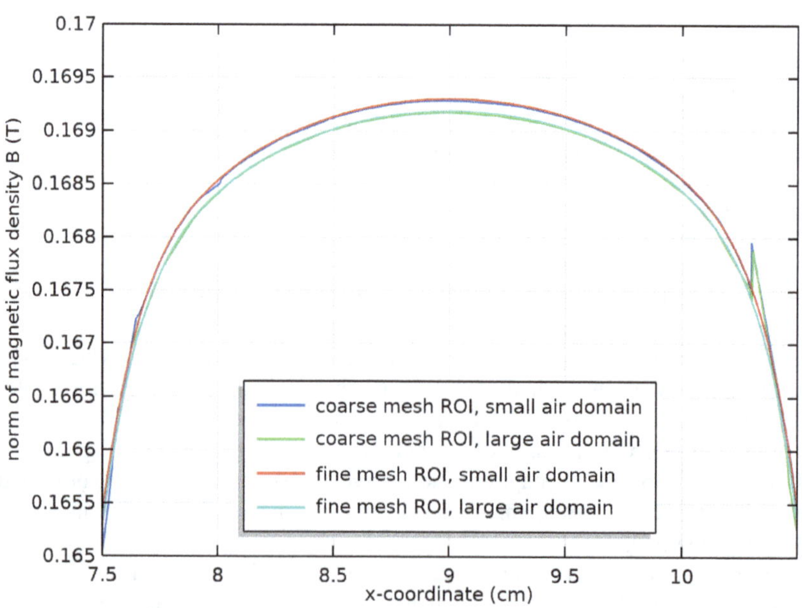

Figure 3: Absolute value of magnetic flux density along a line in the ROI.

3.2 BEM–FEM coupling for the solution of the problem

As demonstrated in the previous sub-section, magnetic field computations in the ROI are sensitive to discretization and size of the air domain. Hence, robustness in an application of the FEM is not guaranteed. To this end, a coupling of the BEM and the FEM is proposed here to improve robustness of magnetic field computations both for the solution of the field problem and for the post-processing.

A coupling of the FEM and the BEM for the solution of magnetic field problems has a long tradition [10]. Latest publications for electromagnetic wave problems [13], acoustics problems [14], or mechanical problems [15] underline that even today a BEM–FEM coupling is competitive and advantageous. However, all recent publications have in common that the post-processing is limited to the computation of field values inside the FEM domain or to only a few characteristic values in the BEM domain.

In the following, the BEM–FEM coupling approach is extended to the post-processing with features comparable to a pure FEM post-processing but with accuracy of BEM. To this end, a propriate pre-processing and solution of the field problem is considered first.

The goal is to compute and to visualize the magnetic field with a high accuracy in the ROI. To overcome interpolation errors caused by the volume discretization of the FEM, the BEM is selected for the ROI. Then, only the artificial boundary of the ROI must be considered to evaluate eqn (10). Normally, the yoke and the pole shoes are discretized using relatively small finite elements. Since the test samples in NMR or ESR applications are positioned in the centre of the ROI, the BEM boundary of the ROI is not placed directly on the surface of the pole shoes but with a small gap to magnetic materials. This enables a layer of finite elements, which increase the element size and hence reduce the number of elements at the BEM domain surface. Then, an efficient solution of the BEM domain based on compression techniques like the ACA or the FMM is possible. A similar approach is used for the air domain. There, also a small FEM mesh reduces the number of boundary elements. In practice, a compromise between computational costs for the BEM with reduced number of elements and for larger FEM domains must be found. Depending on the geometrical configuration of the problem, further parts of the air domain can be solved using the BEM. The proposed splitting is summarized in Fig. 4.

Another advantage of this computational domain splitting is that field singularities, as they occur near edges or corners of the pole shoes, are computed within the FEM domain. There, increasing the number of elements and of integration points is more efficient than in the BEM domain. As a result, relatively smooth boundary values at the BEM domain surfaces are obtained, which can be handled with the techniques presented in [7].

3.3 BEM–FEM coupling for post-processing

For a successful magnet design, a powerful post-processing is indispensable. This includes evaluation of fields in defined points, for instance along a line as shown in Fig. 3, visualization of fields in a plane using colours or arrows, or computation and visualization of field lines or isosurfaces. Both accurate values in form of tables or line plots and a three-dimensional overview of the field characteristics are required for an engineer.

The challenge of a BEM–FEM post-processing is to combine the two completely different working methods in such a way that this coupling and the splitting of the computational domains is invisible for the magnet designer. A classical approach is, to create in the BEM domains finite elements and to compute the field values in the nodes of these elements [16].

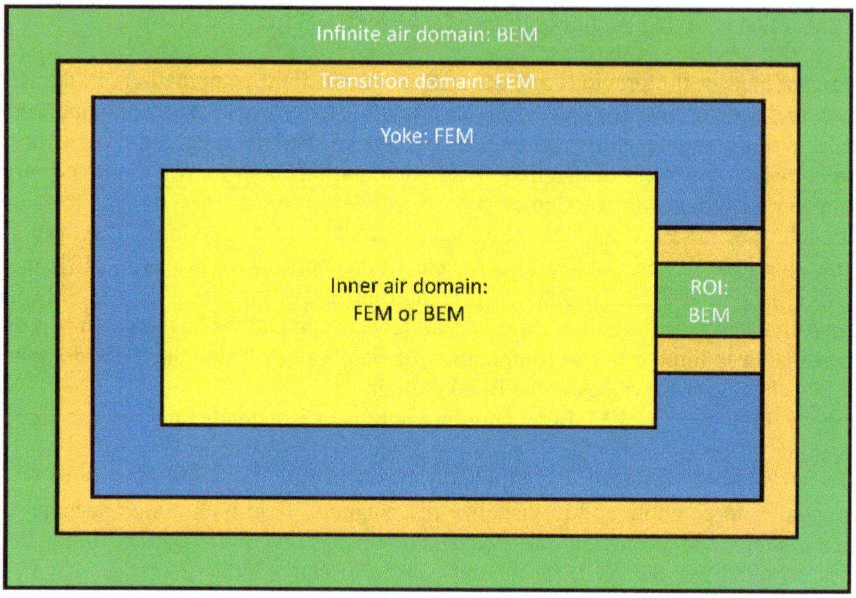

Figure 4: Splitting of computational domains for the BEM or the FEM.

Then, the field values of the FEM and of the BEM are interpolated by standard finite elements in the complete domain and enhanced visualization techniques are applicable [17]. However, that approach requires a large amount of evaluations of eqn (10) along with large computational costs. Hence, an approach, which computes only necessary values, is preferred.

Here, a meshfree post-processing [12] for the BEM is coupled with a classical FEM post-processing. This requires several steps to split and combine the numerical results. First, the designer defines the type of field evaluation. Then, the post-processing software has to select the relevant computational domain automatically. Values and visualization objects in the FEM domain are computed based on grid values and in the BEM domain a meshfree computation is applied to reduce computational costs and to overcome interpolation inaccuracies. Finally, the computed results are composed for virtual reality (VR) or augmented reality (AR) visualization. To this end, a strictly modular field computation software is required.

The software MuPhyN (multiphysics numerics) of the author integrates modules of a fast BEM solver, a meshfree BEM post-processing, and an interface to a FEM solver, here to the commercial implementation COMSOL Multiphysics (Fig. 5).

A key feature of the BEM post-processing module is that the computation of visualization objects like field lines [18] or isosurfaces [19] is strongly coupled with the BEM. This enables efficient and accurate computations. The visualization objects of BEM and FEM are merged in the open source visualization software COVISE (Fig. 6).

To achieve the required accuracy, the BEM must be appropriate parameterized. Since the transition domain between the yoke and the ROI was introduced, no field singularities occur at the BEM domain surface. That means, smooth values of ψ and $\frac{\partial \psi}{\partial n}$ are obtained at the boundary elements. Hence a relatively coarse mesh suffices. On the one hand, a small number

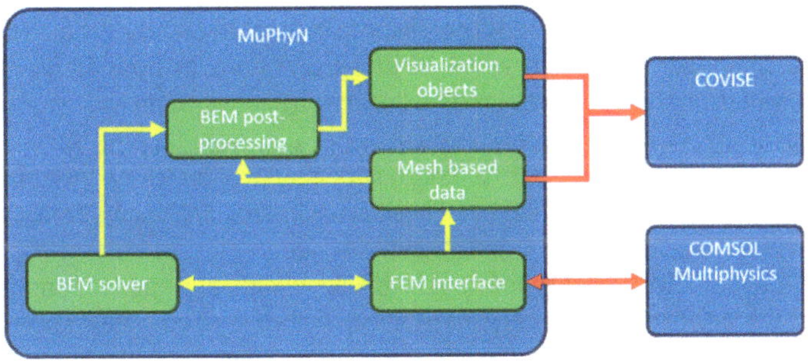

Figure 5: Modular software structure of MuPhyN.

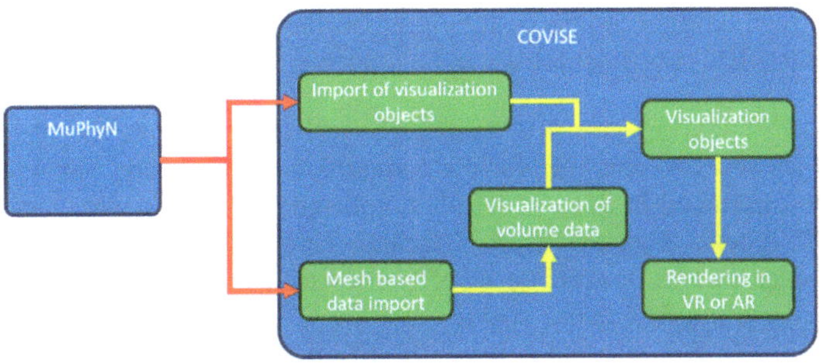

Figure 6: Merging of visualization objects of BEM and FEM.

of boundary elements increases the efficiency of the post-processing in the ROI, on the other hand the relative distance between the evaluation points and the boundary elements can be small. Hence, special treatment of nearly singular and nearly strong singular integrals with sufficiently large number of integration points is necessary [7].

An efficient meshfree BEM post-processing is based on the FMM. The FMM compresses the integrals of eqn (10) and most boundary elements are considered by a series expansion of Green's function in spherical coordinates, e.g. for the single-layer potential part of eqn (10)

$$\psi(\boldsymbol{r}) = \frac{1}{4\pi} \sum_{n=0}^{L} \sum_{m=-n}^{n} r^n Y_n^m(\theta, \phi)\, L_n^m. \tag{13}$$

The order L adjusts the accuracy of the FMM. The local coefficients L_n^m contain information of the boundary values of all boundary elements in the far-field of the evaluation point and $Y_n^m(\theta, \phi)$ are normalized spherical harmonics.

To compute the magnetic field strength $\boldsymbol{H}$ or the magnetic flux density $\boldsymbol{B}$ based on (1), the necessary derivatives can be computed in eqn (13) analytically in Cartesian coordinates. In a similar way, the double-layer part of eqn (10) is considered.

The BEM–FEM post-processing strongly depends on an automatic domain detection method. In the FEM domain, the related finite element of the evaluation point is found with an octree search and an optimization method. The domain detection in the BEM domain is more challenging, since the domain is not discretized [12]. A combination of both approaches determines the relevant domain and computational method automatically and is therefore applicable in algorithms for field line or isosurface computations.

A bidirectional communication between the software modules in Figs 5 and 6 is necessary to evaluate the magnetic fields across BEM and FEM domain boundaries. In the case of isosurface computations, an exchange of node coordinates improves smoothness of the isosurface, especially if a watertight isosurfaces is determined. For accurate field lines, boundaries must be reliably detected to obtain precise coordinates of the intersection point of the field line with the domain surface. These coordinates are used as start coordinates of the field line segment in the next computational domain.

In total, a BEM–FEM post-processing is based on an automatic domain detection, a modular software concept of numerical methods, and a deep integration of numerical methods with post-processing and visualization techniques. For the latter, access to the implementation of the BEM including compression techniques is necessary to modify these methods with respect to efficiency and accuracy [18], [19]. From point of view of a magnet designer, the proposed approach is easily applicable and transparent concerning the computational method. Depending on the user interface of the field computation software, most steps of the BEM–FEM coupling can be hidden.

4 NUMERICAL RESULTS

The proposed BEM–FEM coupling for the computation of magnetic fields has been tested on the magnet introduced in Section 3.1. The yoke and the permanent magnets have been discretized with second order tetrahedral Lagrange elements. A relative fine mesh has been created to consider the non-linear material properties accurately. A relatively coarse mesh for the BEM domain in the ROI and a coarse mesh for the BEM domain of infinite space has been applied (Fig. 7). Between the non-linear FEM domains and the two BEM domains, a FEM domain is added. In total a mesh with 1,232,772 domains and 4,294 boundary elements is obtained.

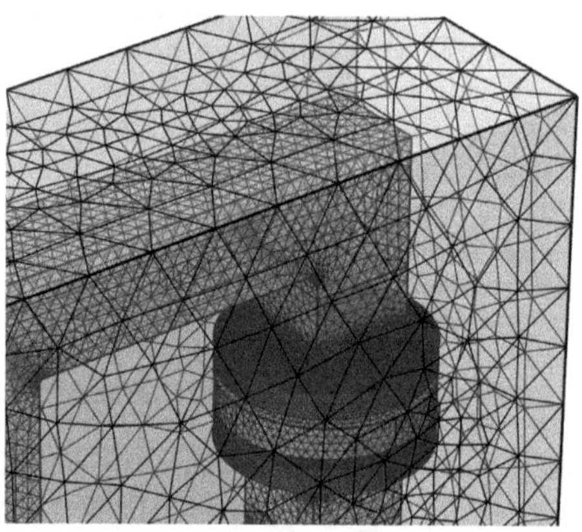

Figure 7: Discretized model of a magnet for a BEM–FEM coupling.

The magnetic properties of the yoke have been modelled based on the non-linear function depicted in Fig. 1. The remanence flux density of the permanent magnets is impressed in the related domains. Since the boundaries of the two BEM domains are in air, continuity of the scalar magnetic potential and of its normal derivative are used as boundary conditions for the BEM–FEM coupling. The ACA has been applied to compress the BEM matrix blocks.

The non-linear problem was solved after three iteration steps of Newton–Raphson method and using BiCGStab solver with direct preconditioner for the underlying linear equation system in 55 minutes on an Intel Xeon CPU E3-1275 v5 with 3.6 GHz clock speed. The memory requirements have been 39 GB.

The norm of the magnetic flux density in a plane through the magnet is depicted in Fig. 8. There, the colour plot has been created using a grid both for the FEM and the BEM domain in the shown plane.

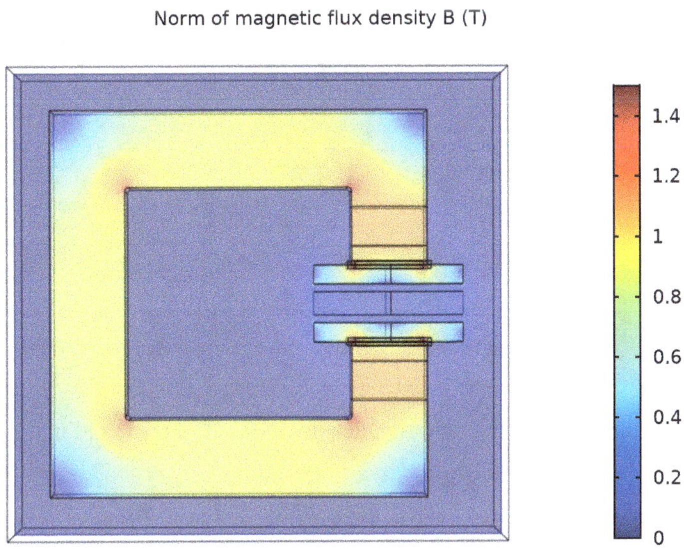

Figure 8: Norm of magnetic flux density computed with BEM and FEM.

The computation of the field lines shown in Fig. 9 has been coordinated by the software MuPhyN. The magnetic flux density in the mesh nodes of the FEM domain was computed in advance and the field lines have been computed using the optimized methods of the FEM solver. In the BEM domain, the field lines haven been obtained using a meshfree BEM post-processing with FMM compression of integrals.

5 CONCLUSIONS

The computation of magnetic fields with a high accuracy, which is required in smart sensor application for NMR or ESR spectroscopy, is implemented by a smart combination and coupling of the finite element method and the boundary element method. The proposed coupling both for the solution of the problem and for the post-processing simplifies the numerical model and ensures reliable numerical results with a high accuracy. In other words, the method is robust and accurate. Thereby, a finite element method in combination with a state-of-the-art boundary element method is very well suited as an easy-to-use and reliable tool for high-end smart sensor design. Mainly the meshfree post-processing of the boundary element method along with an efficient compression technique supports high precision evaluations of magnetic fields in the region of interest.

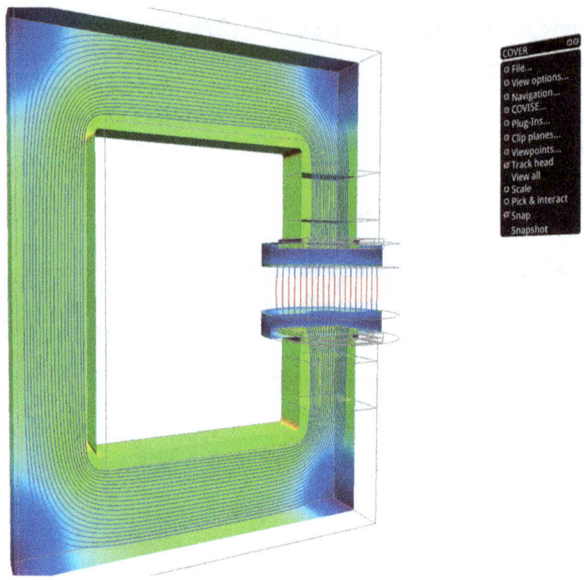

Figure 9: Field lines of the magnetic flux density in the ROI (red) and in the yoke (blue).

REFERENCES

[1] Meijer, G., Makinwa, K. & Pertijs, M., *Smart Sensor Systems: Emerging Technologies and Applications*, Wiley, 2014.

[2] Schlecker, B. et al., VCO-based ESR-on-a-chip as a tool for low-cost, high-sensitivity point-of-care diagnostics. *Proceedings of 2017 IEEE SENSORS*, Glasgow, pp. 1–3, 2017.

[3] Anders, J., Handwerker, J., Ortmanns, M. & Boero, G., A low-power high-sensitivity single-chip receiver for NMR microscopy. *Journal of Magnetic Resonance*, **266**, pp. 41–50, 2016.

[4] Alnajjar, B., Buchau, A., Anders, J. & Blümich, B., An H-shaped low-field permanent magnet for NMR spectroscopy designed using the finite element method. *International Journal of Applied Electromagnetics and Mechanics*, **61**, 2019.

[5] Hafla, W., Buchau, A. & Rucker, W.M., Accuracy improvement in nonlinear magnetostatic field computations with integral equation method and indirect scalar potential formulations. *COMPEL – The International Journal for Computation and Mathematics in Electrical and Electronic Engineering*, **25**(3), pp. 565–571.

[6] Zienkiewicz, O.C. & Zhu, J.Z., The superconvergent patch recovery (SPR) and adaptive finite element refinement. *Computer Methods in Applied Mechanics and Engineering*, **101**, pp. 207–224, 1992.

[7] Huber, C.J., Rieger, W., Haas, M. & Rucker, W.M., The numerical treatment of singular integrals in boundary element calculations. *ACES Journal*, **12**(2), pp. 121–126, 1997.

[8] Bebendorf, M., Approximation of boundary element matrices. *Numerische Mathematik*, **4**, pp. 565–589, 2000.

[9] Greengard, L. & Rokhlin, V., The rapid evaluation of potential fields in three dimensions. *Lecture Notes in Mathematics 1360*, eds C. Anderson & C. Greengard, Springer-Verlag: Berlin, pp. 121–141, 1987.

[10] Buchau, A., Rucker, W.M., Rain, O., Rischmüller, V., Kurz, S. & Rjasanow, S., Comparison between different approaches for fast and efficient 3D BEM computations. *IEEE Transactions on Magnetics*, **39**(3), pp. 1107–1110, 2003.

[11] Buchau, A., Rieger, W. & Rucker, W.M., Fast field computations with the fast multipole method. *COMPEL – The International Journal for Computation and Mathematics in Electrical and Electronic Engineering*, **20**(2), pp. 547–561, 2001.

[12] Buchau, A. & Rucker, W.M., Feasibility of a meshfree post-processing for boundary element methods. *WIT Transactions on Modelling and Simulation*, vol. 61, WIT Press: Southampton and Boston, pp. 327–338, 2015.

[13] Cvetković, M., Dodig, H. & Poljak, D., Numerical comparison of compound and extracted eye models for high frequency dosimetry. *International Journal for Engineering Modelling*, **31**(1–2), 2018.

[14] Zhao, W., Zheng, C., Liu, C. & Chen, H., Minimization of sound radiation in fully coupled structural-acoustic systems using FEM–BEM based topology optimization. *Structural and Multidisciplinary Optimization*, **58**, pp. 115–128, 2018.

[15] Maljaars, P., Kaminski, M. & den Besten, H., Boundary element modelling aspects for the hydro-elastic analysis of flexible marine propellers. *Journal of Marine Science and Engineering*, **67**(6), 2018.

[16] Hafla, W., Weinläder, A., Bardakcioglu, A., Buchau, A. & Rucker, W.M., Efficient post-processing with the integral equation method. *COMPEL – The International Journal for Computation and Mathematics in Electrical and Electronic Engineering*, **26**(3), pp. 873–887, 2007.

[17] Buchau, A., Rucker, W.M., Wössner, U. & Becker, M., Augmented reality in teaching of electrodynamics. *COMPEL – The International Journal for Computation and Mathematics in Electrical and Electronic Engineering*, **28**(4), pp. 948–963, 2009.

[18] Buchau, A. & Rucker, W.M., Meshfree computation of field lines across multiple domains using fast boundary element methods. *IEEE Transactions on Magnetics*, 2015.

[19] Buchau, A. & Rucker, W.M., A meshfree isosurface computation method for boundary element methods. *International Journal of Computational Methods and Experimental Measurements*, **5**(5), pp. 647–658, 2017.

BUCKLING OF PERFORATED PLATES USING THE DUAL RECIPROCITY BOUNDARY ELEMENT METHOD

ROMILDO APARECIDO SOARES JR[1], LEANDRO PALERMO JR[1] & LUIZ CARLOS WROBEL[2,3]
[1]School of Civil Engineering, Architecture and Urban Design, State University of Campinas, Brazil
[2]Department of Mechanical and Aerospace Engineering, Brunel University London, UK
[3]Department of Civil and Environmental Engineering, Pontifical Catholic University of Rio de Janeiro, Brazil

ABSTRACT
The buckling of square perforated plates is analysed with the effect of shear deformation in the bending model. The relationship between the buckling load and the plate thickness is better assessed with this bending model instead of the classical model. The simply supported condition at all sides was adopted in the buckling problem such that opposite sides of the plates were uniformly compressed in one direction. The plates had a thickness to length ratio between 1/1000 and 1/5. This study adopted the dual reciprocity method (DRM) to obtain a formulation without domain integrals. The numerical results obtained were compared with those available in the literature.
Keywords: dual reciprocity method, plate buckling, perforated plates, Reissner plates, boundary elements.

1 INTRODUCTION

In-plane forces affect the bending of plates, and the reduction in plate thickness considerably strengthens the effect of the forces on the plate behavior. Buckling analysis is one of the ways to evaluate the effect of in-plane forces when the plate bending equilibrium includes the effect of geometrical non-linearity. The problem is geometrically nonlinear in the case of large deflections, i.e., the stretching and bending of the plate are coupled [1]. The buckling problem is considered when deflections remain small, and the in-plane forces are only related to in-plane tractions.

Timoshenko and Woinowsky-Krieger [4] presented the basic equation that introduces the effect of geometric non-linearity (GNL) for plate analyses using the classical bending model. Several researchers have shown that the introduction of shear deformation improves the bending model accuracy after the first studies presented by Reissner [2] and Mindlin [3]. The development was extended to bending models considering the effect of shear deformation in several studies that dealt with buckling analyses for thin or moderately thick plates [4]. Dawe and Roufaeil [5] discussed the effect of GNL in the bending of plates with shear deformation and stressed the importance of using the derivatives of rotations as well as the derivatives of deflections in the potential energy density associated with in-plane forces. Mizusawa [6] showed the effect of the derivatives of rotations was greater for certain types of boundary conditions, whereas for others, it was not significant, i.e. the derivatives of deflection would be enough.

The classical bending model has been used in the literature to perform buckling analyses of perforated plates. Levy et al. [7] studied the instability of reinforced perforated plates with a central hole under a uniform compression force. Schlack and Alois [8] computed the critical edge displacement of a simply supported pierced plate under a uniform edge displacement. Yang [9] showed that the buckling load is reduced when a square hole is considered instead of a circular hole. Brown and Yettram [10] studied how the value of the buckling parameter for different load combinations changes with the ratio between the diameter of the hole and the plate side. Shakerley and Brown [11] studied the plate buckling with eccentrically positioned holes. El-Sawy and Nazmy [12] used the finite element method to assess the

www.witpress.com, ISSN 1743-3533 (on-line)
doi:10.2495/BE420081

buckling parameter value for uniaxial loaded plates with different ratios between the diameter of the hole and the plate side. The BEM formulation for plate bending including the effect of shear deformation does not require integration over the thickness, so the problem can be solved as a plane problem. An integral performed on the domain is included in the formulation to allow analyses with the GNL effect. The domain integral can be converted to one performed on the boundary with an additional numerical approach such as the multiple reciprocity method (MRM), the radial integration method (RIM) or the dual reciprocity method (DRM), which is adopted here.

The first time the name DRM was used to the conversion of the domain integral into equivalent boundary integrals was in studies on dynamic and heat transfer problems presented in Nardini and Brebbia [13]. Partridge et al. gave a more detailed explanation of the DRM in a book [14], which included some computer codes used in the method. After the first studies, the DRM has been employed in several BEM formulations including those for plate analyses with the GNL effect as summarized next. Sawaki et al. [15] studied the effect of finite displacements in the bending of thin plates. Elzein and Syngellakis [16] obtained the buckling parameters using the classical bending model. Lin et al. [17] performed studies on the buckling of rectangular and circular plates. Wang et al. [18] presented the solution for the plate-bending problem considering the effect of finite displacements using the von Kärmán equations, similarly to Wen et al. [19] who have also included the effect of shear deformation in the bending model. Purbolaksono and Aliabadi [20] presented a comparison of the results obtained with the DRM versus domain integrations to account for the GNL effect in buckling analyses using the Reissner model. Supriyono and Aliabadi [21] solved plate bending problems with the effect of the shear deformation and included effects of geometrical and physical non-linarites. Purbolaksono and Aliabadi [22] analyzed large deflections in plate bending considering the effect of shear deformation.

The BEM formulation for buckling analyses in Soares and Palermo [23] employed two integrals containing the GNL effect, with one computed in the domain and the other on the boundary. The use of first derivatives of the deflection in kernels of integrals related to the GNL effect, instead of the second derivatives, and the fact that there was no need of relating the derivatives of in-plane forces were the main features of the alternative formulation. This study employs the DRM to replace the domain integral by equivalent boundary integrals in the Soares and Palermo formulation [23]. The numerical implementation employed quadratic shape functions to approximate displacements, plate rotations, distributed shears and moments in the boundary elements as implemented in the study [23], as well as the inverse iteration and Rayleigh quotient to compute the lowest eigenvalue with the corresponding eigenvector. The changes in the value of the buckling parameter according to the plate thickness of perforated plates are studied and compared to those in the literature. The simply supported condition at all sides was adopted in the buckling problem such that opposite sides of the plates were uniformly compressed in one direction. The square plates with a central square hole had a thickness to length ratio between 1/1000 and 1/5.

2 BOUNDARY INTEGRAL EQUATIONS

The buckling analysis considered an isotropic and homogeneous material in the perforated plate under in-plane forces distributed in the domain. The constitutive equations are written next with a unified notation for the Reissner and Mindlin bending models. The Latin indices take on values {1, 2 and 3} and Greek indices take on values {1, 2}

$$M_{\alpha\beta} = D\frac{(1-\nu)}{2}\left(u_{\alpha,\beta} + u_{\beta,\alpha} + \frac{2\nu}{1-\nu}u_{\gamma,\gamma}\delta_{\acute{a}\beta}\right) + \delta_{\alpha\beta}qRE, \tag{1}$$

$$Q_\alpha = D\frac{(1-\nu)}{2}\lambda^2(u_\alpha + u_{3,\alpha}), \tag{2}$$

with

$$\lambda^2 = 12\frac{\kappa^2}{h^2}; \quad RE = \frac{\nu}{\lambda^2(1-\nu)}.$$

D is the flexural rigidity, h is the plate thickness, ν is Poisson's ratio, q is the distributed load on the plate domain and $\delta_{\alpha\beta}$ is the Kronecker delta. The shear parameter κ^2 is equal to 5/6 and $\pi^2/12$ for the Reissner and Mindlin models, respectively. The product qRE in eqn (2) corresponds to the linearly weighted average effect of the normal stress component in the thickness direction, which should be considered in the Reissner model [2] but not in the Mindlin model [3]. This term is null in the buckling problem because the distributed load q is equal to zero.

The displacement boundary integral equations (DBIEs) for the buckling problem used in Soares and Palermo [23] are given next:

$$\frac{1}{2}C_{ij}(x')u_j(x') + \int_\Gamma \left[T_{ij}(x',x)u_j(x) - U_{ij}(x',x)t_j(x)\right]d\Gamma(x) = \cdots$$

$$= \int_\Gamma n_\alpha(x)N_{\alpha\beta}(x)u_{3,\beta}(x)U_{i3}(x',x)d\Gamma(x)$$

$$- \iint_\Omega N_{\alpha\beta}(X)u_{3,\beta}(X)U_{i3,\alpha}(x',X)d\Omega(X), \tag{3}$$

where C_{ij} is an element of the matrix C related to the boundary geometry at the source point, which becomes the identity matrix when a smooth boundary is considered, u_α is the plate rotation in direction α, and u_3 is the plate deflection. U_{ij} represents the rotation (j=1, 2) or the deflection (j=3) due to a unit couple (i=1, 2) or a unit point force (i=3). T_{ij} represents the moment (j=1, 2) or the shear (j=3) due to a unit couple (i=1, 2) or a unit point force (i=3). U_{ij} and T_{ij} are related to the fundamental solution.

According to the development in Soares and Palermo [23], the GNL effect in eqn (3) is introduced with an integral performed on the domain and a boundary integral, which is related to the natural condition in the part of the boundary when the deflection (u_3) is not prescribed. The result in eqn (17) Soares and Palermo [23] showed the boundary integral containing the GNL effect should only be computed along the boundary part with the prescribed deflection. The GNL effect requires the use of the BIE for the gradient of the deflection, which is written next for an internal collocation point:

$$u_{3,\gamma}(X') = \int_\Gamma \{n_\alpha(x)M_{3\alpha\beta,\gamma}(X',x)u_\beta(x) + n_\beta(x)Q_{3\beta,\gamma}(X',x)u_3(x) + \cdots$$

$$\ldots - U_{3\beta,\gamma}(X',x)t_\beta(x) - U_{33,\gamma}(X',x)t_3(x)\}d\Gamma(x)$$

$$- \int_\Gamma n_\alpha(x)N_{\alpha\beta}(x)u_{3,\beta}(x)U_{i3,\gamma}(X',x)d\Gamma(x) + \cdots$$

$$\ldots + \iint_\Omega N_{\alpha\beta}(X)u_{3,\beta}(X)U_{33,\alpha\gamma}(X',X)d\Omega(X). \tag{4}$$

Eqn (4) was written with kernels differentiated with respect to the field point coordinates. The tangential differential operator can be introduced in kernels of integrals in eqn (4) to reduce the singularities resulting from the differentiation as shown in Palermo [24].

The generalized plane stress problem is solved once to obtain the in-plane force distribution used in the buckling analysis. The in-plane force distribution is dependent of the shape and the dimensions of the hole in perforated plates beyond the in-plane tractions distribution on the boundary.

3 APPLICATION OF THE DUAL RECIPROCITY METHOD

The kernel in the domain integral of the DBIEs (eqn (3)) contains the product between the gradient of deflection from the fundamental solution, the in-plane forces tensor and the gradient of the plate deflection. A vector function (b) resulting from the product between the in-plane forces tensor and the gradient of plate deflection can be defined, i.e.:

$$b_\theta(X) = N_{\theta\beta}(X)u_{3,\beta}(X). \tag{5}$$

The DRM is introduced in the approximation of the vector function (b) using the following relation [14]:

$$b_\theta(X) = \textstyle\sum_{m=1}^{N+L} \alpha_\theta^m f^m. \tag{6}$$

The summation in eqn (6) is extended along all points employed in the DRM, i.e. the total number of points placed on the boundary (N) and in the domain (L), f^m and α^m are sets of the approximating functions and weighting coefficients, respectively [14]. Furthermore, the application of the DRM employs the particular solution $\hat{u}_j^m$ [14], which is related to the approximating function f^m according to following equations:

$$\nabla^2(\nabla^2\phi) = \tfrac{1}{D} f(r). \tag{7}$$

$$\nabla^2(\hat{u}_3) = \tfrac{2}{D(1-\upsilon)\lambda^2} f(r) - \nabla^2\phi. \tag{8}$$

$$\hat{u}_\alpha = \phi_{,\alpha}. \tag{9}$$

Eqns (7) to (9) are similar to those used to obtain the fundamental solution for the unit point load under the static condition in Palermo [25] and the dynamic condition in Palermo [26]. The well-known radial basis function (1+r) is considered for f^m. The corresponding particular solution, which is similar to that presented in Wen et al. [27], is given by:

$$f^m = 1 + r \Rightarrow \begin{cases} \hat{u}_\alpha^m = \frac{1}{D}\left(\frac{r^3}{16} + \frac{r^4}{45}\right) r_{,\alpha}, \\ \hat{u}_3^m = \frac{2}{D(1-\upsilon)\lambda^2}\left(\frac{r^2}{4} + \frac{r^3}{9}\right) - \frac{1}{D}\left(\frac{r^4}{64} + \frac{r^5}{225}\right). \end{cases} \tag{10}$$

The distributed shear and bending moments related to the particular solutions are obtained with the constitutive eqns (1) and (2). The generalized tractions are obtained from the distributed shear and bending moments using the following relations:

$$\begin{aligned} \hat{t}_\alpha^m &= \widehat{M}_{\alpha\beta}^m n_\beta \\ \hat{t}_3^m &= \hat{Q}_\beta^m n_\beta. \end{aligned} \tag{11}$$

The gradient of the deflection ($U_{i3,\alpha}$) from the fundamental solution multiplies the terms related to the GNL effect in the DBIEs (eqn (3)). The introduction of the DRM to replace the domain integral requires the use of the BIE for the gradient of the deflections. This equation is written next with the kernel of the domain integral containing the function f^m and the boundary integrals using the displacements and tractions related to the particular solution $\hat{u}_j^m$ [14]:

$$c\hat{u}_{i,\theta}^m(x') = \int_\Gamma \left[n_\alpha(x)M_{i\alpha\beta,\theta}(x',x)\hat{u}_\beta^m(x) + n_\beta(x)Q_{i\beta,\theta}(x',x)\hat{u}_3^m(x) + \cdots\right.$$
$$\left. \ldots - U_{i\beta,\theta}(x',x)\hat{t}_\beta^m(x) - U_{i3,\theta}(x',x)\hat{t}_3^m(x)\right]d\Gamma(x) - \iint_\Omega f^m(X)U_{i3,\theta}(x',X)d\Omega(X). \quad (12)$$

Eqn (12) was written with kernels differentiated with respect to the field coordinates, the scalar c is equal to 1 in case of collocation points located inside the domain and 0.5 in case of points placed on smooth parts of the boundary. The continuity of the first derivative of displacements at x\` is required to apply eqn (12) at points on the boundary. The DBIE for the buckling problem using eqn (12) to introduce the DRM is given by

$$\frac{1}{2}C_{ij}(x')u_j(x') + \int_\Gamma \left[T_{ij}(x',x)u_j(x) - U_{ij}(x',x)t_j(x)\right]d\Gamma(x) = \cdots$$
$$= \int_\Gamma n_\alpha(x)N_{\alpha\beta}(x)u_{3,\beta}(x)U_{i3}(x',x)d\Gamma(x) + \sum_{m=1}^{N+L}\alpha_\theta^m\left\{c\hat{u}_{i,\theta}^m(x') + \cdots\right.$$
$$\ldots\ldots\ldots - \int_\Gamma \left[n_\alpha(x)M_{i\alpha\beta,\theta}(x',x)\hat{u}_\beta^m(x) + n_\beta(x)Q_{i\beta,\theta}(x',x)\hat{u}_3^m(x) + \cdots\right.$$
$$\left.\left.\ldots - U_{i\beta,\theta}(x',x)\hat{t}_\beta^m(x) - U_{i3,\theta}(x',x)\hat{t}_3^m(x)\right]d\Gamma(x)\right\}. \quad (13)$$

The second derivatives of the deflection ($U_{33,\alpha\gamma}$) of the fundamental solution multiply the terms related to the GNL effect in the BIE for the gradient of the deflection (eqn (4)). The introduction of the DRM also requires the BIE for the second derivative of the deflection, as explained to obtain eqn (13). The BIE for the second derivative of the deflection with the kernel of the domain integral containing f^m is given by:

$$c\hat{u}_{3,\gamma\theta}^m(x') + \int_\Gamma \left[n_\alpha(x)M_{3\alpha\beta,\gamma\theta}(x',x)\hat{u}_\beta^m(x) + n_\beta(x)Q_{3\beta,\gamma\theta}(x',x)\hat{u}_3^m(x) + \cdots\right.$$
$$\left.\ldots - U_{3\beta,\gamma\theta}(x',x)\hat{t}_\beta^m(x) - U_{33,\gamma\theta}(x',x)\hat{t}_3^m(x)\right]d\Gamma(x)$$
$$= \iint_\Omega f^m(X)U_{33,\gamma\theta}(x',X)d\Omega(X). \quad (14)$$

Eqn (14) was written with kernels differentiated with respect to the field coordinates, the scalar c is equal to 1 in case of collocation points located inside the domain and 0.5 in case of points placed on smooth parts of the boundary. The continuity of the second derivative of the deflection at x\` is required to apply eqn (14) at points on the boundary. The BIE for the second derivative of the deflection using eqn (14) to introduce the DRM is given by.

$$gu_{3,\gamma}(x') = \int_\Gamma \left\{n_\alpha(x)M_{3\alpha\beta,\gamma}(x',x)u_\beta(x) + n_\beta(x)Q_{3\beta,\gamma}(x',x)u_3(x) + \cdots\right.$$
$$\left.\ldots - U_{3\beta,\gamma}(x',x)t_\beta(x) - U_{33,\gamma}(x',x)t_3(x)\right\}d\Gamma(x)$$
$$- \int_\Gamma n_\alpha(x)N_{\alpha\beta}(x)u_{3,\beta}(x)U_{i3,\gamma}(x',x)d\Gamma(x) + \cdots$$
$$\ldots + \sum_{m=1}^{N+L}\alpha_\theta^m\left\{c\hat{u}_{i,\gamma\theta}^m(x') + \int_\Gamma \left[n_\alpha(x)M_{i\alpha\beta,\gamma\theta}(x',x)\hat{u}_\beta^m(x) + n_\beta(x)Q_{i\beta,\gamma\theta}(x',x)\hat{u}_3^m(x) + ..\right.\right.$$
$$\left.\left.\ldots - U_{i\beta,\gamma\theta}(x',x)\hat{t}_\beta^m(x) - U_{i3,\gamma\theta}(x',x)\hat{t}_3^m(x)\right]d\Gamma(x)\right\}. \quad (15)$$

The scalar g in eqn (15) has the same values explained to the scalar c in eqns (12) and (14).

4 NUMERICAL IMPLEMENTATION

The discretization of the BIEs employed quadratic isoparametric boundary elements and the collocation points were always placed on the boundary. The same mapping function was used for conformal and non-conformal interpolations. The collocation points were placed at positions (-0.67 and 0.0), in the range (1, 1), in the case of continuous elements and at positions (-0.67, 0.0, +0.67) in the case of discontinuous elements, i.e. the collocation points were always shifted inside the boundary elements. The singularity subtraction [28] and the transformation of variable technique [29] were employed for the Cauchy and weak-type singularities, respectively, when integrations were performed on elements containing the collocation points. Hyper- and super-singular integrals in elements containing the collocation points were numerically evaluated using the computer code presented in Gao [30]. The standard Gauss-Legendre scheme was employed for integrations on elements not containing the collocation points.

The DRM considered points distributed in the domain and on the boundary. The points on the boundary were placed at the positions of the collocation points. The first boundary integral on the left-hand side (LHS) of eqn (13) is not related with the DRM, as well as the second boundary integral on the LHS of eqn (15). The discretization of those boundary integrals assumed a constant value along each boundary element for the sum of products between the gradient of the deflection and the in-plane forces, which values were obtained at the central node. The values of derivatives of the deflection computed at the central node used by the DRM were also used in the computation of those boundary integrals.

The eigenvalue analysis used the basic inverse iteration with the Rayleigh quotient [31] as employed in Soares and Palermo [23] and written next:

$$Ax^{(k+1)} = Bx^{k}, \tag{16}$$

$$\lambda_k = \frac{(x^{(k+1)}, x^{k})}{(x^{(k+1)}, x^{(k+1)})}. \tag{17}$$

The vector x^k in eqns (16) to (17) is related to values of the gradient of the deflection at the DRM points. The discretized forms of eqns (13) and (15) were used instead of eqn (16) as done in Soares and Palermo [23]. Starting with an eigenvector x^1 with elements equal to 1.0, the values of the displacements and tractions at the boundary nodes are found with eqn (13); these values are introduced in eqn (15) to obtain the gradient of the deflection (elements of the eigenvector x^2). The elements of α are recalculated at each iteration step according to values of the gradient of the deflection introduced in eqn (5).

After the computation of elements b_θ, the solution of the system of equations given by eqn (6) carries the values of α_θ^m as explained in detail in Partridge et al. [14]. The index θ in elements of b and α requires eqns (5) and (6) to be used twice and one set of α is obtained for each direction θ of the plate.

5 NUMERICAL EXAMPLES

The perforated plate shown in Fig. 1 was studied in El-Sawy and Nazmy [12] where a correction was suggested for the buckling parameters presented in Sabir and Chow [32] for cases of large dimension holes. The results obtained in Soares and Palermo [23] for this problem were very close to those in El-Sawy and Nazmy [12] and were considered here to evaluate the application of the DRM.

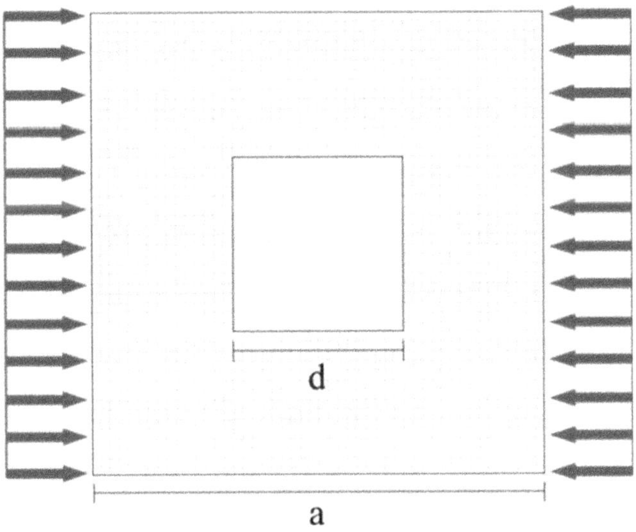

Figure 1: Square plate containing a square hole at the center.

The buckling parameter k is a non-dimensional value related to the critical load of the plate (N_{cr}), the length of the plate side (a) and the flexural rigidity (D), which is obtained according to following expression:

$$k = \frac{a^2 N_{cr}}{\pi^2 D}. \tag{18}$$

The simply supported condition at all sides was adopted in the buckling problem such that opposite sides of the plates were uniformly compressed in one direction. The plates had a thickness to length ratio between 1/1000 and 1/5. The normalized square hole dimensions (d/a) were in the range 0.1 to 0.7. The Young's modulus (E) and Poisson's ratio (ν) were 206.9 GPa and 0.3, respectively. The value of the shear parameter κ^2 was $\pi^2/12$. The comparison between values for the buckling parameter obtained with the DRM and those presented in [23], in which the GNL effect was introduced with domain integration using cells, are presented in Tables 1 to 5.

Table 1: Critical loads for the perforated plate obtained with DRM.

Boundary elements	Domain points	d/a	h/a = 0.001		
			DRM	[23]	Diff. (%)
480	1564	0.1	3.7975	3.7994	-0.05
480	1500	0.2	3.4442	3.4494	-0.15
480	1568	0.3	3.1854	3.1883	-0.09
480	1536	0.4	3.0194	3.0297	-0.34
480	1533	0.5	2.9066	2.9310	-0.84
480	1548	0.6	2.8431	2.8664	-0.82
1280	1632	0.7	2.8168*	2.8449	-1.00

Table 2: Critical loads for the perforated plate obtained with DRM.

Boundary elements	Domain points	d/a	h/a = 0.01		
			DRM	[23]	Diff. (%)
480	1564	0.1	3.7909	3.7932	-0.06
480	1500	0.2	3.4343	3.4412	-0.20
480	1568	0.3	3.1723	3.1782	-0.18
480	1536	0.4	3.0012	3.0153	-0.47
480	1533	0.5	2.8879	2.9096	-0.75
480	1548	0.6	2.8198	2.8352	-0.55
480	1632	0.7	2.8131	2.7998	0.47

Table 3: Critical loads for the perforated plate obtained with DRM.

Boundary elements	Domain points	d/a	h/a = 0.05		
			DRM	[23]	Diff. (%)
480	1564	0.1	3.7284	3.7313	-0.08
480	1500	0.2	3.3659	3.3747	-0.26
480	1568	0.3	3.0957	3.1034	-0.25
480	1536	0.4	2.9028	2.9207	-0.62
480	1533	0.5	2.7565	2.7820	-0.92
480	1548	0.6	2.6412	2.6625	-0.81
480	1632	0.7	2.5730	2.5629	0.39

Table 4: Critical loads for the perforated plate obtained with DRM.

Boundary elements	Domain points	d/a	h/a = 0.10		
			DRM	[23]	Diff. (%)
480	1564	0.1	3.5780	3.5819	-0.11
480	1500	0.2	3.2222	3.2344	-0.38
480	1568	0.3	2.9471	2.9577	-0.36
480	1536	0.4	2.7247	2.7495	-0.91
480	1533	0.5	2.5346	2.5670	-1.28
480	1548	0.6	2.3614	2.3884	-1.14
480	1632	0.7	2.2611	2.2716	-0.47

Table 5: Critical loads for the perforated plate obtained with DRM.

Boundary elements	Domain points	d/a	h/a = 0.20		
			DRM	[23]	Diff. (%)
480	1564	0.1	3.0798	3.1043	-0.79
480	1500	0.2	2.7555	2.8052	-1.81
480	1568	0.3	2.4799	2.5138	-1.37
480	1536	0.4	2.2025	2.2235	-0.95
1280	1533	0.5	1.6960*	1.7185	-1.33
1280	1548	0.6	1.2087*	1.1755	2.75
1280	1632	0.7	0.7813*	0.7263	7.04

6 CONCLUSIONS

The results obtained with the application of the DRM in the formulation presented in Soares and Palermo [23] were close to the solution with the domain integration using cells. Most solutions employed 480 quadratic boundary elements (384 elements on the external boundary and 96 elements along the boundary of the hole) while a few cases required 1280 quadratic boundary elements (1024 elements on the external boundary and 256 elements along the boundary of the hole). A simple radial basis function (1+r) for f^m was used and no shape parameter was adopted. No additional tools were employed to improve the convergence or the integrations over the boundary elements in the numerical implementation of the DRM. The authors believe this DRM formulation for buckling analysis presented accurate results and a consistent behavior with reference to that shown in Soares and Palermo [23].

The differences in the variation of the buckling parameter according to the thickness to plate side ratio shown in Fig. 2 could only be identified when the effect of shear deformation was included in the bending model. The next step in the study of the buckling of perforated plates will be the analysis of circular holes in square and rectangular perforated plates, which will be presented soon.

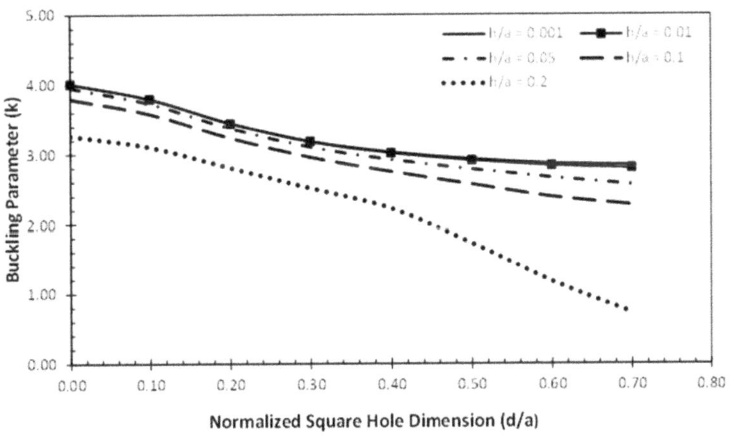

Figure 2: The effect of the hole dimension on the value of the buckling parameter.

ACKNOWLEDGEMENTS

The authors are grateful to CNPq (305997/2016-5) and CAPES (01P03713/2017) for their support in the development of this research.

REFERENCES

[1] Palermo Jr., L., Simões, R. & Mesquita Neto, E., *Finite Displacement Analysis of Plates Subjected to Flexure and Compression Using the Boundary Element Method, International Colloquium on Stability and Ductility of Steel Structures*, Lisboa, 2006.
[2] Reissner, E., The effect of transverse shear deformation on the bending of elastic plates. *Journal of Applied Mechanics*, **12**(2), pp. A66–A77, 1945.
[3] Mindlin, R.D., Influence of rotatory inertia and shear on flexural motions of isotropic elastic plates. *Journal of Applied Mechanics*, **18**, pp. 18–31, 1951.
[4] Timoshenko, S.P. & Woinowsky-Krieger, S., *Theory of Plates and Shells*, McGraw-Hill Book Company: New York, 2nd ed., 1959.
[5] Dawe, D.J. & Roufaeil, O.L., Buckling of rectangular Mindlin plates. *Computers & Structures,* **15**(4), pp. 461–471, 1982.
[6] Mizusawa, T., Buckling of rectangular Mindlin plates with tapered thickness by the spline strip method. *International Journal of Solids & Structures*, **30**(12), pp. 1663–1677, 1993.
[7] Levy, S., Woolley, R.M. & Kroll, W.D., Instability of simply supported square plate with reinforced circular hole in edge compression. *Journal of Research*, pp. 571–577, 1947.
[8] Schlack, J.R. & Alois, L., Elastic stability of pierced square plates. *Experimental Mechanics*, **4**(6), pp. 167–172, 1964.
[9] Yang, H.T., A *Finite Element Formulation for Stability Analysis of Doubly Curved Thin Shell Structures*. PhD thesis, Cornell University, 1969.
[10] Brown, C.J. & Yettram, A.L., The elastic stability of square perforated plates under combinations of bending, shear and direct load. *Thin-Walled Structures*, **4**(3), pp. 239–246, 1986.
[11] Shakerley, T.M. & Brown, C.J., Elastic buckling of plates with eccentrically positioned rectangular perforations. *International Journal of Mechanical Sciences*, **38**(8), pp. 825–838, 1996.
[12] El-Sawy, K.M. & Nazmy, A.S., Effect of aspect ratio on the elastic buckling of uniaxially loaded plates with eccentric holes. *Thin-Walled Structures*, **39**(12), pp. 983–998, 2001.
[13] Nardini, D. & Brebbia, C.A., Boundary integral formulation of mass matrices for dynamic analysis. *Time-dependent and Vibration Problems*, Springer: Berlin and Heidelberg, pp. 191–208, 1985.
[14] Partridge, P.W., Brebbia, C.A. & Wrobel, L.C., *The Dual Reciprocity Boundary Element Method*, Computational Mechanics Publication and Elsevier Applied Science: London and New York, 1992.
[15] Sawaki, Y., Takeuchi, K. & Kamiya, N., Nonlinear bending analysis without domain-cell discretisation. *Engineering Analysis with Boundary Elements*, **7**(3), pp. 130–135, 1990.
[16] Elzein, A. & Syngellakis, S., The dual reciprocity in boundary element formulations of the plate buckling problem. *Engineering Analysis with Boundary Elements*, **9**(2), pp. 175–184, 1992.

[17] Lin, J., Duffield, R.C. & Shih, H., Buckling analysis of elastic plates by boundary element method. *Engineering Analysis with Boundary Elements*, **23**(2), pp. 131–137, 1999.
[18] Wang, W., Ji, X. & Tanaka, M., A dual reciprocity boundary element approach for the problems of large deflection of thin elastic plates. *Computational Mechanics*, **26**(1), pp. 58–65, 2000
[19] Wen, P.H., Aliabadi, M.H. & Young, A., Large deflection analysis of Reissner plate by boundary element method. *Computers & Structures*, **83**(10–11), pp. 870–879, 2005.
[20] Purbolaksono, J. & Aliabadi, M.H., Large deformation of shear deformable plate by boundary element method. *Journal of Engineering Mathematics*, **51**, pp. 211–230, 2005.
[21] Supriyono & Aliabadi, M.H., Analysis of shear deformable plates with combined geometric and material nonlinearities by boundary element method. *International Journal of Solids and Structures*, **44**(3–4), pp. 1038–1059, 2007.
[22] Purbolaksono, J. & Aliabadi, M.H., Stability of Euler's method for evaluating large deformation of shear deformable plates by dual reciprocity boundary element method. *Engineering Analysis with Boundary Elements*, **34**(9), pp. 819–823, 2010.
[23] Soares Jr., R.A. & Palermo Jr., L., Effect of shear deformation on the buckling parameter of perforated and non-perforated plates studied using the boundary element method. *Engineering Analysis with Boundary Elements*, **85**, pp. 57–69, 2017.
[24] Palermo Jr., L., The tangential differential operator applied to a stress boundary integral equation for plate bending including the shear deformation effect. *Engineering Analysis with Boundary Elements*, **36**, pp. 1213–1225, 2012.
[25] Palermo Jr., L., Plate bending analysis using the classical or the Reissner–Mindlin models. *Engineering Analysis with Boundary Elements*, **27**, pp. 603–609, 2003.
[26] Palermo Jr., L., On the harmonic response of plates with the shear deformation effect using the elastodynamic solution in the boundary element method. *Engineering Analysis with Boundary Elements*, **31**, pp. 176–183, 2007.
[27] Wen, P.H., Aliabadi, M.H. & Young, A., Application of dual reciprocity method to plates and shells. *Engineering Analysis with Boundary Elements*, **24**(7–8), pp. 583–590, 2000
[28] Aliabadi, M.H., *The Boundary Element Method: Applications in Solids and Structures*, John Wiley & Sons Ltd., 2002.
[29] Telles, J.C.F., A self-adaptive coordinate transformation for efficient numerical evaluation of general boundary element integrals. *International Journal for Numerical Methods in Engineering*, **24**, pp. 959–973, 1987.
[30] Gao, X., Numerical evaluation of two-dimensional singular boundary integrals – Theory and Fortran code. *Journal of Computational and Applied Mathematics*, **188**, pp. 44–64, 2006.
[31] Wilkinson, J.H., *The Algebraic Eigenvalue Problem*, Oxford University Press and William Clowes & Sons: London, 1972.
[32] Sabir, A.B. & Chow, F.Y., Elastic buckling of flat panels containing circular and square holes. *Proceedings of the International Conference on Instability and Plastic Collapse of Steel Structures*, ed. L.J. Morris, Granada Publishing: London, pp. 311–21, 1983.

BEM–EDM COUPLED ANALYSIS OF MULTI-SCALE PROBLEMS

YONG-TONG ZHENG, XIAO-WEI GAO & HAI-FENG PENG
State Key Laboratory of Structural Analysis for Industrial Equipment, Dalian University of Technology, China

ABSTRACT

In this paper, the element differential method (EDM), a new numerical method proposed recently, is coupled with the boundary element method (BEM), a traditional numerical method, for solving general multi-scale heat conduction problems. The basic algebraic equations in BEM are formulated in terms of temperatures and heat fluxes, which are the same as those in EDM. So, when coupling these two methods, we do not need to transform the variables like the thermal loads into heat fluxes as done with the finite element method (FEM). The key task in the proposed coupled method is to use the temperature consistency condition and the flux equilibrium equation at interface nodes of the two methods to eliminate all BEM nodes except for those on the interfaces. The detailed elimination process is presented in this paper, which can result in the final system of equations without iteration. The coefficient matrix of the final coupled system is sparse, even though a small part is dense. The coupled method inherits the advantage of EDM in flexibility and computational efficiency and the advantage of BEM in the robustness of treating multi-scale problems. A numerical example is given to demonstrate the correctness of this coupled method.

Keywords: coupled analysis, boundary element method, element differential method, multi-scale problems.

1 INTRODUCTION

In mathematics, to solve the type of second-order partial differential equations (PDEs) like heat conduction problems, there are a number of numerical methods available. The frequently used methods are finite element method (FEM) [1]–[7], boundary element method (BEM) [8]–[15] and some of mesh free methods (MFM) [16]–[21]. FEM is the most widely used method due to its flexibility and applicability and most engineering problems can be solved by it. BEM only needs to discretize the boundary of the problem into elements, which not only can reduce the dimension, but also can easily simulate some special cases like crack and stress concentration [8], [12]. However, the final system of BEM has a dense coefficient matrix and the fundamental solutions are derived from the linear problems, which severely limits the application of BEM in large problems, nonlinear and nonhomogeneous problems [12], [15]. Different from mesh-dependent methods, MFM only needs a group of distributed nodes in the computational domain and barely needs elements. Therefore, MFM can easily simulate the problems with irregular geometries. However, MFM has the drawbacks of time consuming to form the global shape functions and difficulty to apply boundary conditions [16], [17].

For multi-scale problems, they have at least a large-scale part and a small-scale part in the computational domain. For example, considering a big block with side length of 50 mm and 0.5 mm coatings around the surface, its large-scale part is block and its small-scale parts are coatings.

When we analyse heat conduction using FEM, both the block and its coating should be discretized into elements. To avoid the ill-formed elements in the coating, the size of each element should be close to the thickness of the coating, which may result in too many elements in modelling the block mentioned above. That is, the thinner the coating, the greater the number of elements.

WIT Transactions on Engineering Sciences, Vol 126, © 2019 WIT Press
www.witpress.com, ISSN 1743-3533 (on-line)
doi:10.2495/BE420091

If we use BEM to solve this type of problem, the scale of different dimensions has less influence on the scale of elements, since we only need to discretize the boundary of the computational domain. However, if the block itself is too big, the total number of nodes will be so many so that the system with dense coefficient matrix to be solved will be very large. Both forming and solving the system are time consuming.

A better way to solve multi-scale problems is using BEM to analyse the coating, and FEM to analyse the block. Then the two systems are coupled together by temperature consistency and flux equilibrium on the interface. However, the basic algebraic equations in BEM are formulated in terms of temperatures and heat fluxes, while those in FEM are formulated in terms of temperatures and thermal loads. So, when coupling these two methods, we should transform the thermal loads into the heat fluxes, which need interpolation and integration [22].

Recently, a new numerical method called element differential method (EDM) is proposed by Gao et al. [23]–[25] and Cui et al. [26]. It is also a flexible method like FEM and can be used in most engineering problems. Crucially, EDM algebraic equations are formulated in terms of temperatures and heat fluxes, the same as with BEM. Therefore, it is easier to couple EDM and BEM together.

In this paper, the theory of EDM and BEM will be introduced briefly, while the coupling process of two methods is mainly illustrated. Finally, an example is given to verify the correctness of this coupled algorithm.

2 GOVERNING EQUATIONS FOR HEAT CONDUCTION PROBLEMS

The main research theme of this paper is multi-scale heat conduction and mechanical problems. Therefore, the governing equations of two problems are considered in the paper.

2.1 Steady heat conduction problems

For transient heat conduction problems, discretization in the time domain is exactly the same as with FEM and will not be considered here.

For steady heat conduction problems with a spatially varying thermal conductivity and heat source, the governing equation can be expressed as

$$\frac{\partial}{\partial x_i}\left(\lambda_{ij}(x)\cdot\frac{\partial T(x)}{\partial x_j}\right)+Q(x)=0 \qquad x\in\Omega, \tag{1}$$

and three types of boundary conditions are shown as follows:

$$T(x)=\overline{T} \qquad \boldsymbol{x}\in\Gamma_1, \tag{2}$$

$$-\lambda_{ij}(x)\cdot\frac{\partial T(x)}{\partial x_j}\cdot n_i(x)=\overline{q} \qquad x\in\Gamma_2, \tag{3}$$

$$-\lambda_{ij}(x)\cdot\frac{\partial T(x)}{\partial x_j}\cdot n_i(x)=h\left(T(x)-T_\infty\right) \qquad x\in\Gamma_3, \tag{4}$$

in which, Ω is the computation domain of problem, Γ is the boundary of Ω, $\Gamma=\Gamma_1+\Gamma_2+\Gamma_3$, $\boldsymbol{x}$ represents coordinate vector, $\boldsymbol{n}$ is the outward normal to the boundary Γ, h is the heat transfer coefficient; $\overline{q}$ and $\overline{T}$ are the prescribed heat flux and temperature on the boundary, respectively; T_∞ is the prescribed environment temperature; λ is the thermal conductivity tensor, T is the temperature, and Q is the heat-generation rate. The repeated subscripts i and j represent the summation over their range which is 2 for 2D and 3 for 3D problems.

3 ELEMENT DIFFERENTIAL METHOD

The main idea of EDM is to transform the physical variables and their first- and second-order partial derivatives in PDE into shape functions and their same order derivatives. Usually, we use the shape function of an isoparametric element [1], [2] because of its advantage in geometry representation and physical variable interpolation. In this section, it is shown briefly how to represents the physical variable derivatives by using shape function derivatives and assemble the final system.

3.1 Computational process from physical variable derivatives to shape function derivatives

For the spatial coordinates and physical variables within the element, we can express them by shape function as:

$$x_i = N_\alpha x_i^\alpha , \tag{5}$$

$$T = N_\alpha T^\alpha , \tag{6}$$

where x_i^α and T^α denote the coordinates and temperatures at the α node over an element, respectively; N_α is the shape function at the node α; the repeated index α represents the summation over all nodes in the element; the subscript i has values 1, 2 and 3 which represent x-, y- and z-directions for 3D problem.

The partial derivatives of physical variables shown in the governing equations can also be expressed by shape function:

$$\frac{\partial T}{\partial x_i} = \frac{\partial N_\alpha}{\partial x_i} T^\alpha , \tag{7}$$

in which $\partial N_\alpha / \partial x_i$, the shape function derivative with respect to global coordinates, can be transformed into them with respect to intrinsic coordinates:

$$\frac{\partial N_\alpha}{\partial x_i} = \frac{\partial N_\alpha}{\partial \xi_j} \cdot \frac{\partial \xi}{\partial x_i} = \frac{\partial N_\alpha}{\partial \xi_j} \cdot \left(\frac{\partial x_i}{\partial \xi_j} \right)^{-1} , \tag{8}$$

where $\partial x_i / \partial \xi_j$ is the Jacobian matrix mapping from the global coordinate system x_i to the intrinsic coordinate system ξ_j , with $(x_1, x_2) = (x, y)$ and $(\xi_1, \xi_2) = (\xi, \eta)$ for 2D problems, and $(x_1, x_2, x_3) = (x, y, z)$ and $(\xi_1, \xi_2, \xi_3) = (\xi, \eta, \varsigma)$ for 3D problems. It should be noticed that $(\partial x_i / \partial \xi_j)^{-1}$ means inversed Jacobian matrix. According to eqn (5), all items in Jacobian matrix can be easily computed using the following expression:

$$[J]_{ij} = \frac{\partial x_i}{\partial \xi_j} = \frac{\partial N_\alpha}{\partial \xi_j} x_i^\alpha . \tag{9}$$

The equations above shows the computational process of the first-order derivatives. Second-order derivatives can be obtained by the same way as shown in the following equations:

$$\frac{\partial^2 T}{\partial x_i \partial x_j} = \frac{\partial^2 N_\alpha}{\partial x_i \partial x_j} T^\alpha , \tag{10}$$

$$\frac{\partial^2 N_\alpha}{\partial x_i \partial x_j} = \frac{\partial}{\partial x_j}\left(\frac{\partial N_\alpha}{\partial x_i}\right) = \frac{\partial}{\partial x_j}\left(\frac{\partial N_\alpha}{\partial \xi_k}\cdot [J]_{ik}^{-1}\right) = \left[[J]_{ik}^{-1}\frac{\partial^2 N_\alpha}{\partial \xi_k \partial \xi_l} + \frac{\partial [J]_{ik}^{-1}}{\partial \xi_l}\frac{\partial N_\alpha}{\partial \xi_k}\right][J]_{jl}^{-1} . \tag{11}$$

3.2 Assembling the system of equations based on governing equations and boundary conditions

When solving the heat conduction or mechanical problem by means of the EDM, the computational domain Ω needs to be discretized into a series of elements that contain some nodes like FEM. The system of equations is established using a collocation scheme at each node. That is, one node matches only one equation and the nodes with different positions in the element have different equations. According to their positions, 3 types of nodes are defined as internal, interface and outer surface nodes. For example, in Fig. 1, there are four 9-node elements in the computational domain. The internal nodes refer to as the nodes located completely within elements like nodes number 7, 9, 17 and 19 in Fig. 1, and the interface nodes refer to as the nodes shared by a number of elements like nodes number 8, 12, 13, 14 and 18 in Fig. 1, whereas the outer surface nodes refer to as the nodes located on the outer boundary of the problem like the rest of the nodes not mentioned above in Fig. 1.

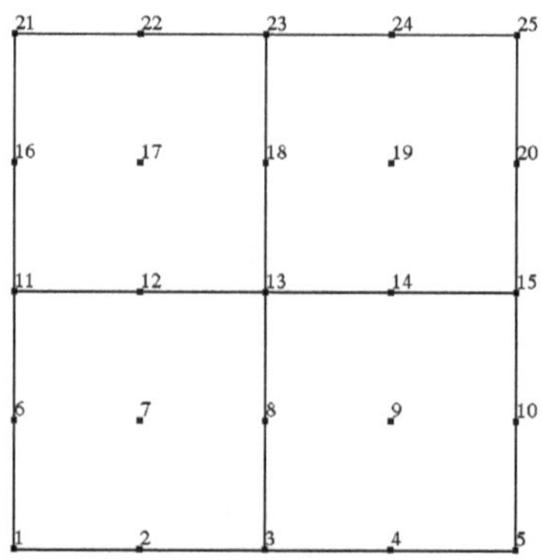

Figure 1: 9-node elements computational domain.

For the internal nodes, eqn (1) should be satisfied. Thus, by substituting eqns (7) and (10) into eqn (1), we can obtain the internal node equation for heat conduction problems:

$$\left[\frac{\partial N_\beta(\xi)}{\partial x_i} k_{ij}^\beta \frac{\partial N_\alpha(\xi)}{\partial x_j} + k_{ij}(\xi)\frac{\partial^2 N_\alpha(\xi)}{\partial x_i \partial x_j}\right] T^\alpha + Q(\xi) = 0 . \tag{12}$$

For the interface nodes, the traction equilibrium condition should be satisfied, i.e.,

$$\sum_{f=1}^{M} q^{f}(\xi_b) = 0, \tag{13}$$

where M is the number of element surface which the interface node is located on. Similarly, substituting eqns (7) and (10) into eqns (13), we can obtain the interface node equation for heat conduction problems (see eqn (14))

$$\sum_{f=1}^{M} -k_{ij}(\xi_b)\frac{\partial N_\alpha(\xi_b)}{\partial x_j} n_i^f(\xi_b) T^\alpha = 0. \tag{14}$$

For the outer surface nodes, the right-hand sides of eqn (13) are not 0, but $\bar{q}$ or $\bar{t_i}$. Therefore, the equation for this kind of nodes become

$$\sum_{f=1}^{M} -k_{ij}(\xi_b)\frac{\partial N_\alpha(\xi_b)}{\partial x_j} n_i^f(\xi_b) T^\alpha = q(\xi_b). \tag{15}$$

Using the above equations, we can list a set of equations involving nodal temperatures and heat fluxes. Multiplying the specified values of temperatures and heat fluxes with their corresponding coefficients, a known vector $\boldsymbol{b}$ can be formed, which is usually moved to the right-hand side of the system of equations. The other terms associated with unknowns are set on the left-hand side of the system of equations. Then, the final system equation can be written as:

$$[A]\{x\} = \{b\}, \tag{16}$$

where $\boldsymbol{x}$ is a vector containing unknowns of each node; $\boldsymbol{A}$ is a sparse coefficient matrix.

By solving eqn (16) for vector $\boldsymbol{x}$, we can obtain all unknowns of the problem. That is all the solution process of heat conduction using EDM.

4 BOUNDARY ELEMENT METHOD

The BEM is a widely known numerical method. So, the BEM theory is introduced briefly in this section.

For heat conduction problems, the boundary integral equation can be expressed as [8]:

$$c(P)T(P) + \int_\Gamma q^*(Q,P)T(Q)d\Gamma(Q) = \int_\Gamma T^*(Q,P)q(Q)d\Gamma(Q), \tag{17}$$

where c=1/2 for smooth boundary points and c=1 for interior points; P and Q represent the source points and field points, respectively; T and q are the temperature and heat flux defined on the boundary Γ; T^* and q^* are the fundamental solutions for heat conduction problems, which are Green function and its derivative [27], [28].

For a BEM computational domain, we can just discretize its boundary into elements. Therefore, a 2D problem needs 1D line elements and a 3D problem needs 2D surface elements. Each node in the elements have their own boundary integral equation. Getting those equations together and moving the temperatures to left-hand side of the system and the heat fluxes to the right-hand side, we can obtain:

$$HT = Gq, \tag{18}$$

where H is the coefficient matrix of temperatures, and G is the coefficient matrix of heat fluxes. To solve the systems, the unknowns are moved to left-hand side of system and the known values to the right-hand side as with EDM. Then the final system is obtained, whose format is like that of eqn (16).

5 COUPLING EDM AND BEM

When solving a multi-scale problem, the large-scale part of the computational domain would be better to be solved by EDM since it is flexible and efficient, while the small-scale part would be better to be solved by BEM because the latter method is robust in treating small scale problems. Then, we can couple these two methods using displacement/temperature consistency and traction/flux equilibrium to solve the multi-scale problems.

In order to couple these two methods, we should list the two systems obtained for their own parts. According to eqn (18), the system of BEM can be written as:

$$H^1_{ss}T^1_s + H^1_{sc}T^1_c = G^1_{ss}q^1_s + G^1_{sc}q^1_c, \tag{19}$$

$$H^1_{cs}T^1_s + H^1_{cc}T^1_c = G^1_{cs}q^1_s + G^1_{cc}q^1_c, \tag{20}$$

where subscript s represents self-boundary nodes; subscript c represents EDM and BEM common interface nodes; the superscript 1 represents that the coefficient matrix or vector belong to BEM. For example, q^1_c is the heat fluxes of EDM and BEM common interface nodes on the BEM computational domain.

The system of EDM can be written as:

$$\begin{bmatrix} A^2_{nn} & H^2_{nc} \\ A^2_{cn} & H^2_{cc} \end{bmatrix} \begin{bmatrix} x^2_n \\ T^2_c \end{bmatrix} = \begin{bmatrix} y^2_n \\ y^2_c + q^2_c \end{bmatrix}, \tag{21}$$

where A represents the coefficient matrix of the unknowns x (unknown displacements and tractions); H represents the coefficient matrix of displacements; y represents the vector forming by multiplying the coefficients with the knowns and adding them; subscript c represents EDM and BEM common interface nodes; subscript n represents the other nodes in the EDM computational domain; the superscript 2 represents that the coefficient matrix or vector belong to EDM; q^2_c is the heat fluxes of EDM and BEM common interface nodes on the EDM computational domain.

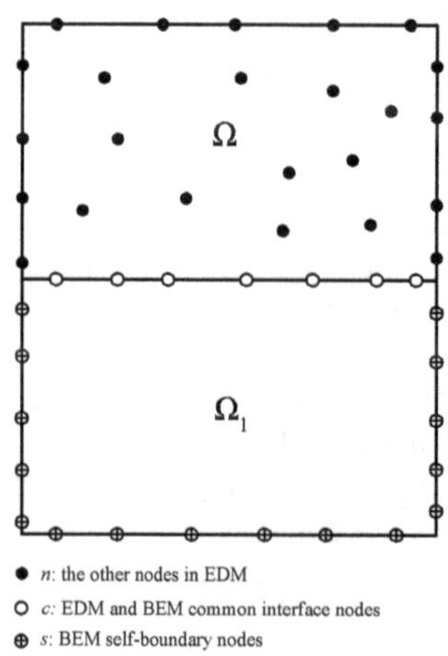

Figure 2: Three kind of nodes in computational domain.

Fig. 2 shows those kinds of nodes above directly. In Fig. 2, Ω is EDM computational domain. Ω_1 is BEM computational domain.

For eqns (19) and (20), the unknowns of self-boundary nodes are moved to left-hand side of the system and the known values to the right-hand side and then the following equation can been obtained:

$$A_{ss}^1 x_s^1 + H_{sc}^1 T_c^1 = y_s^1 + G_{sc}^1 q_c^1 , \tag{22}$$

$$A_{cs}^1 x_s^1 + H_{cc}^1 T_c^1 = y_c^1 + G_{cc}^1 q_c^1 , \tag{23}$$

where A represents the coefficient matrix of unknowns x; y represents the vector formed by multiplying the coefficients with the known values and adding them all.

From eqn (22), we can get:

$$x_s^1 = \left(A_{ss}^1\right)^{-1} \left(y_s^1 + G_{sc}^1 q_c^1 - H_{sc}^1 T_c^1\right). \tag{24}$$

Substituting the above equation into eqn (23), we obtain:

$$q_c^1 = \left(\overline{G_{cc}^1}\right)^{-1} \left(\overline{H_{cc}^1} T_c^1 - \overline{y_c^1}\right), \tag{25}$$

where

$$\begin{cases} \overline{H_{cc}^1} = H_{cc}^1 - H_{cs}^1 \left(H_{ss}^1\right)^{-1} H_{sc}^1 \\ \overline{G_{cc}^1} = G_{cc}^1 - H_{cs}^1 \left(H_{ss}^1\right)^{-1} G_{sc}^1 \\ \overline{y_c^1} = y_c^1 - H_{cs}^1 \left(H_{ss}^1\right)^{-1} y_s^1 . \end{cases} \tag{26}$$

According to displacement consistency condition and traction equilibrium equation, we can get:

$$q_c^1 + q_c^2 = 0 , \tag{27}$$

$$T_c = T_c^1 = T_c^2 . \tag{28}$$

Substituting eqns (25), (27) and (28) into the system of EDM (eqn (21)) yields

$$\begin{bmatrix} A_{nn}^2 & H_{nc}^2 \\ A_{cn}^2 & \left(\overline{G_{cc}^1}\right)^{-1} \overline{H_{cc}^1} + H_{cc}^2 \end{bmatrix} \begin{bmatrix} x_n^2 \\ T_c \end{bmatrix} = \begin{bmatrix} y_s^2 \\ y_c^2 + \left(\overline{G_{cc}^1}\right)^{-1} \overline{y_c^1} \end{bmatrix}. \tag{29}$$

Eqn (29) is the final system of the coupled method. The coefficient matrix $\left(\overline{G_{cc}^1}\right)^{-1} \overline{H_{cc}^1}$ is dense, while the other coefficient matrix is sparse. Usually, in the multi-scale problem, the number of EDM and BEM common interface nodes (c-nodes) is far less than that of the other nodes in EDM (n-nodes).

Therefore, the whole coefficient matrix is sparse even though a small part is dense.

After solving the system, we can use eqns (25) and (24) to calculate q_c^1 and T_s^1. Thus, all the unknowns of computational domain can be obtained.

6 NUMERICAL EXAMPLES

We have already coded the coupled method using FORTRAN code. In this section, an example is computed by the codes to verify the correctness of the proposed method and code.

The problem to be considered is a 2D heat conduction plate with size of 200×100. We divide it into two same parts. The left part is computed by EDM, while the right one is computed by BEM. The EDM part is discretized into 100 9-node elements with 441 nodes, while BEM part is discretized into 40 3-node boundary elements with 80 boundary nodes as shown in Fig. 3. There are 21 common interface nodes in the whole computational domain. The left and right sides are insulated, while the top side is imposed with the temperature condition of 300 K and the bottom side is subjected to heat flux conditions of 1 W/m^2. The heat conductivity of the plate is 1 W/(m·K).

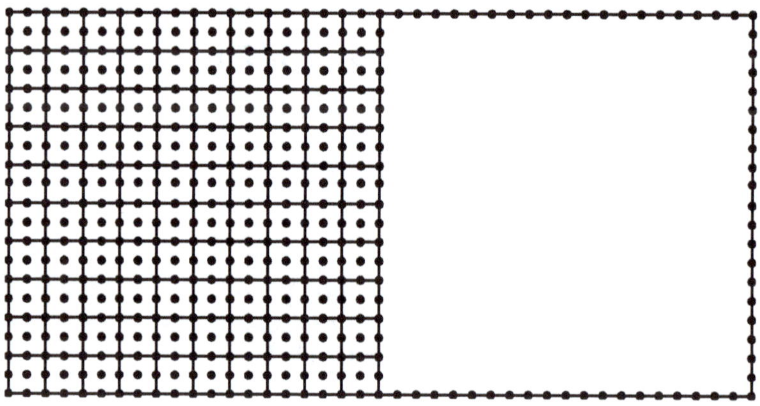

Figure 3: EDM and BEM Mesh over the plate.

After computing, we extract the temperatures of 21 common interface nodes and compare them to the exact solutions (see Fig. 4). From Fig. 4, it can be seen that the computed results are in good agreement with the exact solution. This means that both the coupled method and the FORTRAN code are correct.

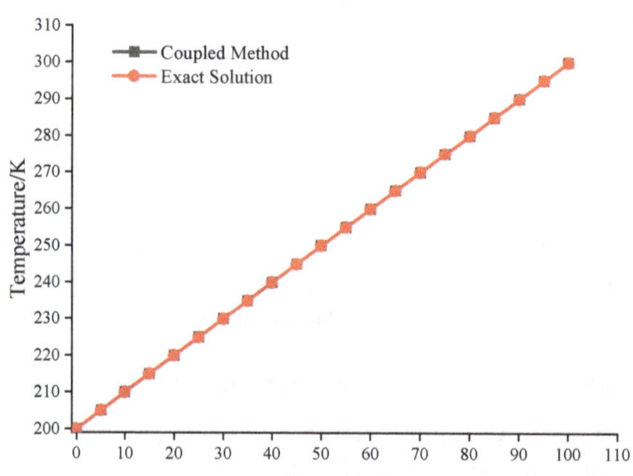

Figure 4: Computed temperatures of 21 common interface nodes and their exact solutions.

7 CONCLUSION

Through the above example, we can conclude that EDM-BEM coupled method can compute the heat conduction problem correctly. With the development of our research, the coupled method will be applied to the solution of more complicated engineering problems.

ACKNOWLEDGEMENT

The support of this investigation by the National Natural Science Foundation of China (grant nos. 11672061, 11702054) is gratefully acknowledged.

REFERENCES

[1] Zienkiewicz, O.C. & Taylor, R.L., *The Finite Element Method*, 6th ed., Elsevier: London, 2005.
[2] Hughes, T.J.R., *The Finite Element Method: Linear Static and Dynamic Finite Element Analysis,* Prentice-Hall: Englewood Cliffs, NJ, 1987.
[3] Belytschko, T., *Nonlinear Finite Elements for Continua and Structures*, John Wiley, 2000.
[4] Quek, G.R. & Liu, S., *The Finite Element Method: A Practical Course*, 2nd ed., Elsevier Science, 2013.
[5] Gunzburger, M.D. & Bochev, P.B., *Least-Squares Finite Element Methods*, Springer: New York, 2009.
[6] Lv, J., Sheng, G., Gao, X. & Zhang, H., Numerical Integration approach based on radial integration method for general 3D polyhedral finite elements. *International Journal of Computational Methods*, **12**(15500265), 2015.
[7] Schroeder, J., Viebahn, N., Balzani, D. & Wriggers, P., A novel mixed finite element for finite anisotropic elasticity; the SKA-element simplified kinematics for anisotropy. *Computer Methods in Applied Mechanics and Engineering*, **310**, pp. 475–494, 2016.
[8] Brebbia, C.A., Boundary Element Techniques: Theory and Applications in Engineering, Springer-Verlag: Berlin and New York, 1984.
[9] Divo, E. & Kassab, A., *Boundary Element Method for Heat Conduction: With Applications in Non-Homogenous Media*, WIT Press, 2003.
[10] Sladek, V., Sladek, J., Tanak, M. & Zhang, C., Transient heat conduction in anisotropic and functionally graded media by local integral equations. *Engineering Analysis with Boundary Elements*, **29**(11), pp. 1047–1065, 2005.
[11] Xiao-Wei Gao, Hai-Feng Peng & Jian Liu, A boundary-domain integral equation method for solving convective heat transfer problems. *International Journal of Heat and Mass Transfer*, **63** pp. 183–190, 2013.
[12] Gao, X.-W., Feng, W.-Z., Zheng, B.-J. & Yang, K., An interface integral equation method for solving general multi-medium mechanics problems. *International Journal for Numerical Methods in Engineering*, **107**(8), pp. 696–720, 2016.
[13] Al-Jawary, M.A. & Wrobel, L.C., Radial integration boundary integral and integro-differential equation methods for two-dimensional heat conduction problems with variable coefficients. *Engineering Analysis with Boundary Elements*, **36**(5), pp. 685–695, 2012.
[14] Liu, Y., *Fast Multipole Boundary Element Method: Theory and Applications in Engineering*, Cambridge University Press: Cambridge, 2009.
[15] Gao, X.W., A meshless BEM for isotropic heat conduction problems with heat generation and spatially varying conductivity. *International Journal for Numerical Methods in Engineering*, **66**(9), pp. 1411–1431, 2006.

[16] Gu, Y.T. & Liu, G.R., A meshfree weak-strong (MWS) form method for time dependent problems. *Computational Mechanics*, **35**(2), pp. 134–145, 2005.

[17] Zhang, X.G., Liu, X.H., Song, K.Z. &. Lu, M.W., Least-squares collocation meshless method. *International Journal for Numerical Methods in Engineering*, **51**(9), pp. 1089–1100, 2001.

[18] Wang, D. & Wu, J., An efficient nesting sub-domain gradient smoothing integration algorithm with quadratic exactness for Galerkin meshfree methods. *Computer Methods in Applied Mechanics and Engineering*, **298**, pp. 485–519, 2016.

[19] Wang, D. & Zhang, H., A consistently coupled isogeometric-meshfree method. *Computer Methods in Applied Mechanics and Engineering*, **268**, pp. 843–870, 2014.

[20] Wen, P.H. & Aliabadi, M.H., An improved meshless collocation method for elastostatic and elastodynamic problems. *Communications in numerical methods in engineering*, **24**(8), pp. 635–651, 2008.

[21] Zheng, B.-J., Gao, X.W., Yang K. & Zhang, C.Z., A novel meshless local Petrov–Galerkin method for dynamic coupled thermoelasticity analysis under thermal and mechanical shock loading. *Engineering Analysis with Boundary Elements*, **60**(SI), pp. 154–161, 2015.

[22] Lin, C.-C., Lawton, E.C., Caliendo, J.A. & Anderson, L.R., An iterative finite element-boundary element algorithm. *Computers and Structures*, **59**(5), pp. 899–909, 1996.

[23] Gao, X.-W., Liu, H.-Y., Xu, B.-B., Cui, M. & Lv, J., Element differential method with the simplest quadrilateral and hexahedron quadratic elements for solving heat conduction problems. *Numerical Heat Transfer Part B-Fundamentals*, **73**(4), pp. 206–224, 2018.

[24] Gao, X.-W. et al., Element differential method for solving general heat conduction problems. *International Journal of Heat and Mass Transfer*, **115**(B), pp. 882–894, 2017.

[25] Gao, X.-W. et al., Element differential method and its application in thermal-mechanical problems. *International Journal for Numerical Methods in Engineering*, **113**(1), pp. 82–108, 2018.

[26] Cui, M., Xu, B.-B., Lv, J., Gao, X.-W. & Zhang, Y., Numerical solution of multi-dimensional transient nonlinear heat conduction problems with heat sources by an extended element differential method. *International Journal of Heat and Mass Transfer*, **126**(A), pp. 1111–1119, 2018.

[27] Brebbia, C.A. & Dominguez, J., *Boundary elements: An introductory course*, Computational Mechanics Publications and McGraw-Hill, 1992.

[28] Yu, B., Zhou, H.-L., Yan, J. & Meng, Z., A differential transformation boundary element method for solving transient heat conduction problems in functionally graded materials. *Numerical Heat Transfer, Part A: Applications*, **70**(3), pp. 293–309, 2016.

A 2.5D BEM-FEM USING A SPECTRAL APPROACH TO STUDY SCATTERED WAVES IN FLUID–SOLID INTERACTION PROBLEMS

FRANCISCO JAVIER CRUZ-MUÑOZ[1], ANTONIO ROMERO[1], PEDRO GALVÍN[1] & ANTÓNIO TADEU[2,3]
[1]Departamento de Mecánica de Medios Continuos y Teoría de Estructuras, Escuela Técnica Superior de Ingeniería, Universidad de Sevilla, Spain
[2]ITeCons – Institute for Research and Technological Development in Construction, Energy, Environment and Sustainability, Portugal
[3]ADAI – LAETA, Department of Civil Engineering, University of Coimbra, Portugal

ABSTRACT

This work presents a two-and-a-half dimensional (2.5D) spectral formulation based on the finite element method (FEM) and the boundary element method (BEM) to study wave propagation in acoustic and elastic waveguides. The analysis involved superposing two dimensional (2D) problems with different longitudinal wavenumbers. A spectral finite element (SFEM) is proposed to represent waveguides in solids with arbitrary cross-section. Moreover, the BEM is extended to its spectral formulation (SBEM) to study unbounded fluid media and acoustic enclosures. Both approaches use Lagrange polynomials as element shape functions at the Legendre–Gauss–Lobatto (LGL) points. The fluid and solid subdomains are coupled by applying the appropriate boundary conditions at the limiting interface. The proposed method is verified by means of a benchmark problem regarding the scattering of waves by an elastic inclusion. The convergence and the computational effort are evaluated for different h-p strategies. Numerical results show good agreement with the reference solution. Finally, the proposed method is used to study the pressure field generated by an array of elastic fluid-filled scatterers immersed in an acoustic medium.
Keywords: SBEM, SFEM, fluid–solid interaction, waveguide, scattered waves, two-and-a-half dimension.

1 INTRODUCTION

Many engineering fields involve time harmonic wave propagation, such as fluid acoustics and solid scattering. Hybrid methods based on the boundary element method (BEM) and the finite element method (FEM) are suitable for studying solid and fluid interaction in unbounded media. Both methods have been used in various works to predict the response of coupled fluid-structure problems. For the low frequency range, the standard formulation using linear elements accurately represents the fluid and solid scattering waves. However, at high frequencies, the solution deteriorates due to so-called pollution effects [1], [2]. Higher element approximations are needed to obtain accurate results.

Different approaches have been proposed in the past to overcome the pollution effect in two and three dimensional problems [3]. The h-refinement method has proved to be useless at high wavenumbers because of the necessary computational effort to represent small wavelengths [4]. Many works suggest using high-order elements to improve the convergence rate of standard formulations. The boundary and finite element methods have thus been extended to their spectral formulations [5], [6] approximating the field variables by high-order interpolation shape functions. Such functions are polynomials over non-uniformly spaced nodes to avoid the Runge phenomenon.

In this work, we propose a two-and-a-half dimensional coupled spectral formulation based on the FEM and the BEM to study wave propagation in fluid and solid waveguides. The proposed method is useful for problems whose geometry and material properties are both uniform in one direction, whereas the excitation exhibits a three dimensional distribution.

WIT Transactions on Engineering Sciences, Vol 126, © 2019 WIT Press
www.witpress.com, ISSN 1743-3533 (on-line)
doi:10.2495/BE420101

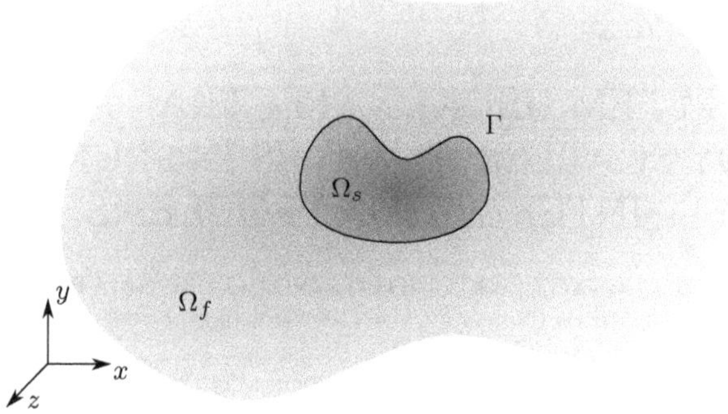

Figure 1: Domain decomposition.

The spectral approach allows the analysis of wave propagation at high wavenumbers, which reduces the pollution effect with low computational effort.

The outline of this work is as follows. First, the fluid–solid interaction problem is formulated in two-and-a-half dimensions. The 2.5D formulation of the BEM and the FEM are extended to their spectral formulations (SBEM and SFEM, respectively). The SBEM is used for the representation of both acoustic enclosures and unbounded domains. Moreover, a solid finite spectral element is proposed to represent elastic waveguides with arbitrary cross-section. The proposed approach is verified with two benchmark problems: the wave propagation in an unbounded acoustic region [7] and the scattering of waves by an elastic inclusion [8]. An h-p convergence analysis is carried out to assess the accuracy of the method. Finally, the proposed method is applied to study the scattered wave field in an unbounded acoustic domain due to an array of elastic fluid-filled cylinders.

2 NUMERICAL MODEL

The spectral formulations of the BEM and FEM are developed to represent an elastic waveguide submerged in an unbounded acoustic region, as shown in Fig. 1. The proposed methodology uses spectral solid elements to represent the elastic waveguide (Ω_s), while the fluid domain (Ω_f) is modelled with spectral boundary elements. Both methods are coupled by imposing the appropriate boundary conditions at the interface Γ: equilibrium of normal pressure and compatibility of normal displacement must be ensured at the fluid–solid interface, with null shear stresses. The 2.5D formulation is handled by defining a characteristic field for the elastic and the acoustic waveguides at point $\mathbf{x} = \mathbf{x}(x, y, z)$ as [9], [10]:

$$\widehat{f}(\mathbf{x}, \kappa_z, \omega) = \widetilde{f}(\widetilde{\mathbf{x}}, \kappa_z, \omega)\, e^{-\iota \kappa_z z}, \tag{1}$$

where $\widetilde{f}(\widetilde{\mathbf{x}}, \kappa_z, \omega)$ is the frequency-wavenumber representation of an unknown variable (e.g. displacement, force or sound pressure), $\widetilde{\mathbf{x}} = \widetilde{\mathbf{x}}(x, y)$ and $\iota = \sqrt{-1}$. The 3D solution is computed as the superposition of 2D solutions with different wavenumber κ_z:

$$f(\mathbf{x}, \omega) = \int_{-\infty}^{+\infty} \widetilde{f}(\widetilde{\mathbf{x}}, \kappa_z, \omega)\, e^{-\iota \kappa_z z} d\kappa_z. \tag{2}$$

2.1 Spectral boundary element method (SBEM)

Next, the SBEM is derived from the standard collocation method, which is fully covered in the literature [11]. The main differences between the two methods concern the field variable approximation and the proposed element integration scheme. In both cases, the boundary integral representation is derived from the Helmholtz equation [11], considering a homogeneous fluid domain denoted by its volume Ω_f and boundary Γ. Although the proposed method is used to obtain the three-dimensional radiated field by an elastic waveguide with an arbitrary cross section, the problem is computed as the superposition of 2D problems with different wavenumber κ_z. Therefore, the boundary Γ is reduced to its cross section Σ at the coordinate plane $z=0$. Thus, the integral representation of the fluid pressure for a point $\widetilde{\mathbf{x}}_i$ located at the boundary Γ in the frequency-wavenumber domain is:

$$c_i(\widetilde{\mathbf{x}}_i)\widetilde{p}_i(\widetilde{\mathbf{x}}_i,\kappa_z,\omega) = -\int_{\Sigma}\left(\iota\rho\omega^2\widetilde{u}_i(\widetilde{\mathbf{x}}_i,\kappa_z,\omega)\widetilde{\Psi}(\widetilde{\mathbf{x}},\kappa_z,\omega;\widetilde{\mathbf{x}}_i)+\widetilde{p}_i(\widetilde{\mathbf{x}}_i,\kappa_z,\omega)\frac{\partial\widetilde{\Psi}(\widetilde{\mathbf{x}},\kappa_z,\omega;\widetilde{\mathbf{x}}_i)}{\partial\mathbf{n}}\right)d\Sigma, \tag{3}$$

where $\widetilde{p}(\widetilde{\mathbf{x}}_i,\kappa_z,\omega)$ and $\widetilde{u}(\widetilde{\mathbf{x}}_i,\kappa_z,\omega)$ are the sound pressure and the particle normal displacement at the cross section Σ of the boundary Γ, respectively. $\widetilde{\Psi}(\widetilde{\mathbf{x}},\kappa_z,\omega;\widetilde{\mathbf{x}}_i)$ is the solution to the Helmholtz equation at point $\widetilde{\mathbf{x}}$ due to a point source located at $\widetilde{\mathbf{x}}_i$ [12]. The integral-free term $c_i(\widetilde{\mathbf{x}}_i)$ only depends on the boundary geometry at the collocation point $\widetilde{\mathbf{x}}_i$.

The boundary is discretised into N elements, with $\Sigma=\bigcup_{j=1}^{N}\Sigma^j$, and the field variables are approximated within the element using a set of interpolation shape functions $\phi(\xi)$. The proposed methodology uses Lagrange polynomials of order p as interpolation shape functions given by:

$$\phi^k=\prod_{j\neq i}\frac{\xi-\xi_i}{\xi_j-\xi_i}, \tag{4}$$

where nodal coordinates ξ_i are found at the Lobatto–Gauss–Legendre (LGL) points. The use of a family of orthogonal polynomial ($\phi^i(\xi_j)=\delta_{ij}$), such as Lagrange polynomials, as interpolation shape functions, mitigates the appearance of the Runge phenomenon, which can result in a highly ill-conditioned problem for high-order elements [13].

Once the sound pressure and particle normal displacement are approximated within the element, the former integrals in eqn (3) become:

$$\widetilde{g}_i^j\left(\widetilde{\mathbf{x}},\kappa_z,\omega;\widetilde{\mathbf{x}}_i\right)=\int_{\Sigma^j}\phi\widetilde{\Psi}\left(\widetilde{\mathbf{x}},\kappa_z,\omega;\widetilde{\mathbf{x}}_i\right)d\Sigma, \tag{5}$$

$$\widetilde{h}_i^j\left(\widetilde{\mathbf{x}},\kappa_z,\omega;\widetilde{\mathbf{x}}_i\right)=\int_{\Sigma^j}\phi\frac{\partial\widetilde{\Psi}\left(\widetilde{\mathbf{x}},\kappa_z,\omega;\widetilde{\mathbf{x}}_i\right)}{\partial n}d\Sigma. \tag{6}$$

The spatial integrations in eqns (5) and (6) are numerically evaluated using a standard Gauss-Legendre quadrature with $(p+1)$ integration points whenever the collocation point does not belong to the integration element. However, the integrals become singular when the collocation point falls on the integration element. These integrals are numerically evaluated by element subdivision, isolating the collocation point at the integration endpoints

$$\int_{\Sigma^j} \widetilde{f}\left(\widetilde{\mathbf{x}}, \kappa_z, \omega; \widetilde{\mathbf{x}}_i\right) d\Sigma = \int_{-1}^{1} \widetilde{f}\left(\widetilde{\mathbf{x}}, \kappa_z, \omega; \widetilde{\mathbf{x}}_i\right) |J^j(\xi)| d\xi$$
$$= \int_{-1}^{\xi_p} \widetilde{f}\left(\widetilde{\mathbf{x}}, \kappa_z, \omega; \widetilde{\mathbf{x}}_i\right) |J^j(\xi)| d\xi + \int_{\xi_p}^{1} \widetilde{f}\left(\widetilde{\mathbf{x}}, \kappa_z, \omega; \widetilde{\mathbf{x}}_i\right) |J^j(\xi)| d\xi, \tag{7}$$

with ξ_p being the natural coordinate of the collocation point $\widetilde{\mathbf{x}}_i$, $\widetilde{f}\left(\widetilde{\mathbf{x}}, \kappa_z, \omega; \widetilde{\mathbf{x}}_i\right)$ the kernel of integrals in eqns (5) and (6), and $|J^j(\xi)|$ the Jacobian of the transformation to the natural coordinate system for element Σ^j. These integrals are numerically evaluated using a smooth polynomial transformation as described in Reference [14]. Accurate results are given in the next section for a number of integration points $10(p+1)$.

Finally, the boundary element matrices are assembled into a single system of equations relating sound pressure and particle normal displacement at nodal points:

$$\widetilde{\mathbf{H}}\left(\widetilde{\mathbf{x}}, \kappa_z, \omega\right) \widetilde{p}\left(\widetilde{\mathbf{x}}, \kappa_z, \omega\right) = \widetilde{\mathbf{G}}\left(\widetilde{\mathbf{x}}, \kappa_z, \omega\right) \widetilde{u}\left(\widetilde{\mathbf{x}}, \kappa_z, \omega\right), \tag{8}$$

where $\widetilde{p}\left(\widetilde{x}, \kappa_z, \omega\right)$ and $\widetilde{u}\left(\widetilde{\mathbf{x}}, \kappa_z, \omega\right)$ collect the nodal pressure and the normal displacement to the boundary, and $\widetilde{\mathbf{H}}\left(\widetilde{\mathbf{x}}, \kappa_z, \omega\right)$ and $\widetilde{\mathbf{G}}\left(\widetilde{\mathbf{x}}, \kappa_z, \omega\right)$ are the fully populated non-symmetrical boundary element system matrices.

2.2 Spectral finite element method (SFEM)

The spectral finite element formulation is based on the virtual work principle in the frequency-wavenumber domain, which states [15]:

$$-\omega^2 \int_{\Omega_s} \delta\widehat{\mathbf{u}}^T(\mathbf{x}, \kappa_z, \omega) \rho_s \widehat{\mathbf{u}}(\mathbf{x}, \kappa_z, \omega) d\Omega + \int_{\Omega_s} \delta\widehat{\boldsymbol{\varepsilon}}^T(\mathbf{x}, \kappa_z, \omega) \widehat{\boldsymbol{\sigma}}(\mathbf{x}, \kappa_z, \omega) d\Omega$$
$$= \int_{\Omega_s} \delta\widehat{\mathbf{u}}^T(\mathbf{x}, \kappa_z, \omega) \rho_s \widehat{\mathbf{b}}(\mathbf{x}, \kappa_z, \omega) d\Omega + \int_{\Gamma} \delta\widehat{\mathbf{u}}^T(\mathbf{x}, \kappa_z, \omega) \widehat{\mathbf{q}}(\mathbf{x}, \kappa_z, \omega) d\Gamma, \tag{9}$$

where $\widehat{\mathbf{u}}(\mathbf{x}, \kappa_z, \omega)$ is the displacement vector, $\widehat{\boldsymbol{\varepsilon}}(\mathbf{x}, \kappa_z, \omega)$ and $\widehat{\boldsymbol{\sigma}}(\mathbf{x}, \kappa_z, \omega)$ are respectively the strain and stress tensors, $\rho_s \widehat{\mathbf{b}}(\mathbf{x}, \kappa_z, \omega)$ is the body force in the domain Ω_s, ρ_s is the solid density, and $\widehat{\mathbf{q}}(\mathbf{x}, \kappa_z, \omega)$ is the nodal force vector. A variable preceded by δ denotes a compatible variation of the displacement or the strain field.

The stress tensor is related to the strain tensor through the constitutive relation $\widehat{\boldsymbol{\sigma}} = \mathbf{C}\widehat{\boldsymbol{\varepsilon}}$, where, in the case of a linear isotropic material, the constitutive matrix $\mathbf{C}$ depends on the Young's modulus E and the Poisson ratio ν.

The dependence of the strain tensor on the longitudinal coordinate z can be removed considering the displacement definition in the frequency-wavenumber domain according to eqn (1):

$$\widehat{\mathbf{u}}\left(\mathbf{x}, \kappa_z, \omega\right) = \widetilde{\mathbf{u}}\left(\widetilde{\mathbf{x}}, \kappa_z, \omega\right) e^{-\iota \kappa_z z}. \tag{10}$$

Then, the strain tensor can be derived from the displacement vector $\widehat{\mathbf{u}}$ as:

$$\widehat{\boldsymbol{\varepsilon}} = \mathbf{L}_1 \widehat{\mathbf{u}} + \mathbf{L}_2 \frac{\partial \widehat{\mathbf{u}}}{\partial z} = \mathbf{L}_1 \widehat{\mathbf{u}} - \iota \kappa_z \mathbf{L}_2 \widehat{\mathbf{u}}, \tag{11}$$

where the differential operators are:

$$\mathbf{L}_1 = \begin{bmatrix} \frac{\partial}{\partial x} & 0 & 0 \\ 0 & \frac{\partial}{\partial y} & 0 \\ 0 & 0 & 0 \\ \frac{\partial}{\partial y} & \frac{\partial}{\partial x} & 0 \\ 0 & 0 & \frac{\partial}{\partial y} \\ 0 & 0 & \frac{\partial}{\partial x} \end{bmatrix}, \qquad \mathbf{L}_2 = \begin{bmatrix} 0 & 0 & 0 \\ 0 & 0 & 0 \\ 0 & 0 & 1 \\ 0 & 0 & 0 \\ 0 & 1 & 0 \\ 1 & 0 & 0 \end{bmatrix}. \tag{12}$$

The 2.5D formulation is derived assuming that solid material properties and geometry are both homogeneous in the longitudinal direction. Therefore, eqn (9) is further elaborated by eliminating the dependency on the coordinate z rewriting the volume integrals over the section A_s of the domain integration Ω_s by the coordinate plane $z = 0$:

$$-\omega^2 \int_{A_s} \delta\widetilde{\mathbf{u}}^T \rho_s \widetilde{\mathbf{u}} dA + \int_{A_s} \delta\widetilde{\varepsilon}^T \widetilde{\sigma} dA = \int_{A_s} \delta\widetilde{\mathbf{u}}^T \rho_s \widetilde{\mathbf{b}} dA + \int_{\Sigma} \delta\widetilde{\mathbf{u}}^T \widetilde{\mathbf{q}} d\Sigma, \tag{13}$$

where Σ is defined as the boundary Γ at $z = 0$.

The solid subdomain is discretised into elements, where the approximated displacement vector $\widetilde{\mathbf{u}}$ is defined as:

$$\widetilde{\mathbf{u}} = \varphi \widetilde{\mathbf{u}}^e, \tag{14}$$

where φ are the two-dimensional shape functions obtained from the one-dimensional function $\phi(\xi)$ as:

$$\varphi^k(\xi, \eta) = \phi^i(\xi)\phi^j(\eta), \qquad k = (i-1)(p+1) + j. \tag{15}$$

The proposed spectral finite element is defined from local nodal coordinates $(\xi, \eta) \in ([-1, 1] \times [-1, 1])$ at the LGL points.

Once the element approximation is defined, the strain tensor is expressed in terms of the nodal displacement as:

$$\widetilde{\varepsilon}^e = \mathbf{L}_1 \varphi \widetilde{\mathbf{u}}^e - \iota\kappa_z \mathbf{L}_2 \varphi \widetilde{\mathbf{u}}^e = \mathbf{B}_1 \widetilde{\mathbf{u}}^e - \iota\kappa_z \mathbf{B}_2 \widetilde{\mathbf{u}}^e, \tag{16}$$

where $\mathbf{B}_1 = \mathbf{L}_1\varphi$ and $\mathbf{B}_2 = \mathbf{L}_2\varphi$.

Introducing eqns (14) and (16) into eqn (13) and operating, the virtual work principle can be rewritten for each element as:

$$\begin{aligned} &-\omega^2 \int_{A_s^e} \delta\widetilde{\mathbf{u}}^{eT} \varphi^T \rho_s \varphi \widetilde{\mathbf{u}}^e dA + \int_{A_s^e} \delta\widetilde{\mathbf{u}}^{eT} \mathbf{B}_1^T \mathbf{C} \mathbf{B}_1 \widetilde{\mathbf{u}}^e dA \\ &- \iota\kappa_z \int_{A_s^e} \delta\widetilde{\mathbf{u}}^{eT} \left(\mathbf{B}_1^T \mathbf{C} \mathbf{B}_2 + \mathbf{B}_2^T \mathbf{C} \mathbf{B}_1\right) \widetilde{\mathbf{u}}^e dA - \kappa_z^2 \int_{A_s^e} \delta\widetilde{\mathbf{u}}^{eT} \mathbf{B}_2^T \mathbf{C} \mathbf{B}_2 \widetilde{\mathbf{u}}^e dA \\ &= \int_{A_s^e} \delta\widetilde{\mathbf{u}}^{eT} \varphi^T \rho_s \widetilde{\mathbf{b}}^e dA + \int_{\Sigma^e} \delta\widetilde{\mathbf{u}}^{eT} \varphi^T \widetilde{\mathbf{q}}^e d\Sigma, \end{aligned} \tag{17}$$

where A_s^e is the cross-sectional area of the three dimensional element in the z direction and Σ^e its boundary. The following expression is obtained taking into account that eqn (17) must be satisfied for any compatible displacement $\delta\widetilde{\mathbf{u}}^e$:

$$\left[-\omega^2 \mathbf{M}^e + \mathbf{K}_0^e - \iota\kappa_z \mathbf{K}_1^e - \kappa_z^2 \mathbf{K}_2^e\right] \widetilde{\mathbf{u}}^e = \widetilde{\mathbf{f}}^e, \tag{18}$$

where $\mathbf{M}^e$ is the element mass matrix, $\mathbf{K}_0^e$, $\mathbf{K}_1^e$ and $\mathbf{K}_2^e$ are the stiffness element matrices and $\widetilde{\mathbf{f}}^e$ is a vector that collects the forces applied to the element. These matrices are defined by:

$$\mathbf{M}^e = \int_{A_s^e} \varphi^T \rho_s \varphi dA, \tag{19}$$

$$\mathbf{K}_0^e = \int_{A_s^e} \mathbf{B}_1^T \mathbf{C} \mathbf{B}_1 dA, \tag{20}$$

$$\mathbf{K}_1^e = \int_{A_s^e} \left(\mathbf{B}_1^T \mathbf{C} \mathbf{B}_2 - \mathbf{B}_2^T \mathbf{C} \mathbf{B}_1 \right) dA, \tag{21}$$

$$\mathbf{K}_2^e = \int_{A_s^e} \mathbf{B}_2^T \mathbf{C} \mathbf{B}_2 dA, \tag{22}$$

and the external load vector $\widetilde{\mathbf{f}}^e$ as:

$$\widetilde{\mathbf{f}}^e(\widetilde{\mathbf{x}}, \kappa_z, \omega) = \int_{A_s^e} \varphi^T \rho_s \widetilde{\mathbf{b}}^e(\widetilde{\mathbf{x}}, \kappa_z, \omega) dA + \int_{\Sigma^e} \varphi^T \widetilde{\mathbf{q}}^e(\widetilde{\mathbf{x}}, \kappa_z, \omega) d\Sigma, \tag{23}$$

which contains the contributions of both body forces and surface loads.

Then, eqn (18) can be written as:

$$\widetilde{\mathbf{K}}\widetilde{\mathbf{u}} = \widetilde{\mathbf{f}}, \tag{24}$$

where $\widetilde{\mathbf{K}} = \left[-\omega^2 \mathbf{M} + \mathbf{K}_0 - \iota \kappa_z \mathbf{K}_1 - \kappa_z^2 \mathbf{K}_2 \right]$ is the dynamic stiffness matrix obtained in the frequency-wavenumber domain from the assembled mass $\mathbf{M}$ matrix, and stiffness $\mathbf{K}_0$, $\mathbf{K}_1$ and $\mathbf{K}_2$ matrices.

The mass $\mathbf{M}^e$ matrix and stiffness $\mathbf{K}_0^e$, $\mathbf{K}_1^e$ and $\mathbf{K}_2^e$ matrices are numerically integrated using a LGL quadrature of order p. Therefore, the nodal element coordinates coincide with the element integration points producing a diagonal mass $\mathbf{M}^e$ matrix and stiffness $\mathbf{K}_2^e$ matrices.

2.3 SBEM-SFEM coupling

Eqns (8) and (24) are coupled by imposing equilibrium of forces, compatibility of normal displacement and null shear stresses at the interface Γ. These equations are assembled into a single comprehensive system, together with the equilibrium and compatibility conditions.

The load vector $\widetilde{\mathbf{f}}_f$ at the fluid–solid interface is obtained by integrating the fluid pressure field $\widetilde{p}_f$ over the boundary Γ:

$$\widetilde{\mathbf{f}}_f = -\int_{\Gamma} \varphi^T \mathbf{n} \phi \widetilde{p}_f d\Gamma = -\mathbf{R}^T \widetilde{p}_f, \tag{25}$$

where $\mathbf{n}$ is the outward normal vector along Γ, and $\mathbf{R}$ is the coupling fluid–solid matrix which relates force at the solid subdomain and pressure at the boundary. Substituting eqn (25) into eqn (24) yields:

$$\begin{bmatrix} \widetilde{\mathbf{K}}_{ss} & \widetilde{\mathbf{K}}_{sf} \\ \widetilde{\mathbf{K}}_{fs} & \widetilde{\mathbf{K}}_{ff} \end{bmatrix} \begin{bmatrix} \widetilde{\mathbf{u}}_s \\ \widetilde{\mathbf{u}}_f \end{bmatrix} = \begin{bmatrix} \widetilde{\mathbf{f}}_s \\ -\mathbf{R}^T \widetilde{p}_f \end{bmatrix}, \tag{26}$$

where subscript f indicates degrees of freedom belonging to Γ and s stands for the rest of solid degrees of freedom.

Finally, the coupling of eqns (8) and (26) is carried out with the imposition of compatibility of normal displacement at the boundary Γ, and null shear-stresses. Both

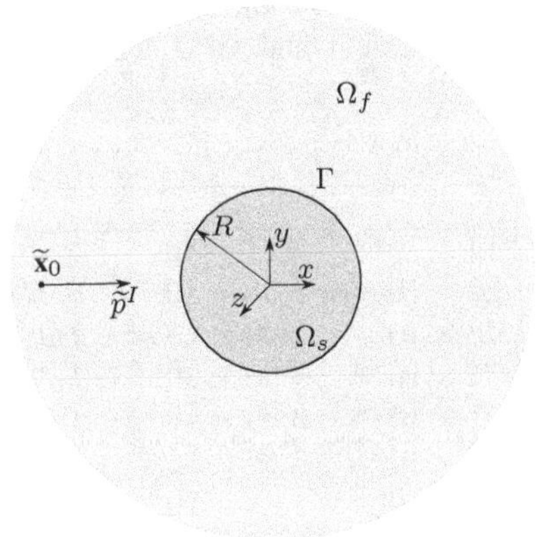

Figure 2: Elastic inclusion in an unbounded acoustic region.

systems of equations are assembled into an overall system:

$$\begin{bmatrix} \widetilde{\mathbf{K}}_{ss} & \widetilde{\mathbf{K}}_{sf} & \mathbf{0} \\ \widetilde{\mathbf{K}}_{fs} & \widetilde{\mathbf{K}}_{ff} & \mathbf{R}^T \\ \mathbf{0} & -\widetilde{\mathbf{G}}\mathbf{N}^T & \widetilde{\mathbf{H}} \end{bmatrix} \begin{bmatrix} \widetilde{\mathbf{u}}_s \\ \widetilde{\mathbf{u}}_f \\ \widetilde{p}_f \end{bmatrix} = \begin{bmatrix} \widetilde{\mathbf{f}}_s \\ \widetilde{\mathbf{f}}_f \\ \mathbf{0} \end{bmatrix}, \tag{27}$$

where $\mathbf{N}$ is a matrix containing the outward normal vector $\mathbf{n}$ at the boundary. Eqn (27) is solved for each frequency-wavenumber step to compute the coupled fluid–solid response.

3 MODEL VERIFICATION

The proposed coupled spectral boundary and finite element method was verified on a benchmark problem involving a cylindrical elastic waveguide in a homogeneous unbounded fluid medium. The inclusion had a radius $r = 0.6$ m. The solid properties were compression wave propagation velocity $c_p = 2630$ m/s, shear wave propagation velocity $c_s = 1416$ m/s, and density $\rho_s = 2250\ \text{kg/m}^3$, while the fluid properties were sound wave propagation velocity $c_f = 1500$ m/s and density $\rho_f = 1000\ \text{kg/m}^3$. The proposed SFEM represented the solid inclusion, while the SBEM was used to model the fluid domain.

The elastic inclusion was subjected to a dilatational point source located in the fluid at distance $\widetilde{\mathbf{x}}_0 = (-3.5, 0)$ from the cavity centre (Fig. 2).

The problem solution was computed for a fluid wavenumber $\kappa_f = 5.236$ rad/m over a grid of receivers using a density of six points per wavelength. A constant value of $\kappa_z = 1.0$ rad/m was assumed for comparison purposes. The analytical solution to this problem can be found in Reference [8]. The numerical results were compared with the reference solution and the scaled L_2-error ϵ_2 was used to asses the accuracy of the proposed method.

Three discretisations were investigated. The characteristic element sizes were $1/h = \{2.12, 4.24, 8.48\}\ \text{m}^{-1}$, resulting in $\kappa_f h = \{2.47, 1.23, 0.62\}$. The element order p was set to obtain nodal densities per wavelength $d_\lambda = \{10, 15, 20\}$, which was limited to $d_\lambda = 20$ for the finest discretisation because of the element size.

Table 1 summarizes the computed scaled L_2-error ϵ_2 and the CPU time for the different h-p discretisations used in this problem. The computed results presented a good agreement with the reference solution, as the error varied from $\epsilon_2 \approx 1 \times 10^{-3}$ to $\epsilon_2 \approx 1 \times 10^{-5}$. The smallest error was found for the medium size discretisation with an acceptable computational

Table 1: Summary of scaled L_2-error ϵ_2 of the total fluid pressure, horizontal and longitudinal solid displacement and CPU time, computed for the different hp discretisations.

$1/h[\mathrm{m}^{-1}]$	p	d_λ	ϵ_2^p	$\epsilon_2^{u_x}$	$\epsilon_2^{u_z}$	CPU time [s]
	2	5	2.14×10^{-1}	2.24×10^{-1}	5.76×10^{-2}	0.526
	3	7.5	1.22×10^{-2}	1.58×10^{-2}	8.07×10^{-3}	1.051
2.12	4	10	1.78×10^{-3}	2.08×10^{-3}	1.08×10^{-3}	1.713
	6	15	5.71×10^{-4}	4.90×10^{-4}	1.03×10^{-3}	4.379
	8	20	5.01×10^{-4}	3.99×10^{-4}	1.00×10^{-4}	11.36
	2	10	3.93×10^{-3}	5.50×10^{-3}	3.92×10^{-3}	0.899
4.24	3	15	1.29×10^{-4}	3.08×10^{-4}	2.37×10^{-4}	2.018
	4	20	6.0×10^{-5}	2.75×10^{-5}	7.55×10^{-5}	3.976
8.48	2	20	1.67×10^{-4}	4.20×10^{-4}	3.24×10^{-4}	2.449

time, while the finest mesh had a higher error in a similar time. On the other hand, the coarsest mesh needed a higher computational effort to reach an acceptable error because of the higher element order approximation. Therefore, a hp-refinement is more efficient in terms of error and computational effort than a single h-refinement or p-refinement.

4 NUMERICAL EXAMPLE

Finally, the capabilities of the proposed method are explored with a numerical example. The problem solved herein was a study of the wave field scattered by an array of elastic cylinders submerged in an inviscid fluid medium (Fig. 3). The elastic cylinders had an external radius $r = 0.4$ m and a thickness $t = 0.05$ m. The centres of the scatterers were equally spaced $s = 1.2$ m. The material was assumed to be PVC, with compression wave propagation velocity $c_p = 2143$ m/s, shear wave propagation velocity $c_s = 875$ m/s, and density $\rho = 1400$ kg/m^3. The outer fluid was water, with sound wave propagation velocity $c_f = 1500$ m/s and density $\rho_f = 1000$ kg/m^3. Three configurations were considered based on the interior of the scatterers: $i)$ water-filled; $ii)$ air-filled, with sound propagation velocity 340 m/s and density 1.225 kg/m^3; and $iii)$ state of vacuum.

The array of scatterers was subjected to a dilatational point source located in the fluid at position $\tilde{\mathbf{x}}_0 = (-4.0, 0)$ (Fig. 3). The incident wave field $\tilde{p}^I$ was defined as in the previous section. The problem response of the fluid-structure system was studied for a frequency $f = 6400$ Hz.

The scatterers were modelled with spectral finite elements while outer and inner fluid subdomains were represented with the SBEM. The discretisation was chosen to enable a nodal density per wavelength of $d_\lambda = 12$ in the outer fluid, while the scatterers and the inner fluid meshes were defined to match the outer fluid mesh. The outer fluid was modelled with 34 elements, so the cylinders and the inner fluid were modelled again with 34 elements. The element order was set to $p = 4$ in all cases. This discretisation resulted in a nodal density $d_\lambda = 12$ with and $\kappa_f h = 2$.

The 3D solution was computed as the superposition of 2D problems with different longitudinal wavenumbers by means of eqn (2). The 3D solution can be computed in a set of regularly spaced points along the longitudinal direction where $\Delta z = 2\pi/\max\{\kappa_z\}$. The maximum wavenumber $\max\{\kappa_z\} = 160$ rad/m was chosen to compute the solution enabling six points in a longitudinal wavelength, $\Delta z = 2\pi/\max\{\kappa_z\} = 0.0393$ m. The

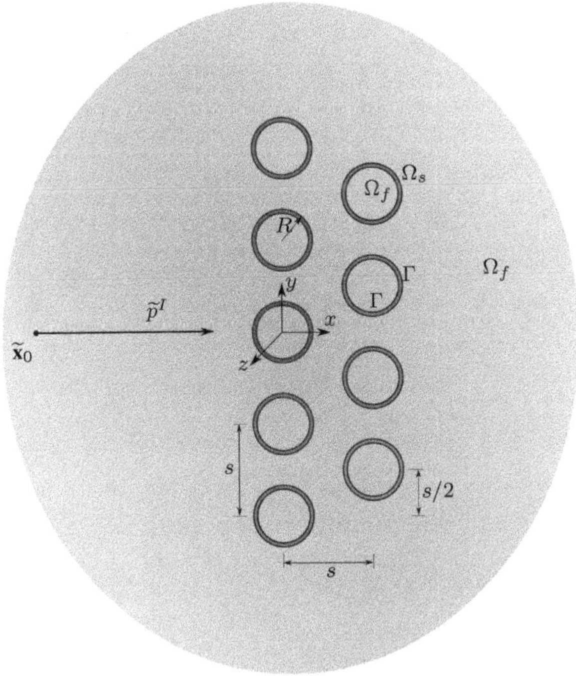

Figure 3: Configuration of scatterers.

wavenumber sampling $\Delta\kappa_z = 0.5178$ rad/m made it possible to compute the solution for a maximum distance $L = \pi/\Delta\kappa_z = 6.0672$ m from the source.

First, the behaviour of a cross section is analysed. Fig. 4 shows the total pressure field in the outer fluid and the scattered pressure field inside the cylinders at a vertical plane $z = 0$ for the three studied cases. The total pressure fields are quite similar in the three cases. The maximum pressure was found at the source point and the scatterer configurations produced a shadowed zone, with a lower pressure amplitude. The vibration of the internal surface of the cylinders caused a scattered wave field inside them in the fluid-filled cases. When the cylinders were filled with water (Fig. 4(a)), the scattered pressure field shows a distribution in accordance with the incident field, similar wavelength, amplitude, phase and direction, as the fluid properties were the same. The fluid-filled case (Fig. 4(b)) exhibited a much more complex pattern, with lower amplitude and wavelength and any concordance with the incident field in direction or phase due to the difference in the properties of inner and outer fluids. The vacuum and air-filled systems presented the best attenuation, as the water-filled system allowed the waves to propagate through the scatterers.

Fig. 5 shows the 3D representation of the sound pressure at the surface of the scatterers and the total pressure field in the fluid domain for the water-filled example. The scattered pressure field at a vertical plane $z = 6.0672$ m inside the cylinders is also shown. Only half of the model is represented for clarity's sake, because of problem symmetry. The pressure field had a spherical distribution centred at the source, which was progressively decreasing as the distance to the source increased. Fig. 6 shows the deformed shape of the solid scatterers external surface and displays the normal displacement to the boundary in the colour scale. The displacement field also had a spherical distribution centered at the source, which decreased

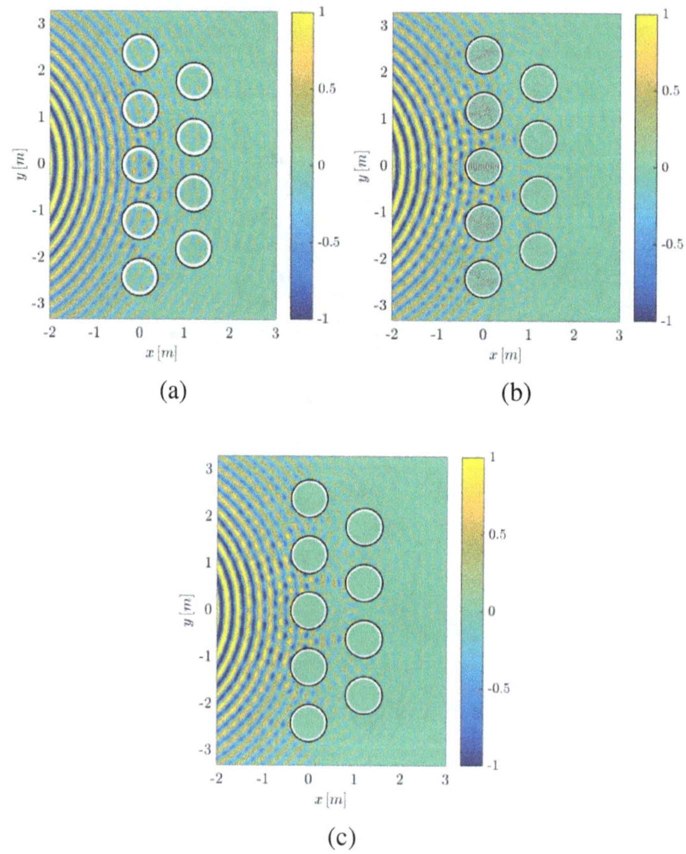

Figure 4: Real part of the pressure field at cross-section of the scatterers system at the vertical plane $z = 0$ for the (a) water-filled cylinders, (b) air-filled cylinders and (c) vacuum cylinders. Sound pressure in water medium was normalised to 0.15 Pa, and to 5×10^{-5} Pa in air medium.

as the distance from the source increased. Maximum amplitudes of both the pressure and displacement were found at the plane $z = 0$, where the source was located and the incident field was higher.

This example has shown that the proposed method can be used to study the wave propagation in a fluid-filled elastic waveguides immersed in an unbounded acoustic medium. The proposed method was able to represent different fluid properties for the unbounded fluid and the acoustic enclosures.

5 CONCLUSIONS

This work has proposed a spectral element formulation based on the BEM and the FEM to study fluid and solid wave propagation. The presented methodology looks at 3D problems whose materials and geometric properties remain homogeneous in one direction. Solid subdomains were modelled with the SFEM, whereas the unbounded fluid media and acoustic enclosures were represented by the SBEM. The coupling of both methods was carried out by

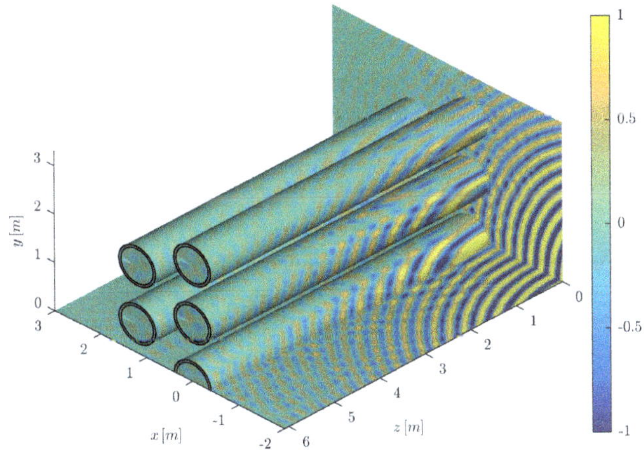

Figure 5: Real part of the pressure field (normalised to 0.15 Pa) in an unbounded fluid by a system of water-filled elastic scatterers.

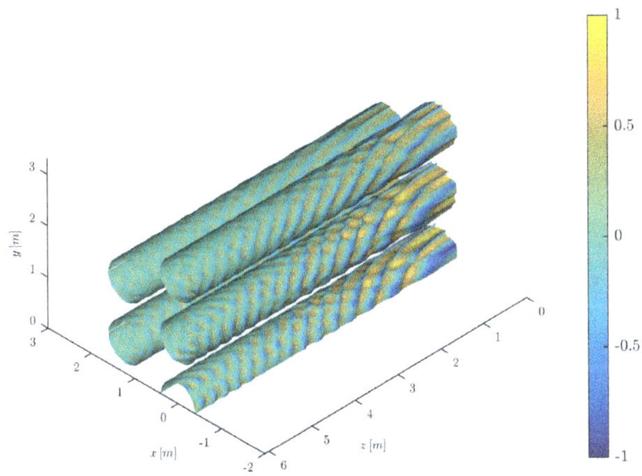

Figure 6: Real part of normal displacement at external surface (normalised to 1.2×10^{-12} m) in a system of water-filled elastic scatterers, represented over the deformed shape.

imposing the appropriate boundary conditions at the interface to study fluid-structure wave propagation.

The model was verified with a benchmark problem having an analytical solution. The coupled SBEM-SFEM was verified with a problem concerning a cylindrical elastic waveguide in an unbounded fluid medium. Numerical results show good agreement with the analytical solution. An h-p analysis shows that an optimal solution in terms of accuracy and computational effort can be obtained through an hp-refinement. Discretisations with a nodal density per wavelength of ten are enough to achieve reasonable accuracy. The accuracy improves by increasing the nodal density per wavelength up to twenty.

Finally, the coupled methodology was used to study the wave field radiated by a three dimensional fluid-filled elastic scatterers system submerged in an unbounded acoustic medium. The sound pressure field and the scatterers' surface displacements were studied. The proposed method made it possible to represent different fluids for the unbounded medium and the acoustic enclosures.

ACKNOWLEDGEMENTS

The research work presented herein was supported by the Spanish Ministry of the Economy and Competitiveness (Ministerio de Economía y Competitividad) through research project BIA2016-75042-C2-1-R. It was also supported by project POCI-01-0247-FEDER-017759 (SmartCore – Desenvolvimento de painéis multifuncionais de elevada performance), funded by the Operational Program for Competitiveness and Internationalization (POCI) of Portugal 2020, with the support of the European Regional Development Fund (FEDER). Financial support is gratefully acknowledged. The authors also wish to acknowledge the support provided by the Andalusian Scientific Computing Centre (CICA).

REFERENCES

[1] Babuška, I., Ihlenburg, F., Paik, E.T. & Sauter, S.A., A generalized finite element method for solving the Helmholtz equation in two dimensions with minimal pollution. *Computer Methods in Applied Mechanics and Engineering*, **128**(3), pp. 325–359, 1995.

[2] Ihlenburg, F., Babuška, I. & Sauter, S., Reliability of finite element methods for the numerical computation of waves. *Advances in Engineering Software*, **28**(7), pp. 417–424, 1997.

[3] Oberai, A.A. & Pinsky, P.M., A numerical comparison of finite element methods for the Helmholtz equation. *Journal of Computational Acoustics*, **08**(01), pp. 211–221, 2000.

[4] Willberg, C., Duczek, S., Vivar Perez, J.M., Schmicker, D. & Gabbert, U., Comparison of different higher order finite element schemes for the simulation of Lamb waves. *Computer Methods in Applied Mechanics and Engineering*, **241–244**, pp. 246–261, 2012.

[5] Patera, A.T., A spectral element method for fluid dynamics: Laminar flow in a channel expansion. *Journal of Computational Physics*, **54**(3), pp. 468–488, 1984.

[6] Ostachowicz, W., Kudela, P., Krawczuk, M. & Zak, A., *Guided Waves in Structures for SHM: The Time–domain Spectral Element Method*, Wiley, 2012.

[7] Cerrato, A., Rodríguez-Tembleque, L., González, J.A. & Ferri Aliabadi, M.H., A coupled finite and boundary spectral element method for linear water-wave propagation problems. *Applied Mathematical Modelling*, **48**, pp. 1–20, 2017.

[8] Pereira, A., Tadeu, A. & António, J., Influence of the cross-section geometry of a cylindrical solid submerged in an acoustic medium on wave propagation. *Wave Motion*, **36**(1), pp. 23–39, 2002.

[9] Gavrić, L., Computation of propagative waves in free rail using a finite element technique. *Journal of Sound and Vibration*, **185**(3), pp. 531–543, 1995.

[10] Gavrić, L., Finite element computation of dispersion properties of thin-walled waveguides. *Journal of Sound and Vibration*, **173**(1), pp. 113–124, 1994.

[11] Domínguez, J., *Boundary elements in dynamics*, Computational Mechanics Publications and Elsevier Applied Science, 1993.

[12] Tadeu, A. & Godinho, L., Three-dimensional wave scattering by a fixed cylindrical inclusion submerged in a fluid medium. *Engineering Analysis with Boundary Elements*, **23**(9), pp. 745–755, 1999.

[13] Fornberg, B. & Zuev, J., The Runge phenomenon and spatially variable shape parameters in RBF interpolation. *Computers and Mathematics with Applications*, **54**(3), pp. 379–398, 2007.
[14] Wu, T.W., *Boundary Element Acoustics Fundamentals and Computer Codes*, WIT Press, 2000.
[15] Zienkiewicz, O., *The Finite Element Method*, 3rd edn, McGraw-Hill, 1986.

COUPLED FINITE AND BOUNDARY ELEMENT METHOD FOR SOLVING MAGNETIC HYSTERESIS PROBLEMS

ILYA STUPAKOV, MIKHAIL ROYAK & NATALIA KONDRATYEVA
Novosibirsk State Technical University, Russia

ABSTRACT

This paper is devoted to effective solving of magnetic hysteresis problems in 3D modeling of accelerator magnets. It is proposed to use boundary elements and a reduced scalar magnetic potential in the air and finite elements, in addition to a full scalar magnetic potential in ferromagnetic materials. There is no need to mesh the air domain by using that approach. Also, it allows nonlinear and inhomogeneous magnetization in ferromagnets to be taken into account. Numerical simulation is carried out using three different models of hysteresis: the Preisach model, the model based on dry friction and the Vinokurov model. To verify the accuracy of the developed approach, the simulation results are compared with the magnetic field measurements for the deflecting accelerator magnet.

Keywords: finite elements, boundary elements, magnetic field, hysteresis, total and reduced potential.

1 INTRODUCTION

Designing acceleration magnets requires effective solving of magnetostatic problems. The most often used method to solve such problems is the finite element method. Employing the boundary element method to solve magnetostatic problems is proposed in two studies by Andjelic et al. [1], [2]. However, even though the boundary element method significantly simplifies the mesh construction, it can only account for nonlinear properties of ferromagnetic materials in ineffective way, while taking nonlinearity into account is important for obtaining an accurate solution of real problems. Thus, the coupled approach with the use of both finite and boundary elements is gaining popularity in recent years [3], [4]. Most notably, this approach is used in CERN and is described in Russenschuck [5]. However, the approach presented in this book employs vector magnet potential. This is less effective for solving various problems than the approach based on two scalar potentials – total and reduced. This idea was first proposed in Simkin and Trowbridge [6], but the algorithm used for combining these potentials was overcomplicated which led to low popularity of the method. In this paper, we propose a simplified formulation with two potentials for the coupled finite and boundary element method.

One more difference from the existing works is the employment of magnetization instead of magnetic permeability. This enables us to account for hysteresis effects in ferromagnetic materials, which significantly improves the accuracy of modeling in some situations. This paper presents a comparison of computational results using several hysteresis models with the magnetic field measurements obtained from a physics experiment.

2 MATHEMATICAL MODEL

Let us consider quasistatic magnetic problems, i.e. the problems where eddy currents and displacement currents are not taken into account. Those problems are described by a special case of Maxwell's equations:

$$\nabla \times \mathbf{H} = \mathbf{J}, \tag{1}$$

$$\nabla \cdot \mathbf{B} = \mathbf{0}, \tag{2}$$

where **H** is magnetic field intensity, **B** is magnetic induction, and **J** is external current density.

Instead of a traditional relation between **H** and **B**, i.e. $\mathbf{B} = \mu\mathbf{H}$, let us use a more general one:

$$\mathbf{B} = \mu_0(\mathbf{H} + \mathbf{M}), \tag{3}$$

where μ_0 is magnetic permeability of vacuum, and **M** is magnetization. Let us consider that the computational domain Ω is simply connected, and its boundary $\partial\Omega$ can be divided into two parts Γ_H and Γ_B, on which homogeneous boundary conditions are specified:

$$\mathbf{H} \times \mathbf{n}|_{\Gamma_H} = 0, \tag{4}$$

$$\mathbf{B} \cdot \mathbf{n}|_{\Gamma_B} = 0, \tag{5}$$

where **n** is an external normal to the domain. Let us seek **H** in the form of

$$\mathbf{H} = \mathbf{H}_{ext} - \nabla u, \tag{6}$$

where u is a scalar magnetic potential, and $\mathbf{H}_{ext}$ is the external field that should satisfy the condition of eqn (7):

$$\nabla \times \mathbf{H}_{ext} = \mathbf{J}. \tag{7}$$

A solution of the magnetostatic equations in an unbounded homogeneous domain can be used as the external field. It can be obtained in an explicit form of the Biot–Savart Law [7]:

$$\mathbf{H}_{ext}(\mathbf{x}) = \frac{1}{4\pi}\int_{\Omega} \frac{\mathbf{J}\times(\mathbf{x}-\mathbf{y})}{|\mathbf{x}-\mathbf{y}|^3} d\mathbf{y}. \tag{8}$$

Note that the field $\mathbf{H}_{ext}$ presented in the way above also satisfies eqn (9):

$$\nabla \cdot \mathbf{H}_{ext} = 0. \tag{9}$$

Since the domain Ω is simply connected, the existence of the scalar potential u satisfying eqn (9) follows from $\nabla \times (\mathbf{H} - \mathbf{H}_{ext}) = 0$ according to Helmholtz's theorem. Thus, eqn (1) is fulfilled and one needs to seek for potential satisfying eqn (2). The obtained scalar potential is called a reduced potential because to calculate the magnetic field one needs to add the external field $\mathbf{H}_{ext}$ to it. In case **M** is dependent on field **H**, this could be inconvenient because of significant computational costs required to obtain field $\mathbf{H}_{ext}$. To avoid this, the formulation with coupled total and reduced scalar potentials can be used. Let us show that an analogous formulation can be obtained in a simpler way than that described in Simkin and Trowbridge [6]. To do that, it is enough to modify the external field $\mathbf{H}_{ext}$. Let us divide the computational domain Ω into two parts: domain Ω_0 with no magnetic materials ($\mathbf{M} = \mathbf{0}$), and domain Ω_M with no currents ($\mathbf{J} = \mathbf{0}$). Let Γ_I be the interface between these domains. Fig. 1 shows the described division.

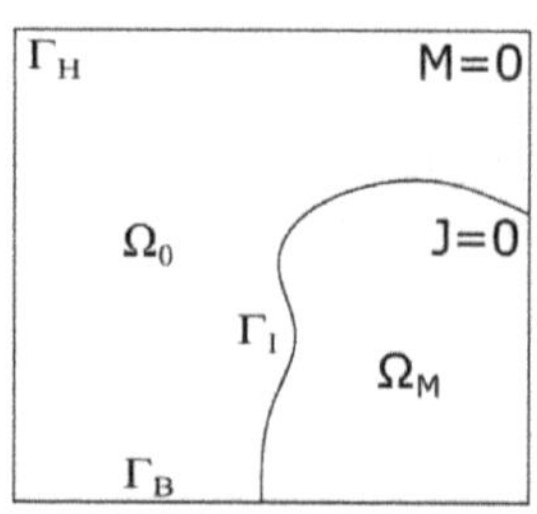

Figure 1: The division of the computational domain.

Let us require that the modified external field $\widetilde{\mathbf{H}}_{ext}$ not only satisfies eqn (7) but also vanishes in the domain Ω_M. Thus, field $\mathbf{H}$ in this domain is defined by the scalar potential only, i.e. the potential becomes total in this domain.

To modify the external field in this way, it is sufficient to ensure that eqn (7) is fulfilled in the domain Ω_0 and the tangential part of the external field vanishes on the interface Γ_I. This can be done if the external field is represented in the following way:

$$\widetilde{\mathbf{H}}_{ext} = \mathbf{H}_{ext} - \nabla h, \tag{10}$$

where function h should satisfy the condition

$$\nabla h \times \mathbf{n}|_{\Gamma_I} = \mathbf{H}_{ext} \times \mathbf{n}|_{\Gamma_I}. \tag{11}$$

Lemma 1: A continuous function h satisfying eqn (11) exists if, and only if, the integral of the currents bounded by a closed contour C is equal to zero for every C lying in Γ_I.

Proof: From the theorem about magnetic field circulation (integral representation of eqn (7)), it follows that a circulation of $\mathbf{H}_{ext}$ around any contour is equal to the integral of the currents bounded by that contour. Considering this, the lemma is equivalent to the following: for the existence of the function h it is necessary and sufficient that the circulation of $\mathbf{H}_{ext}$ around any contour lying in Γ_I was equal to zero. Necessity follows from the circulation of the gradient of a continuous function being equal to zero.

Sufficiency can be proved constructively. Firstly, let us consider a case when the boundary Γ_I is connected. Let us select some point on the boundary where h will be assumed to vanish. From the circulation of $\mathbf{H}_{ext}$ being equal to zero it follows that the work of the vector field $\mathbf{H}_{ext}$ is independent from the integration path. Thus, the value of the function h in an arbitrary point can be defined as the work of $\mathbf{H}_{ext}$ from this point to the point where h is equal to zero. In case the boundary is not connected, h could be analogously defined on each connected component. Outside of the surface Γ_I, the function h can be extended into the domain Ω_0 in any continuous way.

Also note that the sufficient condition for the existence of h is Ω_M being simply connected. However, this condition is not necessary, e.g., Fig. 2 shows a domain Ω_M for which conditions of Lemma 1 are fulfilled, thus, function h exists. Note that the conditions of existence of function h almost do not reduce the applicability of the proposed method. This is due to the fact, that if the contour around the current exists inside the magnetic material, it is usually the case when the effects of eddy currents are significant and should be taken into account.

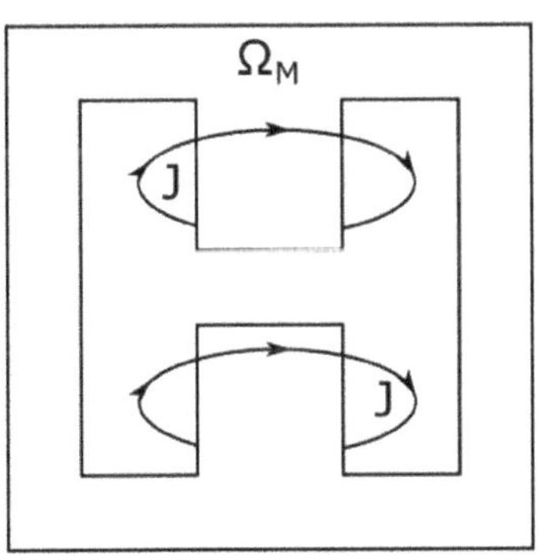

Figure 2: Example of not simply connected domain Ω_M.

In practice, instead of calculating the work, function h can be obtained as a solution of a best approximation problem

$$h = \arg\min_{t} ||\mathbf{H}_{ext} - \nabla t||_{L_2(\Gamma_I)}. \tag{12}$$

3 VARIATIONAL FORMULATION

To obtain a solution of the magnetostatic problem, one need to seek potential u that satisfies eqn (2) in the domains Ω_0 and Ω_M. In addition, the continuity of the normal part of $\mathbf{B}$ should be ensured on the interface between the domains. Firstly, let us consider the domain Ω_M. In this domain, the potential is total, and eqn (2) takes the form

$$\nabla \cdot (-\nabla u + \mathbf{M}) = 0. \tag{13}$$

The normal part of the field $\mathbf{B}$ on the boundary of the domain Ω_M can be presented in the form

$$\frac{1}{\mu_0}\mathbf{B} \cdot \mathbf{n} = -\frac{\partial u}{\partial n} + \mathbf{M} \cdot \mathbf{n}. \tag{14}$$

In this domain, we use the finite element method for the approximation of the potential. To do so, we rewrite eqn (13) in the weak form using the Galerkin method

$$(\nabla \cdot (-\nabla u + \mathbf{M}), v)_{\Omega_M} = 0, \tag{15}$$

where $(\cdot,\cdot)_X$ denotes an inner product of the space $L_2(X)$, $u, v \in H_0^1(\Omega_M)$, and H_0^1 is a Sobolev space which is defined as follows:

$$H_0^1(\Omega_M) = \{f | f \in L_2(\Omega_M), \nabla f \in L_2(\Omega_M), f|_{\Gamma_H} = 0\}. \tag{16}$$

We integrate eqn (15) by parts

$$\left(-\frac{\partial u}{\partial n} + \mathbf{M} \cdot \mathbf{n}, v\right)_{\partial\Omega_M} = (-\nabla u + \mathbf{M}, \nabla v)_{\Omega_M}. \tag{17}$$

We replace the left part with the normal part of $\mathbf{B}$, according to eqn (14), and employ the boundary conditions. After that, we have

$$\left(\frac{1}{\mu_0}\mathbf{B} \cdot \mathbf{n}, v\right)_{\Gamma_I} = (\mathbf{M}, \nabla v)_{\Omega_M} - (\nabla u, \nabla v)_{\Omega_M}. \tag{18}$$

Secondly, let us consider the domain Ω_0. In this domain, the potential is reduced. If we extend function h in a way that it is harmonic, then eqn (2) in this domain is reduced to the Laplace's equation

$$\Delta u = 0. \tag{19}$$

The normal part of the field $\mathbf{B}$ on the boundary of the domain Ω_0 can be presented in the form

$$\frac{1}{\mu_0}\mathbf{B} \cdot \mathbf{n} = \mathbf{H}_{ext} \cdot \mathbf{n} - \frac{\partial h}{\partial n} - \frac{\partial u}{\partial n}. \tag{20}$$

We use the boundary element method for the approximation of the potential in this domain. The approach used is based on Steinbach [8]. We will employ the fact that the solution of the Laplace's equation can be presented in the form

$$u(y) = \int_{\partial\Omega_0} U^*(\mathbf{x},\mathbf{y}) \frac{\partial}{\partial n} u(\mathbf{x}) dS - \int_{\partial\Omega_0} \frac{\partial}{\partial n_x} U^*(\mathbf{x},\mathbf{y}) u(x) dS_x, \tag{21}$$

where $U^*(\mathbf{x},\mathbf{y})$ is the fundamental solution of the Laplace's equation. From this equation, we can obtain the boundary integral equations

$$u = \mathrm{V}\frac{\partial u}{\partial n} + \left(\frac{1}{2}\mathrm{I} - \mathrm{K}\right)u, \tag{22}$$

$$\frac{\partial u}{\partial n} = \left(\frac{1}{2}\mathrm{I} + \mathrm{K}'\right)\frac{\partial u}{\partial n} + \mathrm{D}u, \tag{23}$$

where V is a single layer potential operator, K is a double layer potential operator, and D is a hypersingular integral operator.

After substituting the flows in the second equation by the expression from the first equation, we can obtain the equation for a relation between u and $\frac{\partial u}{\partial n}$:

$$\frac{\partial u}{\partial n} = \left(\mathrm{D} + \left(\frac{1}{2}\mathrm{I} + \mathrm{K}'\right)\mathrm{V}^{-1}\left(\frac{1}{2}\mathrm{I} + \mathrm{K}\right)\right)u = \mathrm{S}u, \tag{24}$$

where S is the Steklov–Poincare operator.

We rewrite eqn (20) in the weak form using the Galerkin method

$$\left(\frac{1}{\mu_0}\mathbf{B}\cdot\mathbf{n}, v\right)_{\partial\Omega_0} = (\mathbf{H}_{ext}\cdot\mathbf{n}, v)_{\partial\Omega_0} - \left(\frac{\partial h}{\partial n}, v\right)_{\partial\Omega_0} - \left(\frac{\partial u}{\partial n}, v\right)_{\partial\Omega_0}, \tag{25}$$

where $u, v \in H_0^{1/2}(\partial\Omega_0)$, and $H_0^{1/2}$ is a Sobolev space which is defined as follows:

$$H_0^{1/2}(\partial\Omega_0) = \{f \mid \exists g \in H^1(\Omega_0): g|_{\partial\Omega_0} = f, f|_{\Gamma_H} = 0\}. \tag{26}$$

Using the Steklov–Poincare operator, this equation can be transformed into the following form of eqn (27):

$$\left(\frac{1}{\mu_0}\mathbf{B}\cdot\mathbf{n}, v\right)_{\Gamma_I} = (\mathbf{H}_{ext}\cdot\mathbf{n}, v)_{\Gamma_I} - (\mathrm{S}h, v)_{\Gamma_I} - (\mathrm{S}u, v)_{\partial\Omega_0}. \tag{27}$$

To obtain the final equation, one needs to equate the normal components of **B**:

$$(\mathrm{S}u, v)_{\partial\Omega_0} + (\nabla u, \nabla v)_{\Omega_M} = (\mathbf{H}_{ext}\cdot\mathbf{n}, v)_{\Gamma_I} - (\mathrm{S}h, v)_{\Gamma_I} + (\mathbf{M}, \nabla v)_{\Omega_M}, \tag{28}$$

where $u, v \in H_0^1(\Omega_M) \cup H_0^{1/2}(\partial\Omega_0)$.

4 DISCRETIZATION INTO SYSTEM OF EQUATIONS

To seek u using the coupled finite and boundary element method, let us introduce a finite dimensional subspace V_h of the space $H_0^1(\Omega_M) \cup H_0^{1/2}(\partial\Omega_0)$. Let the functions $\psi_i \in V_h$ form a basis of the space V_h. Then the projections of the potential u and the function h onto the space V_h can be presented in the following form:

$$u = \sum_j q_j \psi_j, \ \ h = \sum_j r_j \psi_j, \tag{29}$$

where q_j and r_j denote weights of basis functions. Then we can rewrite eqn (28) as a system of equations:

$$(\mathbf{S}^{BEM} + \mathbf{S}^{FEM})\mathbf{q} = \mathbf{f}^H - \mathbf{S}^{BEM}\mathbf{r} + \mathbf{f}^M, \tag{30}$$

where matrix $\mathbf{S}^{BEM}$ is an approximation of Steklov–Poincare boundary element operator, $\mathbf{S}^{FEM}$ is a finite element stiffness matrix,

$$f_i^H = (\mathbf{H}_{ext}\cdot\mathbf{n}, \psi_i)_{\Gamma_I}, \tag{31}$$

$$f_i^M = (\mathbf{M}, \nabla\psi_i)_{\Omega_M}. \tag{32}$$

Let us consider the calculation of the contributions from the magnetization f_i^M. Let the domain Ω_M be divided into finite elements Ω_k. We assume the magnetization is constant on elements Ω_k and also is dependent on the mean value of the field $\mathbf{H}$ on this element

$$\widehat{\mathbf{H}}_k = \frac{1}{|\Omega_k|}\int_{\Omega_k} \mathbf{H}\, d\Omega, \tag{33}$$

where $|\Omega_k|$ is the volume of the element Ω_k. If we denote

$$\mathbf{G}_j^k = \int_{\Omega_k} \nabla\psi_j d\Omega, \tag{34}$$

then the mean value of $\mathbf{H}$ can be calculated using eqn (35):

$$\widehat{\mathbf{H}}_k = -\frac{1}{|\Omega_k|}\sum_j \mathbf{G}_j^k\, q_j. \tag{35}$$

Then the contributions from the magnetization f_i^M can be written as

$$f_i^M = \sum_k \mathbf{M}\,(\widehat{\mathbf{H}}_k)\cdot \mathbf{G}_i^k. \tag{36}$$

Eqn (30) is nonlinear, because contributions f_i^M are dependent on the vector of weights $\mathbf{q}$. Let us define a residual vector, whose norm should be minimized

$$\mathbf{R}(\mathbf{q}) = (\mathbf{S}^{BEM} + \mathbf{S}^{FEM})\mathbf{q} - \mathbf{f}^H + \mathbf{S}^{BEM}\mathbf{r} - \mathbf{f}^M(\mathbf{q}). \tag{37}$$

To minimize the residual, we use the Gauss–Newton method

$$\mathbf{J}\mathbf{p}_m = -\mathbf{R}(\mathbf{q}_{m-1}), \tag{38}$$

$$\lambda_m = \arg\min_{\lambda} \|\mathbf{R}(\mathbf{q}_{m-1} + \lambda\mathbf{p}_m)\|, \tag{39}$$

$$\mathbf{q}_m = \mathbf{q}_{m-1} + \lambda_m \mathbf{p}_m, \tag{40}$$

where m is a sequence number of the iteration over nonlinearity, and a vector of zeros can be used for $\mathbf{q}_0$. Matrix $\mathbf{J}$ is calculated using the weights from the previous iteration via eqn (41):

$$\mathbf{J} = (\mathbf{S}^{BEM} + \mathbf{S}^{FEM}) - \frac{\partial \mathbf{f}^M}{\partial \mathbf{q}}, \tag{41}$$

where the following eqn (42) can be used to calculate the derivatives of f_i^M

$$\frac{\partial f_i^M}{\partial q_j} = -\frac{1}{|\Omega_k|}\sum_k (\mathbf{G}_i^k)^T \frac{\partial \mathbf{M}}{\partial \mathbf{H}} \mathbf{G}_j^k. \tag{42}$$

Note that if the tensor $\frac{\partial \mathbf{M}}{\partial \mathbf{H}}$ is symmetric, the matrix $\mathbf{J}$ will also be symmetric.

5 HYSTERESIS MODELS

To solve the classic problem of magnetostatics, one can employ $\mathbf{M}(\mathbf{H}) = (\mu - 1)\mathbf{H}$. However, if hysteresis effects should be accounted for, one needs to use a hysteresis model, which describes the relation $\mathbf{M}(\mathbf{H})$. There are a lot of different models and none of them are unquestionably the best [9], [10]. In this paper, we consider the three following models.

5.1 Preisach model

The Preisach model is one of the most well-known models of hysteresis [12]. Let us use the Preisach model in the form with a one-dimensional distribution for which the Everett function can be analytically calculated [13], [14]. A one-dimensional distribution is given by the following expression:

$$\varphi(x) = \frac{ae^{-bx}}{(1+ce^{-bx})^2}, \tag{43}$$

where a, b and c are model parameters. The magnetization in this model can be expressed as

$$M(H) = \int\int \varphi(-h_1)\varphi(h_2)\gamma_H(h_1, h_2)dh_1dh_2, \tag{44}$$

where $\gamma_H(h_1, h_2)$ can be $+1$ or -1 depending on the history of H. This expression can be represented as a combination of the Everett function values in some points. The Everett function can be obtained analytically by calculating the following integrals:

$$E(x,y) = \int_x^y \int_x^{h_2} \varphi(-h_1)\varphi(h_2)dh_1dh_2. \tag{45}$$

To obtain a vector model, let us use the method proposed in Mayergoyz [15]:

$$\mathbf{M(H)} = \sum_k w_k\, \mathbf{r}_k M(\mathbf{H}\cdot\mathbf{r}_k), \tag{46}$$

where w_k and $\mathbf{r}_k$ represent the weights and points of some unit sphere quadrature, respectively.

5.2 Vinokurov model

The Vinokurov model was proposed in Vinokurov et al. [16]. Let us use this model in the form of eqn (47):

$$M(H) = \sum_k w_k\, M_s\Big(1 - \frac{1}{\sqrt{1+2\frac{\chi_0}{M_s}|H_{e,k}|}}\Big)\frac{H_{e,k}}{|H_{e,k}|}, \tag{47}$$

where w_k, M_s and χ_0 are model parameters and $H_{e,k}$ can be obtained using the following update rule:

$$H_{e,k}^{new} = \begin{cases} Hcos(\theta_k) - H_{c,k} & |Hcos(\theta_k) - H_{c,k}| > H_{e,k}, \\ Hcos(\theta_k) + H_{c,k} & |Hcos(\theta_k) + H_{c,k}| < H_{e,k}, \\ H_{e,k} & otherwise, \end{cases} \tag{48}$$

where θ_k and $H_{c,k}$ are model parameters. The vector variant of Vinokurov model will be used in eqn (46).

5.3 Dry friction model

The main idea of the dry friction model was proposed by Bergqvist [17]. The difference of this model from two previous models is that the Bergqvist's model is a vector model by definition.

Let us use the modification described in Henrotte et al. [18] and Henrotte and Hameyer [19]. In this model, the reversible component of **H** is expressed as

$$\mathbf{H}_r = \sum_k w_k\, \mathbf{H}_{r,k}, \tag{49}$$

and $\mathbf{H}_{r,k}$ can be obtained by the following update rule:

$$\mathbf{H}_{r,k}^{new} = \begin{cases} \mathbf{H}_{r,k} & |\mathbf{H} - \mathbf{H}_{r,k}| \leq r_k, \\ \mathbf{H} - r_k \frac{\mathbf{H}-\mathbf{H}_{r,k}}{|\mathbf{H}-\mathbf{H}_{r,k}|} & otherwise, \end{cases} \tag{50}$$

where w_k and r_k are model parameters. The magnetization can be calculated by the following expression:

$$\mathbf{M}(\mathbf{H}) = M_s L\left(\frac{|\mathbf{H}_r|}{a}\right)\frac{\mathbf{H}_r}{|\mathbf{H}_r|}, \tag{51}$$

where

$$L(x) = \coth(x) - \frac{1}{x}, \tag{52}$$

is the Langevin function, M_s and a is a model parameter.

6 NUMERICAL EXPERIMENTS

To compare these three models, let us use the experimental measurements for a deflecting accelerator magnet described in Royak et al. [11]. The 3D model of this magnet with current coils is shown in Fig. 3.

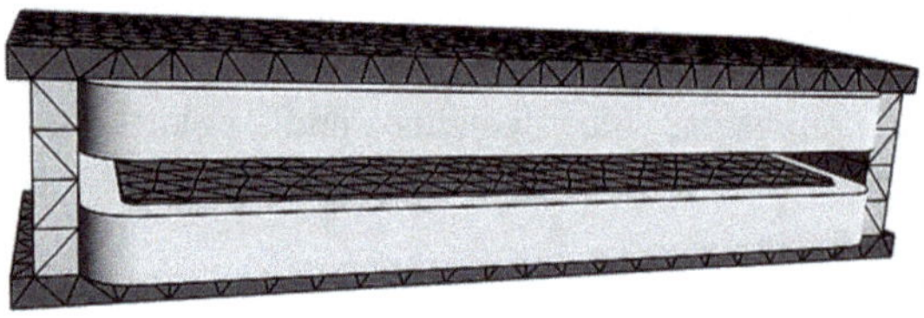

Figure 3: The 3D model of the deflecting accelerator magnet with current coils.

This magnet consists of two materials. For the first material, constructional steel 3, there are some measurements described in Vinokurov et al. [16] that allow us to obtain hysteresis model parameters of the material. The second material is used for the magnet pole. The material is low carbon steel that is close to the pure iron in the chemical composition and magnetic properties. However, numerical simulation was complicated due to exhaustion of samples of this steel; for this reason, we had to determine the hysteresis model parameters by three-dimensional modeling. To select model parameters, a symmetric variation in the current was chosen. The current variation is presented in Fig. 4.

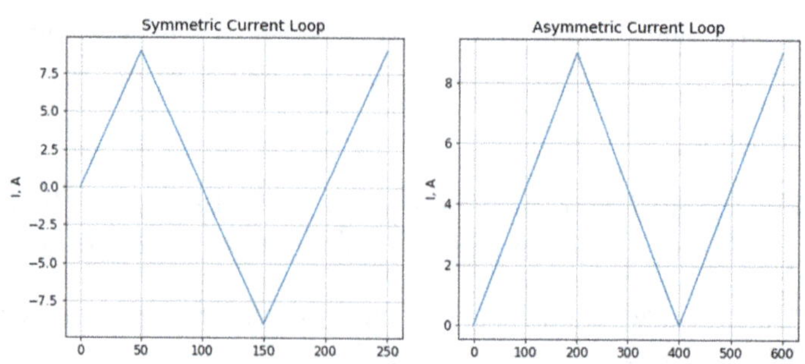

Figure 4: Current loops.

Then we will compare models both on the symmetric variation in the current and on an asymmetric variation in the current. The hysteresis loops of all three models and measurements for the symmetric current are presented in Fig. 5.

The parameters of pole material for all three models are found by three-dimensional modeling for this loop. As seen, all three models have slight differences from the experimental measurements.

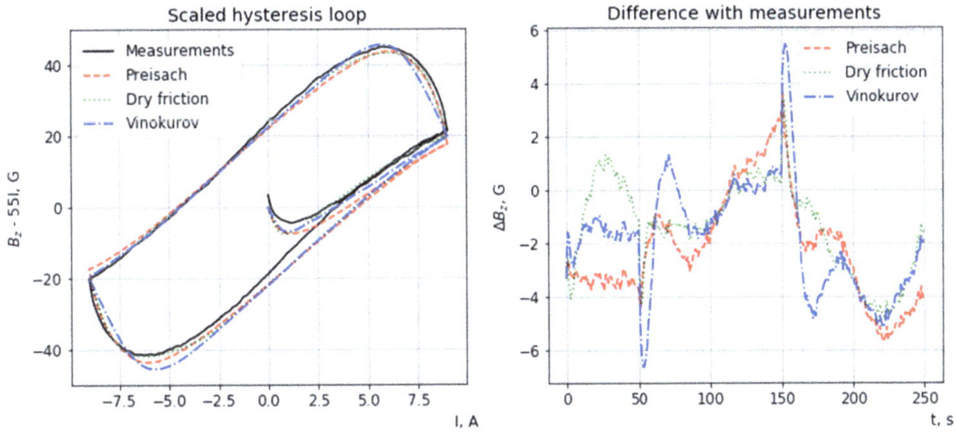

Figure 5: Result on symmetric loop.

The hysteresis loops of all three models and measurements for the asymmetric current are presented in Fig. 6. The parameters of pole material for all three models are the same as for the previous modeling for the symmetric loop. As seen, all three models also have slight differences from the experimental measurements.

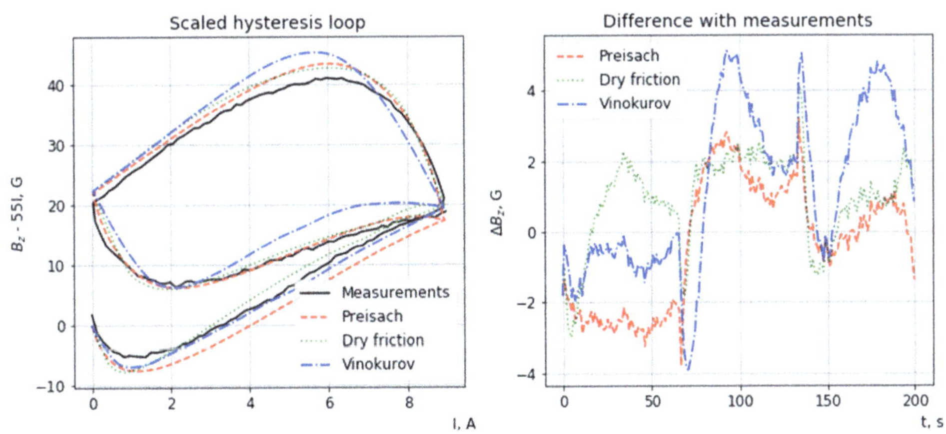

Figure 6: Result on asymmetric loop.

Finally, let us present all results to compare in Table 1. The "time" column in this table represents the time of the whole loop simulation for a model.

Table 1: Models comparison.

Model	Symmetric loop		Asymmetric loop	
	error, G	*time, s*	*error, G*	*time, s*
Preisach	4.26	141	3.08	84
Vinokurov	4.17	154	4.00	129
Dry friction	3.16	115	2.70	103

7 CONCLUSION

The results of the numerical experiments show that the proposed approach based on the coupled finite and boundary element method is viable and can be used to solve real physics problems. The promising outlook of the approach is also supported by the fact that even on the coarse mesh used in the experiments, the difference between computational results and the magnetic field measurements is around 1%, which is comparable to the error of the measurements.

All three hysteresis models have slight differences from the measurements for both current loops, so we cannot choose the best model by the accuracy criteria. The simulation times for all models are not significantly different, so time cannot be chosen as a criterion either. But the number of history parameters that corresponds to the computer memory consumption differs significantly. For the results presented in this paper, the minimal memory (2 vectors per finite element) was used by the dry friction model and other models used much more memory (2 float numbers for each sphere quadrature node per element for the Vinokurov model and not less than 3 floats for each sphere quadrature node per element for the Preisach model).

ACKNOWLEDGEMENT

The research was fully supported by the Russian Science Foundation grant (project 17-71-10203).

REFERENCES

[1] Andjelic, Z., Of, G., Steinbach, O. & Urthaler, P., Boundary element methods for magnetostatic field problems: A critical view. *Computing and Visualization in Science*, **14**(3), pp. 117–130, 2011.

[2] Andjelic, Z., Of, G., Steinbach, O. & Urthaler, P., Fast boundary element methods for industrial applications in magnetostatics. *Fast Boundary Element Methods in Engineering and Industrial Applications*, Springer: Berlin and Heidelberg, pp. 111–143, 2012.

[3] Pusch, D. & Ostrowski, J., Robust FEM/BEM coupling for magnetostatics on multiconnected domains. *IEEE Transactions on Magnetics*, **46**(8), pp. 3177–3180, 2010.

[4] Bruckner, F. et al., 3D FEM–BEM-coupling method to solve magnetostatic Maxwell equations. *Journal of Magnetism and Magnetic Materials*, **324**(10), pp. 1862–1866, 2012.

[5] Russenschuck, S., *Field Computation for Accelerator Magnets: Analytical and Numerical Methods for Electromagnetic Design and Optimization*, John Wiley & Sons, 2011.

[6] Simkin, J. & Trowbridge, C.W., Three-dimensional nonlinear electromagnetic field computations, using scalar potentials. *IEE Proceedings B-Electric Power Applications*, IET, vol. 127, no. 6, pp. 368–374, 1980.

[7] Stupakov, I.M., Royak, M.E. & Kondratyeva, N.S., The method for calculating magnetic field induced by current coils. *13th International Scientific-Technical Conference on Actual Problems of Electronics Instrument Engineering* (APEIE), vol. 2, pp. 347–350, 2016.

[8] Steinbach, O., Numerical approximation methods for elliptic boundary value problems: finite and boundary elements. *Springer Science & Business Media*, 2007.

[9] Liorzou, F., Phelps, B. & Atherton, D.L., Macroscopic models of magnetization. *IEEE Transactions on Magnetics*, **36**(2), pp. 418–428, 2000.

[10] Benabou, A., Clénet, S. & Piriou, F., Comparison of Preisach and Jiles–Atherton models to take into account hysteresis phenomenon for finite element analysis. *Journal of Magnetism and Magnetic Materials*, **261**(1–2), pp. 139–160, 2003.

[11] Royak, M.E. et al., Application of a new model of residual magnetization of iron for calculating the deflecting magnet of an accelerator. *Technical Physics Letters*, **43**(10), pp. 924–927, 2017.

[12] Preisach, F., On the magnetic aftereffect. *IEEE Transactions on Magnetics*, **53**(3), pp. 1–11, 2017.

[13] Szabó, Z., Tugyi, I., Kádár, G. & Füzi, J., Identification procedures for scalar Preisach model. *Physica B: Condensed Matter*, **343**(1–4), pp. 142–147, 2004.

[14] Fuzi, J., Analytical approximation of Preisach distribution functions. *IEEE Transactions on Magnetics*, **39**(3), pp. 1357–1360, 2003.

[15] Mayergoyz, I.D., *Mathematical Models of Hysteresis and their Applications*, Academic Press, 2003.

[16] Vinokurov, N.A. et al., Allowing for hysteresis in the calculation of fields in the elements of accelerator magnetic systems. *Technical Physics Letters*, **42**(7), pp. 708–711, 2016.

[17] Bergqvist, A., Magnetic vector hysteresis model with dry friction-like pinning. *Physica B: Condensed Matter*, **233**(4), pp. 342–347, 1997.

[18] Henrotte, F., Nicolet, A. & Hameyer, K., An energy-based vector hysteresis model for ferromagnetic materials. *COMPEL – The International Journal for Computation and Mathematics in Electrical and Electronic Engineering*, **25**(1), pp. 71–80, 2006.

[19] Henrotte, F. & Hameyer, K., A dynamical vector hysteresis model based on an energy approach. *IEEE Transactions on Magnetics*, **42**(4), pp. 899–902, 2006.

COUPLING OF THE BOUNDARY ELEMENT METHOD WITH A HYBRID METHOD FOR INVERSE STRESS ANALYSIS OF PIPELINES

CIBELE CORNEJO JACINTO PORTELA[1,2], AMANDA JAREK[3,4] & LUIZ ALKIMIN DE LACERDA[2,3,4]
[1]ITeCons – Institute for Research and Technological Development in Construction, Energy, Environment and Sustainability, Portugal
[2]Postgraduate Program in Numerical Methods in Engineering, UFPR – Federal University of Paraná, Brazil
[3]LACTEC – Institute of Technology for Development, Civil Engineering, Brazil
[4]Department of Civil Engineering, UFPR – Federal University of Paraná, Brazil

ABSTRACT

The aim of this work is to develop a numerical tool for structural inverse analysis to obtain boundary conditions and their location. The inverse analysis is performed through an optimization process having as restrictions the information obtained from a direct analysis, representing data which could be obtained from surface monitoring of pipelines. Thus, the numerical models comprehend the solution of two problems: the direct one in which the boundary element method (BEM) is used to obtain the displacements and tractions in the structure, and the inverse problem, in which two optimization models were implemented. The first approach uses the genetic algorithm (GA) method associated with the BEM (GA-BEM), minimizing the residuals between the calculated and the "monitored" relative surface displacements. The second approach is a hybrid method (HM) in which the GA is associated with the Newton–Raphson method for performing the inverse analysis with the BEM (HM-BEM). All tests demonstrated the efficiency of the numerical tools for structural inverse analysis, with the HM approach showing lower computational cost and better accuracy.

Keywords: boundary element method, inverse analysis, genetic algorithm, Newton–Raphson method, elasticity, hybrid method.

1 INTRODUCTION

Transportation of oil and gas is an important issue in the oil industry where the main means of transportation are the pipelines. With the increase of demand and the use of oil and its derivatives, there has been a significant increase in pipeline projects requiring larger diameters and smaller wall thicknesses operating under high pressure. Due to the variable geotechnical conditions, the risks of rupture cannot be neglected. It is ideal to minimize these risks with structural or geotechnical improvements in order to prevent accidents while obtaining lower cost of production. Therefore, it is very important to provide proper tools to the maintenance engineers to detect and avoid conditions which can potentially result in structural failure, and consequently, generating environmental damages and/or economic and human losses. For this reason, analyses of pipelines were performed by many researchers up to present days and a wide range of analytical and numerical models can be found in literature [1]–[3], characterized by different assumptions for pipelines behaviour.

In this context, the present work aimed the development of two computational codes, in Fortran 95 programming language, for the analysis of three dimensional linear elastic problems. The purpose of these tools is the application of structural inverse analysis in deformed ducts to retrieve their boundary conditions. Inverse problems constitute a large field of research in science and engineering, since in many applications it is necessary to identify parameters such as boundary unknown conditions, geometry or material properties from experimental measurements.

WIT Transactions on Engineering Sciences, Vol 126, © 2019 WIT Press
www.witpress.com, ISSN 1743-3533 (on-line)
doi:10.2495/BE420121

In this work, a first numerical tool was built employing the boundary element method (BEM) to obtain the displacements and stresses distributed in the structure. This initial implementation was used for direct analyses and production of synthetic data and also as an auxiliary program for the inverse algorithms.

The first inverse analysis was performed with BEM and a genetic algorithm (GA). This numerical tool was called GA-BEM. There are many works in the technical literature, in several areas of application that have been successful in solving optimization problems using this combination [4]–[6]. For instance, Katsifarakis and Petala [4] presented numerical results for the maximization of the rate of extraction of underground water without saline intrusion for management in coastal aquifers. Miltiadou and Wrobel [5] used this coupling technique for the identification of polarization curves in cathodic protection systems. Another reference is the work of Kita and Tanie [6] for shape optimization of structures.

The GA's tools have the disadvantage of generally requiring a high computational effort when compared to gradient methods. Thus, a second numerical tool for inverse analysis was elaborated, using the same BEM code with a hybrid method (HM), which conjugates GA with the Newton–Raphson gradient method (NR). This numerical tool was called HM-BEM. It is known that the NR method implies in lower computational cost, but requires a good initial value to ensure convergence. The work of Feliz [7] is an example that used the proposal of a hybrid system to study power flux convergence problem in energy systems. The system was composed by GA method responsible to provide an initial solution from a group of possibilities, followed by the NR method to achieve the final result. In this work, a combination of GA and NR is evaluated for the inverse analysis of deformed structures searching the identification of load boundary condition magnitude and location.

2 BOUNDARY ELEMENT METHOD FOR ELASTICITY

Most of the physical phenomena occurring in nature can be described by differential equations where initial conditions or boundary conditions are imposed. There are different methods for solving these problems.

Analytical solutions to mathematical problems that satisfy both differential equations and boundary conditions can be obtained only for some very simple problems. Therefore, numerical methods are essential if complex problems are under investigation. The BEM is one possible approach for the solution of three-dimensional problems [8]–[9].

2.1 Fundamental solution

For the BEM, a solution to the governing equation is required, and is called the fundamental solution. Considering a homogenous three-dimensional linear elastic infinite domain under a concentrated unit load, the fundamental Kelvin displacement and traction solutions in 3D domains are [9]:

$$U_{ij}(X',x) = \frac{1}{16\pi G(1-\upsilon)r}\left[(3-4\upsilon)\delta_{ij} + r_i r_j\right], \quad (1)$$

$$T_{ij}(X',x) = \frac{-1}{8\pi(1-\upsilon)r^2}\left[\frac{\partial r}{\partial n}\left[(1-2\upsilon)\delta_{ij} + 3r_i r_j\right] + (1-2\upsilon)\left(n_i r_j - n_j r_i\right)\right], \quad (2)$$

where $U_{ij}(X',x)$: displacement at any field point x, in the direction j, when the unit load is applied in a collocation point X' in the direction i; $T_{ij}(X',x)$: traction at any field point x, in the direction j, when the unit load is applied in a collocation point X' in the direction i; r: distance from collocation point X' to field point x; n: normal direction at x, in the boundary; G: Shear modulus; υ: Poisson's ratio; δ: Kronecker delta.

2.2 Boundary integral equation

The Somigliana's identity brings the displacement value at an internal point X' in terms of the values of displacements u_j and tractions t_j on the boundary surface Γ, and of the body forces b_j. Considering $x \in \Gamma$, $X \in \Omega$ and $X' \in \Omega$. Somigliana's identity is expressed by:

$$u_i(X') = \int_\Gamma U_{ij}(X',x)t_j(x)d\Gamma - \int_\Gamma T_{ij}(X',x)\ u_j(x)d\Gamma + \\ + \int_\Omega U_{ij}(X',X)\ b_j(X)\ d\Omega. \tag{3}$$

The direct boundary element formulation relating boundary tractions and displacements can be obtained from eqn (3) by considering a limiting process as the internal point X' approaches the boundary, becoming x'. The resulting expression is the displacement boundary integral equation, written as:

$$C_{ij}(x')u_j(x') + \int_\Gamma T_{ij}(x',x)u_j(x)d\Gamma = \int_\Gamma U_{ij}(x',x)t_j(x)d\Gamma + \\ + \int_\Omega U_{ij}(x',X)\ b_j(X)\ d\Omega, \tag{4}$$

where the free term coefficient $Cij(x')$ depends on the boundary geometry at the collocation point x'.

Discretizing eqn (4), neglecting body forces and applying it to all nodal points in the boundary, one can assemble a system of equations in the format [10]:

$$\boldsymbol{Hu} = \boldsymbol{Gt}, \tag{5}$$

where $\boldsymbol{H}$ and $\boldsymbol{G}$ are coefficient matrices, which depend on the discretized geometry and material properties, and $\boldsymbol{u}$ and $\boldsymbol{t}$ are the vectors of nodal values of displacements and tractions in the boundary, respectively.

If there is only one unknown value at each nodal point, the system can be rearranged by placing all the unknowns on the left-hand side and all known variables to the right-hand side, forming:

$$\boldsymbol{Ax} = \boldsymbol{F}, \tag{6}$$

where all unknown nodal values are located in vector $\boldsymbol{x}$ and $\boldsymbol{F}$ is found by multiplying the corresponding $\boldsymbol{H}$ and $\boldsymbol{G}$ columns by the known boundary data.

Once this solution is complete, all boundary variables are known and the solution at any domain point can be calculated.

3 INVERSE PROBLEMS

When solving a direct problem in elasticity, the boundary conditions, the geometry and the material properties are all known. Stresses, strains and displacements along the body are the usual results. On the other hand, in an inverse problem, boundary parameters, geometry and material characteristics, may not be known explicitly. One way to solve an inverse problem is to consider it as an optimization problem, searching to maximize or minimize an objective function.

Genetic algorithms and the Newton–Raphson method are examples of techniques which can be used to solve an optimization problem. Each technique is briefly described in this section pointing out how they are coupled to the BEM. Moreover, a strategy for their combination as a hybrid method is also shown, bringing their advantages into a single solution method.

3.1 Genetic algorithms

Genetic algorithms are numerical optimization algorithms based on the natural process of biological evolution (Darwinism) [11]. From the theory of Darwin, the principle of selection privileges the individuals who are more apt, with more possibility of survival and, therefore, with higher probability of reproduction. The larger the number of descendants, the bigger the chances of their genetic codes remain for the next generations. These genetic codes constitute the identity of each individual and are represented in the chromosomes. These are the principles that inspire the construction of GA's, which seek a better solution to a given problem through the evolution of populations of solutions encoded by means of artificial chromosomes [12].

In GA's, a chromosome is a structure that can be seen as a solution point in the search space of the problem to be optimized. These chromosomes are then subjected to an evolutionary process. After several generations, the population evolves in order to obtain more fit individuals.

According to Coley [12], GA's are initialized with a population of possible solutions that are usually randomly scattered in the search space and are converted to real variables. After obtaining the first population, the value of an objective function for each generated variable is calculated. Through this function, also called the fitness function, it is possible to evaluate the quality of each solution for the maximum or minimum problem.

In the present work, GA is implemented to determine the minimizers of the objective function that is represented by:

$$Z_k = \sum_{i=1}^{N_v} \left(\frac{x_{expi} - x_{calci}}{x_{calci}}\right)^2, k = 1,2 \dots N_p\ , \tag{7}$$

where Z_k: value of the objective function for each individual; x_{expi}: experimentally measured relative displacement between two points of the structure under analysis. x_{calci}: numerically calculated relative displacement between two points, with the BEM; N_v: number of measurements; N_p: number of individuals.

These chromosomes are then subjected to an evolutionary process with the application of Selection and Genetic Operators. A typical genetic algorithm uses two operators to "guide" the population over a series of generations to the overall optimum: crossover and mutation [13].

Selection is applied on the population in a manner similar to the natural selection found in biological systems. Using the fitness function, weaker individuals are eliminated, and the more fit individuals have higher chance of promoting the information they have for the next generation. This process does not totally exclude individuals less able to maintain the diversity of the population.

The crossover allows the exchange of information from the solutions in a manner similar to that used by natural organisms submitted to reproduction. This operator promotes a blend of the characteristics of the "parents", who were previously selected based on their fitness.

The mutation randomly changes the bit value of an individual. Its purpose is to maintain the genetic diversity of the population by ensuring that the probability of reaching any point in the search space will not be zero. Its application creates new characteristic values that does not exist or that appeared in small numbers in the population.

After selection, crossover and mutation have been applied in the initial population, a new population is formed, and the generation counter is increased by one. These processes of selection, crossover and mutation continue until a certain number of generations or until some form of convergence criterion is fulfilled.

3.2 Newton–Raphson method

The Newton–Raphson method consists of a procedure for calculating the root of a function $f(x)$. The basic idea is to successively apply a formula starting from an initial approximation x_0 and generating a sequence of approximations that will converge to the root of the function [14]. Generally, for a single function, one has:

$$x_{i+1} = x_i - \frac{f(x_i)}{f'(x_i)}. \tag{8}$$

But this method can also be applied to a function that represents a system of equations as:

$$\boldsymbol{F}(\boldsymbol{x}) = 0, \tag{9}$$

where $\boldsymbol{x} = [x_1, x_2, \ldots, x_n]^T$.

Using the Taylor Series of $\boldsymbol{F}(\boldsymbol{x})$ about the point $\boldsymbol{x}_0$, where $\boldsymbol{x}_0$ is an approximation for the solution of $\boldsymbol{F}(\boldsymbol{x})=0$ and keeping only the first order terms:

$$\boldsymbol{F}(\boldsymbol{x}) \approx \boldsymbol{F}(\boldsymbol{x}_0) + (\boldsymbol{x} - \boldsymbol{x}_0)\boldsymbol{F}'(\boldsymbol{x}_0). \tag{10}$$

The Jacobian Matrix is formed by the partial derivatives of $\boldsymbol{F}$ components:

$$\boldsymbol{J_F} = \begin{pmatrix} \frac{\partial f_1}{\partial x_1} & \cdots & \frac{\partial f_1}{\partial x_n} \\ \vdots & \cdots & \vdots \\ \frac{\partial f_m}{\partial x_1} & \cdots & \frac{\partial f_m}{\partial x_n} \end{pmatrix}. \tag{11}$$

And it is possible to rewrite eqn (10) as:

$$\boldsymbol{F}(\boldsymbol{x}) = \boldsymbol{F}(\boldsymbol{x}_0) + (\boldsymbol{x} - \boldsymbol{x}_0)\boldsymbol{J}_F(\boldsymbol{x}_0). \tag{12}$$

Finally, the NR formula for iteratively solving problems involving systems of equations is:

$$\boldsymbol{x}_{i+1} = \boldsymbol{x}_i - \boldsymbol{J}_F^{-1}\boldsymbol{F}(\boldsymbol{x}_i). \tag{13}$$

3.3 Model formed by the hybrid method and the boundary element method (HM-BEM)

The scheme in Fig. 1 briefly illustrates the operation of the HM for inverse analysis with the BEM. First, an initial population (possible solutions of the problem) is created and represented by binary vectors. Then, these binary vectors are transformed into real numbers. For each generation, all individuals are submitted to the BEM. According to the results, each individual of this population is evaluated by the fitness function and classified according to its quality.

The best individual of this population is used as the starting point for the NR and the new individual is created and converted to a binary vector. The previous population along with this new individual are subjected to genetic operators and a new population is created. This cycle is repeated until a stopping criterion is satisfied.

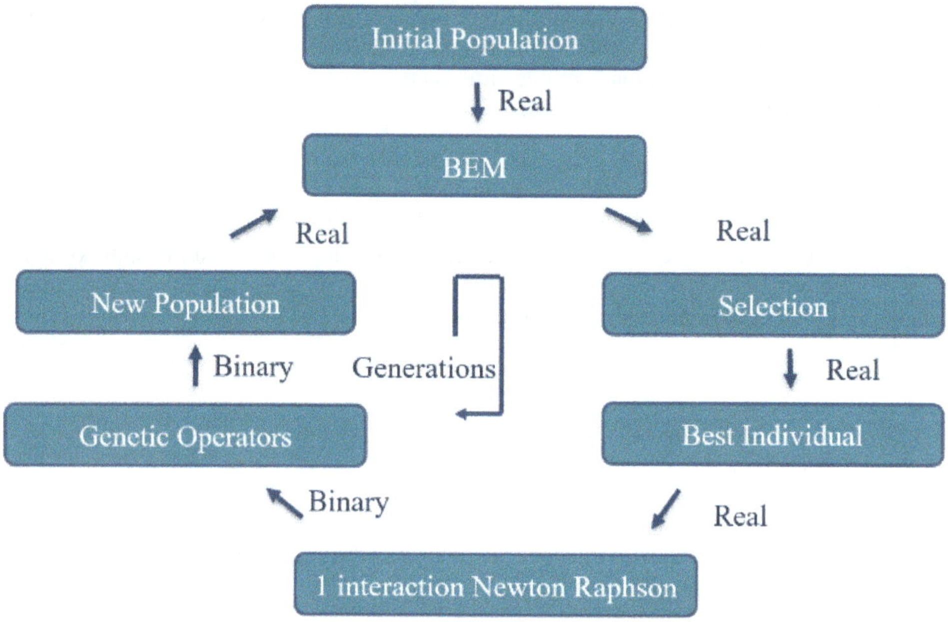

Figure 1: Diagram of the hybrid method associated with the BEM.

4 NUMERICAL ANALYSES

Two problems are proposed in order to test the hybrid strategy. In both cases body forces are neglected and a steel duct is modelled in 3D with the following material properties and geometry:

- Young's modulus: 200 GPa;
- Poisson's ratio: 0.32;
- Length: 7.0 m;
- External diameter: 0.3 m;
- Internal diameter: 0.26 m.

In each example, a direct analysis is initially performed in order to obtain synthetic experimental data. Afterwards, an inverse analysis is conducted for obtaining unknown parameters.

The inverse analysis is performed twice using the GA-BEM and the HM-BEM optimization techniques for comparative purposes.

4.1 First numerical test

This example consists of a fixed ended duct with a central vertical load represented by traction on an element in the central region, as shown in Fig. 2. Two parameters are selected to be unknowns in the inverse analysis:

- The magnitude of the central vertical traction (t_y);
- The traction t_y location (n).

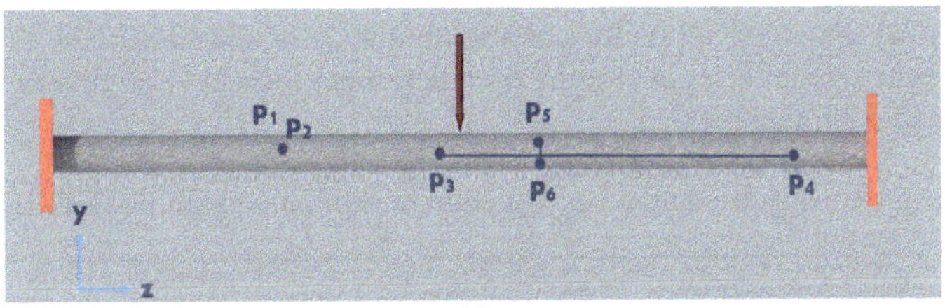

(a)

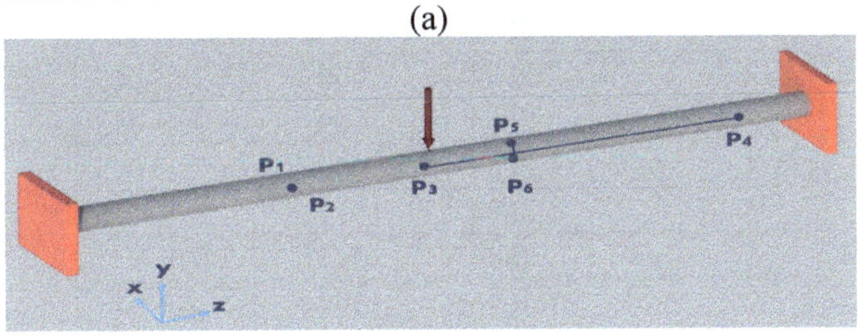

(b)

Figure 2: Model study of the first numerical example. (a) Front view; (b) Perspective view.

For searching the traction location parameter (*n*), a matrix ***M***(16x32) is defined representing all elements of the external surface of the duct. In this sense, searching for this location (*n*) requires identifying, two variables of matrix ***M***: column (k_1) and row (k_2).

4.1.1 Direct analysis

As mentioned previously, the direct problem is initially solved. The boundary was discretized into 1056 constant quadrilateral boundary elements. The problem is solved by assuming a vertical $t_y =$ -900*k*Pa; located at the element $n = 529$, corresponding to $k_1 = 2$ and $k_2 = 19$.

The values of displacements at points P_1 to P_6 located on the boundary, depicted in Fig. 2, are used to calculate their relative distances. These distances represent the experimental data, which are necessary to solve the inverse problems.

4.1.2 Inverse analysis

The following value intervals are assumed for the unknown parameters t_y, k_1, k_2: [-2MPa, 0MPa] [1, 32] and [1, 16], respectively. They constitute the GA searching limits, in which its GA operators and parameters are shown in Table 1.

Table 1: Genetic parameters.

Population Size	200
Number of generations	100
Selection scheme	Tournament selection
Crossover scheme	Segmented crossover
Mutation scheme	Bit-flipping mutation
Other operators	Elitism
Crossover probability	90%
Mutation probability	1%
Binary chromosomes	23 bits

Table 2 shows the best individual result obtained by the GA-BEM inverse analysis formulation in each generation, including its magnitude, location and fitness function value (FFV) calculated from eqn (7).

Table 2: Best individual of each generation of first example (GA-BEM).

Generation	t_y(MPa)	location (n)	FFV
1	-1.2871	563	2.37E-013
2	-1.2884	1009	3.78E-015
3	-1.2884	1009	3.78E-015
4	-1.2884	1009	3.78E-015
5	-1.2884	1011	3.67E-015
6	-1.2884	1011	3.67E-015
7	-1.3096	1009	1.36E-015
8	-1.3096	1009	1.36E-015
9	-1.3323	1011	8.21E-017
10	-1.3323	1011	8.21E-017
11	-0.9043	529	6.26E-017
12	-0.9043	529	6.26E-017
13	-0.8965	529	3.95E-017
14	-0.8965	529	3.95E-017
15	-0.8965	529	3.95E-017
16	-0.8965	529	3.95E-017
17	-0.8965	529	3.95E-017
18	-0.8965	529	3.95E-017
19	-0.8965	529	3.95E-017
20	-0.9004	529	6.56E-019
:	:	:	:
81	-0.8999	529	6.03E-019
:	:	:	:
100	-0.8999	529	6.03E-019

A stopping criterion for fitness function value from eqn (7), of the best individual was adopted with tolerance 10^{-26}.

Table 3 shows the results obtained with the HM-BEM inverse approach. Once again, the best individual data parameter of each generation is presented. It must be emphasized that the NR is only applied for searching the traction and not its location on the boundary. Clearly, generation after generation the best result either remains the same or improves. This is due to the process called elitism employed in the algorithms, where the best individual is kept in the new population of the next generation. Fig. 3 shows the fitness function values of the best individual along the generations, with both GA-BEM and HM-BEM inverse analysis formulations. Clearly, both results converge towards the correct solution. In the GA-BEM analysis, the boundary traction magnitude relative error was 4.74E-03% in the last generation, while in the MH-BEM analysis it was 2.22E-06% in the 11th generation. In both cases the correct loading element was found.

Table 3: Best individual of each generation of first example (HM-BEM).

Generation	t_y(MPa)	location (n)	FFV
1	-1.2871	563	2.37E-013
2	-1.2884	1009	3.78E-015
3	-1.2884	1009	3.78E-015
4	-1.2884	1009	3.78E-015
5	-1.2884	1011	3.67E-015
6	-1.2884	1011	3.67E-015
7	-1.3096	1009	1.36E-015
8	-1.3096	1009	1.36E-015
9	-1.3323	1011	8.21E-017
10	-1.3323	1011	8.21E-017
11	-0.9000	529	1.78E-027

Fig. 3 shows the fitness function values of the best individual along the generations, with both GA-BEM and HM-BEM inverse analysis formulations. Clearly, both results converge towards the correct solution. In the GA-BEM analysis, the boundary traction magnitude relative error was 4.74E-03% in the last generation, while in the MH-BEM analysis it was 2.22E-06% in the 11th generation. In both cases the correct loading element was found.

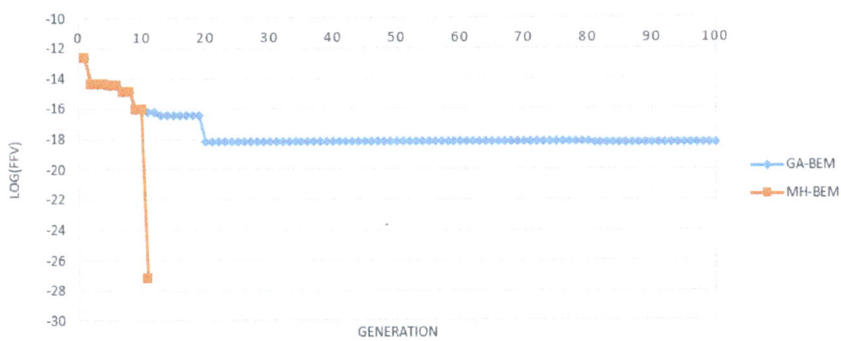

Figure 3: Variation of the function fitness value of the best individual in log 10 of each generation with the use of the GA-BEM and HM-BEM in the first example.

4.2 Second numerical test

This second numerical test, illustrated in Fig. 4 consists of a fixed ended steel duct, loaded with the action of three orthogonal concentrated forces at one point, represented by 3 tractions (t_x, t_y, t_z) applied on one element.

This example had as its objective the search of four unknown parameters:

- The magnitude of t_x;
- The magnitude of t_y;
- The magnitude of t_z;
- The location which (t_x, t_y, t_z) are acting (n).

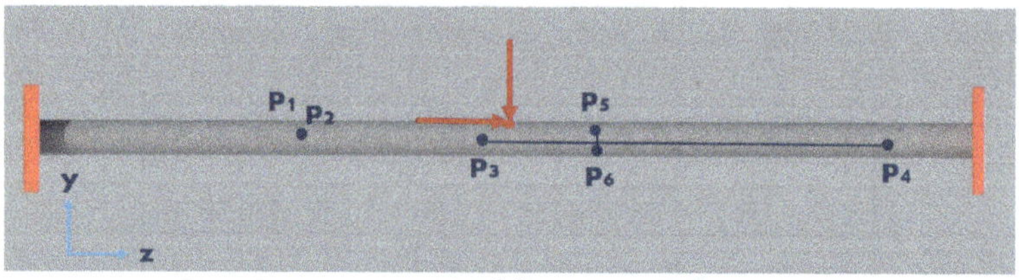

(a)

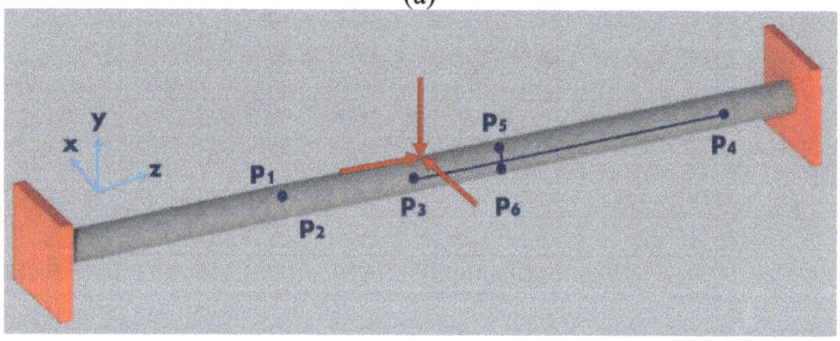

(b)

Figure 4: Model study of the second numerical example. (a) Front view; (b) Perspective view.

In the HM-BEM analysis, once the correct loading element location was found, the NR method lead the iterative process towards the expected result.

4.2.1 Direct analysis

The direct problem for 3D domain has been solved for this problem with the same mesh with 1056 constant quadrilateral boundary elements of the previous example. The direct problem is solved by assuming the parameters boundary tractions as t_x = 3MPa, t_y= -9MPa and t_z= 14MPa, located at the element number n=529, corresponding to k_1 = 2 and k_2 = 19.

4.2.2 Inverse analysis

The genetic operators and parameters to GA of both program (GA-BEM and HM-BEM) are shown in Table 4. Table 5 presents the result of the best individual for each generation using GA-BEM.

Table 4: Genetic parameters.

Population size	200
Number of generations	200
Selection scheme	Tournament selection
Crossover scheme	Segmented crossover
Mutation scheme	Bit-flipping mutation
Other operators	Elitism
Crossover probability	90%
Mutation probability	1%
Binary chromosomes	51 bits

Table 5: Best individual of each generation of second example (GA-BEM).

Generation	t_x(MPa)	t_y(MPa)	t_z(MPa)	location (n)	FFV
1	3.6037	-6.9511	13.5237	529	1.37E-011
2	3.6037	-6.9511	13.5237	529	1.37E-011
3	3.6037	-6.9511	13.5237	529	1.37E-011
4	3.6037	-6.9511	13.5237	529	1.37E-011
5	3.2960	-7.2636	13.5201	529	9.04E-012
6	3.6476	-9.7637	13.5201	529	2.53E-012
7	3.6476	-9.7637	13.5201	529	2.53E-012
:	:	:	:	:	:
200	2.9994	-8.9957	13.9998	529	5.21E-019

The search space is made up of 2^{51} possibilities of answers, for these 4 search variables. It is verified, by the analysis of the results, a convergence to the expected result over the generations even for 4 variables, demonstrating the validation of the algorithm GA-BEM. The program generated a result with relative errors of 1.83E-02%, 4.74E-03% and 8.72E-04% for the three tractions parameters t_x, t_y and t_z respectively, and found the exact location of these tractions action.

For the HM a stopping criterion was chosen for the fitness value of the best individual with tolerance 10^{-26}, as in the previous example. The results generated are shown in Table 6.

Table 6: Best individual of each generation of second example (HM-BEM).

Generation	t_x(MPa)	t_y(MPa)	t_z(MPa)	location (n)	FFV
1	3.0000	-9.0000	14.0000	529	1.37E-028

It is verified that in the first generation the stop criterion was satisfied, generating results with relative errors of 1E-06%, 7.77E-07% and 8.57E-08% to tx, ty and tz, respectively. This is because the GA found in the first generation the correct actuation element and, with the application of the NR, with only one iteration, generated results with a good precision. It was found, then, that in this example too, the inverse analysis with the HM-BEM was more efficient and accurate than with the use of the GA-BEM.

5 CONCLUSION

The contribution of this work is the association of the BEM with optimization formulations obtaining robust and efficient techniques for the search of unknown parameters applied to elasticity problems. More specifically, BEM was combined with genetic algorithms and the Newton–Raphson method to produce such formulations. Two numerical tests were carried out and the developed tools, GA-BEM and MH-BEM showed good approximations in their results. However, the use of MH-BEM presented better results with a lower computational cost, showing a good potential for use in structural analysis.

REFERENCES

[1] Czinkota, S., Structural analysis of pipeline stresses created by line lowering. *Canadian Journal of Civil Engineering*, **14**(6), pp. 719–727, 1987.

[2] Safarudin, M. & Setiawan, J.D., Structural analysis for in-service gas pipeline lowering using numerical method. *JMATEC Web of Conferences*, vol. 159, 2018.

[3] Magura, M. & Brodniansky, J., Structural analysis and of braking block sleeves on transit gas-pipeline. *Procedia Engineering*, **40**, pp. 257–261, 2012.

[4] Katsifarakis, K.L. & Petala, Z., Combined use of genetic algorithms and boundary elements to optimize coastal aquifers management. *Journal of Hydrology*, **327**, pp. 200–207, 2006.

[5] Miltiadou, P. & Wrobel, L.C., A BEM-based genetic algorithm for identification of polarization curves in cathodic protection systems. *Journal for Numerical Methods in Engineering*, **54**, pp. 159–174, 2002.

[6] Kita, E. & Tanie, H., Shape optimization of continuum structures by genetic algorithm and boundary element method. *Engineering Analysis with Boundary Elements*, **19**, pp. 129–136, 1997.

[7] Feliz, E.J.O., *Aplicação de Algoritmos Genéticos para Convergência de Fluxo de Potência em Sistemas de Energia.* Master's dissertation, Universidade do Vale do Itajaí: São José, 2009.

[8] Aliabadi, M.H., *The Boundary Element Method: Applications in Solids and Structure*, vol. 2, John Wiley & Sons: England, 2002.

[9] Brebbia, C.A. & Dominguez, J., *Boundary Elements an Introductory Course*, McGraw-Hill: Southampton, 1992.

[10] Brebbia, C.A., Telles, J.C.F. & Wrobel, L.C., *Boundary Element Techniques*, Springer-Verlag: Berlin and New York, 1984.

[11] Davis, L., *Handbook of Genetic Algorithms*, Van Nostrand Reinhold: New York, 1991.

[12] Coley, D.A., *An Introduction to Genetic Algorithms for Scientists and Engineers*, World Scientific: Exeter, 1999.

[13] Goldberg, D.E., *Genetic Algorithms in Search, Optimization, and Machine Learning*, Addison-Wesley: Alabama, 1989.

[14] Federico, D.S. & Hurtado, A.N., *Métodos Numéricos Aplicados a la Ingeniería*, Patria: México, 2014.

ADAPTIVE RBF-FD METHOD FOR POISSON'S EQUATION

JURE SLAK[1,2] & GREGOR KOSEC[1]
[1]Parallel and Distributed Systems Laboratory, Jožef Stefan Institute, Slovenia
[2]Faculty of Mathematics and Physics, University of Ljubljana, Slovenia

ABSTRACT
Solutions to many physical problems governed by partial differential equations (PDEs) often vary significantly in magnitude throughout the problem domain. Although in some special cases the areas with high error are known in advance, in general the error distribution is unknown beforehand. Adaptive techniques for solving PDEs are a standard way of dealing with this problem, where problematic regions are iteratively refined. A step further is automatic adaptivity, where problematic regions are chosen automatically using an error indicator and then refined, until a certain error threshold is reached. In this paper, we apply a recently published technique for automatic adaptivity for strong form meshless methods and solve the Poisson equation and its generalisations, using the popular RBF-FD method. Both 2D and 3D cases are considered, comparing uniform and adaptive refinement, illustrating the advantages of fully automatic adaptivity.
Keywords: adaptivity, mesh-free methods, RBF-FD, Poisson equation, Helmholtz equation, PDEs.

1 INTRODUCTION

Adaptive modification of partial differential equation (PDE) discretizations is needed in problems with variable physical behaviour or precision requirements throughout the domain. The techniques used in traditional methods, such as the Finite Element Method (FEM), are well developed [1]. Besides the procedure for the solution of the PDE, two more parts are needed for fully automatic adaptivity: an indicator of the solution quality and a refinement strategy. In the following paragraphs we offer a brief review of all three elements in a strong form meshless method context.

Local strong form methods were developed as a cheaper alternative to global collocation methods and are usually presented as generalisations of Finite Difference Method to scattered grids. The most direct generalisation is the Finite Point Method, which uses monomials to compute the stencil weights [2]. Due to stability issues, Radial Basis Functions (RBFs) have been suggested and successfully used instead of monomials [3]. To ensure consistency of the approximations and eliminate the influence of the shape parameter, Polyharmonic splines, augmented with monomials were researched and successfully applied to various problems [12], which we will also use for our solution procedure.

The most widely used error indicator was developed by Zienkiewicz and Zhu [4] for FEM and has since been generalized to meshless methods. The main idea of the indicator is to approximate the error of the solution as the difference between the original solution and the so-called recovered solution, obtained by appropriate post-computation. In strong form meshless methods, it has already been used with the Finite Point Method [5] and RBF-FD [11]. A different type of error indicator is available for least-squares based meshless methods, which use the residual of the least squares approximations to estimate the error [6]. Additionally, several ad hoc error indicators have been used, mostly using the variability of the solution to determine refinement regions, such as Davydov and Oanh [7] and Kosec and Šarler [8]. This method will be used in the present work as well, for simplicity.

After the problematic regions have been identified, the (de)refinement can be done in multiple ways. Either we can modify the approximation, resulting in p-refinement schemes, or the discretization itself can be modified. This can be done either by moving the nodes

WIT Transactions on Engineering Sciences, Vol 126, © 2019 WIT Press
www.witpress.com, ISSN 1743-3533 (on-line)
doi:10.2495/BE420131

(r-refinement), used in, for example, Afshar et al. [6], or by placing more nodes in regions with high estimated error and removing nodes from low-error regions (h-refinement), used in, for example, Kosec and Šarler [8] and Oanh et al. [11].

Another variation in h-refinement strategies is whether the existing discretisation is only modified and partially reused, or generated completely anew. The former can cause problems with conditioning, causing the need for specially selected stencils [11], which is why we opted for the latter option.

Similar methodology to the one described above, has been recently used to solve linearly elastic contact problems [10]. In this work we apply it to the Poisson equation and its generalizations. The rest of the paper is organized as follows: in Section 2 the RBF-FD method is presented, in Section 3 the adaptive methodology is described, and in Section 4 the results of the numerical experiments are presented.

2 RBF-GENERATED FINITE DIFFERENCES

Consider an elliptic boundary value problem:

$$\mathcal{L}u = f \text{ in } \Omega, \tag{1}$$

$$u = u_0 \text{ on } \partial\Omega, \tag{2}$$

which will be used as a sample theoretical problem throughout this paper. Domain Ω is discretized by placing N nodes p_i in and on the boundary of the domain. Nodes are positioned by algorithm presented in [13], which can distribute nodes according to any continuous spacing function h. Each node p_i is assigned a *stencil* of n nodes, $S(p_i) = (p_{j_1}, \dots, p_{j_n})$. The operator $\mathcal{L}$ is approximated for each point p_i in the domain interior. This is done by assuming an approximation of form

$$(\mathcal{L}u)(p_i) \approx \textstyle\sum_{k=1}^{n} w_{i,k} u(p_{j_k}). \tag{3}$$

Equality in the above approximation is desired for a certain set of functions: if node p_i are arranged in a grid and equality is enforced for monomials, standard finite difference approximation is obtained. To derive the "RBF-generated finite differences", RBFs are used to compute the weights $w_{i,k}$.

Polyharmonic splines will be used in this work, which are radial functions of form $\phi(r) = r^k$, for an odd integer k. Radial basis functions are often thought of as being centred in one node, denoting $\phi_j(p) = \phi(\|p - p_j\|)$. To obtain RBF-FD weights, radial functions centred on the stencil nodes are used as the basis. Substituting ϕ_{j_ℓ} for u in eqn (3) and enforcing equality, we arrive at the following linear system:

$$\begin{bmatrix} \phi(\|p_{j_1} - p_{j_1}\|) & \cdots & \phi(\|p_{j_n} - p_{j_1}\|) \\ \vdots & \ddots & \vdots \\ \phi(\|p_{j_1} - p_{j_n}\|) & \cdots & \phi(\|p_{j_n} - p_{j_n}\|) \end{bmatrix} \begin{bmatrix} w_{i,1} \\ \vdots \\ w_{i,n} \end{bmatrix} = \begin{bmatrix} \left(\mathcal{L}\phi(\|\cdot - p_{j_1}\|)\right)(p_i) \\ \vdots \\ \left(\mathcal{L}\phi(\|\cdot - p_{j_n}\|)\right)(p_i) \end{bmatrix}, \tag{4}$$

written compactly as $Aw = b$. Additional consistency constraints can be enforced to ensure convergence [12], augmenting the above matrix into

$$\begin{bmatrix} A & Q^T \\ Q & 0 \end{bmatrix} \begin{bmatrix} w \\ \lambda \end{bmatrix} = \begin{bmatrix} b \\ c \end{bmatrix}, \tag{5}$$

where the second equation $Q\lambda = c$ is expanded as

$$\begin{bmatrix} q_1(p_{j_1}) & \cdots & q_1(p_{j_n}) \\ \vdots & \ddots & \vdots \\ q_\ell(p_{j_1}) & \cdots & q_\ell(p_{j_n}) \end{bmatrix} \begin{bmatrix} \lambda_1 \\ \vdots \\ \lambda_\ell \end{bmatrix} = \begin{bmatrix} (\mathcal{L}q_1)(p_i) \\ \vdots \\ (\mathcal{L}q_\ell)(p_i) \end{bmatrix}, \tag{6}$$

with q_l representing monomials up to and including some order m. Values λ are discarded after computation.

With the stencil weights $w_{i,k}$ computed for each node p_i the PDE $(\mathcal{L}u)(p_i) = f(p_i)$ can be approximated with a linear algebraic equation $\sum_{k=1}^{n} w_{i,k}u(p_{j_k}) = f(p_i)$. These equations can be assembled in a sparse matrix and after including the boundary conditions, can be solved for unknowns $u(p_i)$ to obtain the numerical solution.

3 ADAPTIVE METHODOLOGY

The adaptive methodology in this paper behaves similarly to "remeshing" used commonly in FEM and has been successfully used to solve contact problems from linear elasticity [7]. Some initial (possibly variable) nodal spacing $h^0(p)$ is chosen, as well as its lower and upper bounds $h_L(p)$ and $h_U(p)$, respectively. Domain Ω is filled with nodes, conforming to h^0 and the solution u^0 is obtained.

An error indicator is employed to determine which nodes should be (de)refined and the nodal density h^0 is altered appropriately. This adaptive cycle below is repeated until the convergence criterion is met. The procedure on j-th iteration is written in more detail below:

1. Fill Ω with nodes conforming to h^j.
2. Solve the problem to obtain u^j.
3. Compute the error indicator values ε_i^j for each node p_i.
4. If the mean of ε_i^j is below some tolerance ε return u^j as the solution and stop.
5. Adapt h^j to obtain h^{j+1}.

More details on steps 3 and 5 are given in the next two sections.

3.1 Error indicator

The error indicator used in this paper is designed to detect the variability of the solution around node p_i. First, the recovered solution $\tilde{u}_i$ is computed as $\tilde{u}_i = \frac{1}{n}\sum_{k=1}^{n} u_k$ and then the indicator ε_i is computed as $\varepsilon_i = \sqrt{\frac{1}{n}\sum_{k=1}^{n} e_i^2}$, where local errors e_i are computed as $e_i = |u_i - \tilde{u}_i|$. Despite its simplicity, this error indicator turned out to work well enough for our purposes.

3.2 Nodal spacing adaptation

The existing nodal spacing function h^j is evaluated at nodes p_i to obtain values $h_{i,j} = h^j(p_i)$. These values $h_{i,j}$ are modified by a *density factor* f_i as

$$h_{i,j+1} = \min\left(max\left(h_{i,j}/f_i\,, h_L(p_i)\right), h_U(p_i)\right), \tag{7}$$

where factor f_i is computed as

$$f_i = \begin{cases} 1 + \frac{\eta - \varepsilon_i}{\eta - m}\left(\frac{1}{\beta} - 1\right); & \varepsilon_i \le \eta, \\ 1; & \eta \le \varepsilon_i \le \varepsilon, \\ 1 + \frac{\varepsilon_i - \varepsilon}{M - \varepsilon}(\alpha - 1); & \varepsilon_i \ge \varepsilon, \end{cases} \tag{8}$$

and α represents the refine aggressiveness, β the derefine aggressiveness, ε the refinement threshold, η the derenfinement threshold, and $m = \min_i \varepsilon_i$ is the minimal and $M = \max_i \varepsilon_i$ is the maximal value of the error indicator. Note that setting $\alpha = 1$ or $\beta = 1$ disables refinement and derefinement, respectively.

Lower refinement bound h_L is often not necessary and can be set to ∞, while upper bound h_U is often set to the initial density h_0. Unless otherwise stated this will be the case in this paper as well.

After the values $h_{i,j+1}$ have been computed, the function h^{j+1} is obtained using modified Shepard's scattered interpolation of pairs $(p_i, h_{i,j+1})$. The 15 closest neighbouring nodes were used in this paper.

4 RESULTS

We test the fully adaptive solution procedure described in the previous section on some standard test problems, taken from Mitchell [9].

4.1 L-shaped domain

The first test case is the "L domain" test case. Laplace's equation

$$\nabla^2 u = 0 \tag{9}$$

is solved on domain $\Omega = [-1,1]^2 \setminus ([0,1] \times [-1,0])$ with a known solution u given in polar coordinates as $u = r^{\frac{2}{3}} \sin\frac{2}{3}\theta$ and Dirichlet boundary conditions obtained from u.

RBF-FD method with Polyharmonic splines augmented with monomials up to and including 2nd order was used to approximate the differential operators. The stencils for each node were chosen by simply selecting the closest $n = 15$ nodes.

The resulting sparse system was solved using the Intel® MKL Pardiso sparse solver. Both uniform and fully adaptive refinement was tested. The adaptive procedure was run with $\alpha = 3$, $\varepsilon = 10^{-2}$, $\beta = 1$ and $\eta = 0$.

The error was between the exact solution u and numerical solution $\hat{u}$ was measured in three different norms

$$e_1 = \frac{\|u - \hat{u}\|_1}{\|u\|_1}, \quad \|u\|_1 = \frac{1}{N}\sum_{p \in G} |u(p)|, \tag{10}$$

$$e_2 = \frac{\|u - \hat{u}\|_2}{\|u\|_2}, \quad \|u\|_2 = \sqrt{\frac{1}{N}\sum_{p \in G} |u(p)|^2}, \tag{11}$$

$$e_\infty = \frac{\|u - \hat{u}\|_\infty}{\|u\|_\infty}, \quad \|u\|_\infty = \max_{p \in G} |u(p)|, \tag{12}$$

where G is a set of independent points arranged in a dense uniform grid over the whole domain.

The errors e_1, e_2 and e_∞ for uniform refinement are shown in Fig. 1. Fig. 2 shows the errors under adaptive refinement. The adaptive refinement finished in 6 iterations. We can see that better rates and lower errors are obtained using this technique.

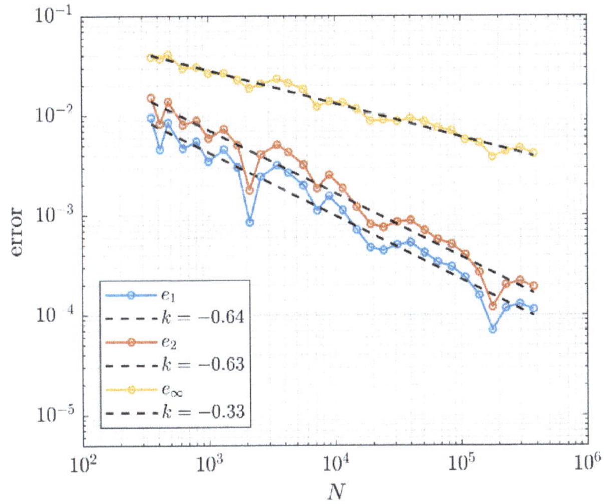

Figure 1: Errors when solving the "L domain" problem using uniform refinement.

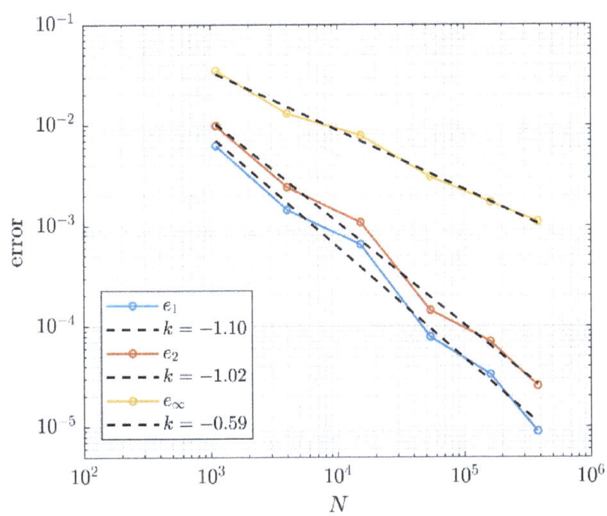

Figure 2: Errors when solving the "L domain" problem using adaptive refinement.

Additionally, local node errors and nodal densities were observed during the course of the adaptive iteration. The local errors on each iteration are computed for each node as $\epsilon_i = |\hat{u}(p_i) - u(p_i)|$. The local nodal separation distance is computed as the average distance $\bar{d}_i$ from each node to its closest neighbours, $\bar{d}_i = \frac{1}{3}(d_{i,1} + d_{i,2} + d_{i,3})$. From $\bar{d}_i$ we can compute the relative node density ρ_i as

$$\rho_i = -\log_2\left(\frac{\bar{d}_i}{\max_j \bar{d}_j}\right). \tag{13}$$

Value $\rho_i = 0$ means that node p_i has maximal separation distance (least dense) while e.g. $\rho_i = 3$ means that nodes around p_i are 8 times as dense as in the coarsest part. Values ϵ_i and ρ_i corresponding to the solutions, whose errors were computed in Fig. 2, are shown in Fig. 3. It can be seen that the node density adapts according to the error and begins to concentrate in the inner corner.

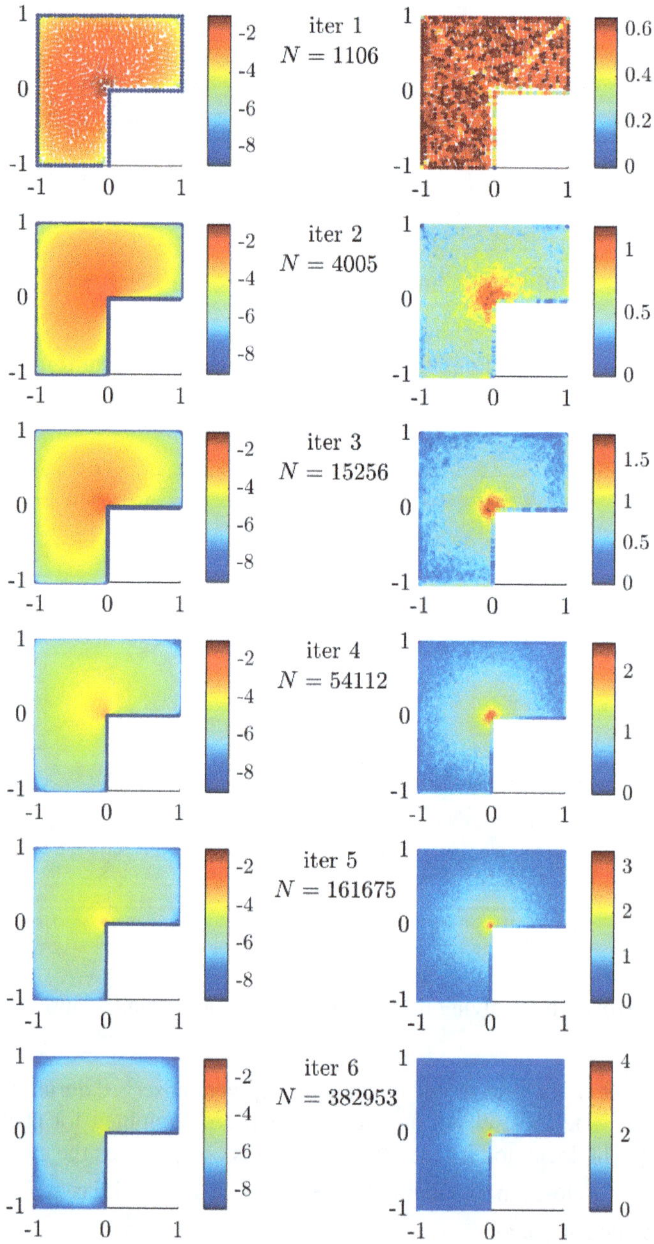

Figure 3: Plots of logarithms of errors ($\log \epsilon_i$, left column) and nodal densities (ρ_i, right column) over the course of the adaptive iteration.

4.2 2D Helmholtz equation

Next, we investigate the behaviour of the method on a Helmholtz equation:

$$-\nabla^2 u - \frac{1}{(\alpha+r)^4} u = f \tag{14}$$

with an oscillatory solution $u(r) = \sin\left(\frac{1}{\alpha+r}\right)$, where the right-hand side f is computed from the solution u.Value $\alpha = \frac{1}{5\pi}$ was used in all computations. The same numerical setup as in Section 4.1 was used to obtain the numerical results, except that the tolerance ϵ was set to 10^{-3}. Fig. 4 shows the errors during uniform and adaptive refinement. We can see that no satisfactory solution can be obtained using uniform refinement and that there is no indication of convergence. The adaptive solution procedure manages to find a satisfactory solution and exhibits better convergence properties.

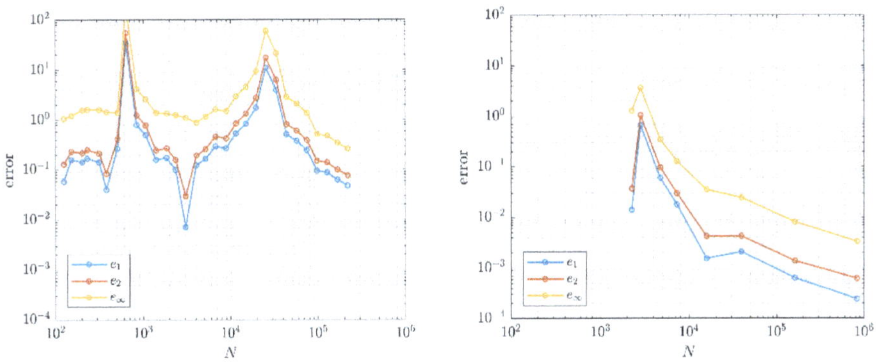

Figure 4: Errors during uniform refinement (left) and adaptive refinement (right) when solving the 2D Helmholtz equation.

The initial jump in error has been observed before [10], [11] and is usually caused by too small initial node density. The solution in the last iteration is shown in Fig. 5, along with the obtained nodal density function. Node separation distance in the densest part was around 400-times smaller than in the coarsest part.

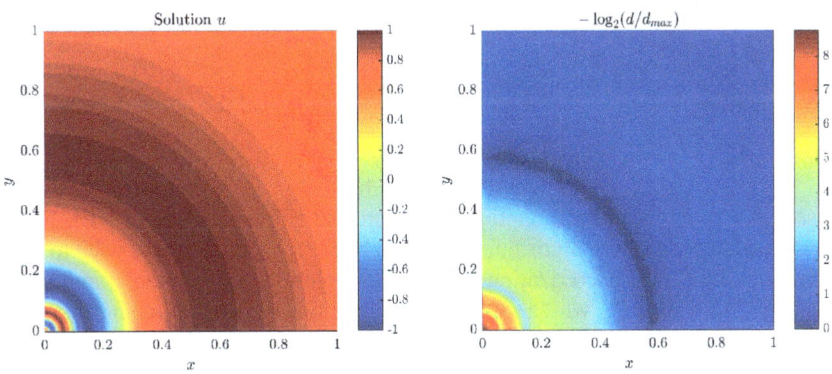

Figure 5: Solution u_i of the 2D Helmholtz equation (left) and the nodal density ρ_i (right).

4.3 3D Helmholtz equation

The same problem as in Section 4.2 is used to assess the behaviour of the method in 3D as well. The same numerical setup as in 2D was used, except that the number of nearest neighbours used to the stencil increased to $n = 25$. Uniform refinement behaves similarly to its 2D counterpart. The errors during the adaptive iteration are shown in Fig. 6. The obtained solution and nodal density in the last iteration are shown in Fig. 7.

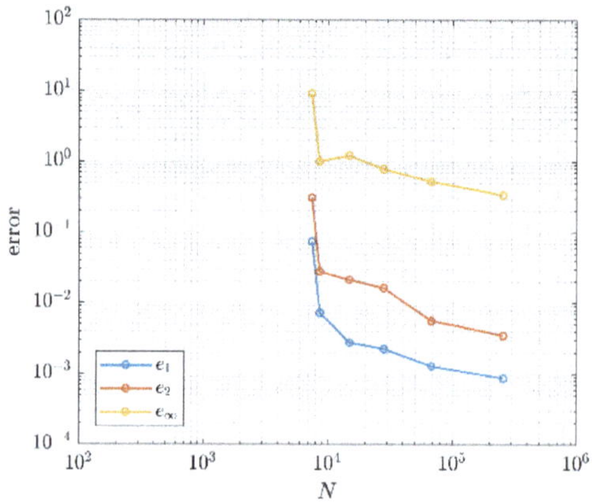

Figure 6: Errors during adaptive solution of the 3D Helmholtz equation.

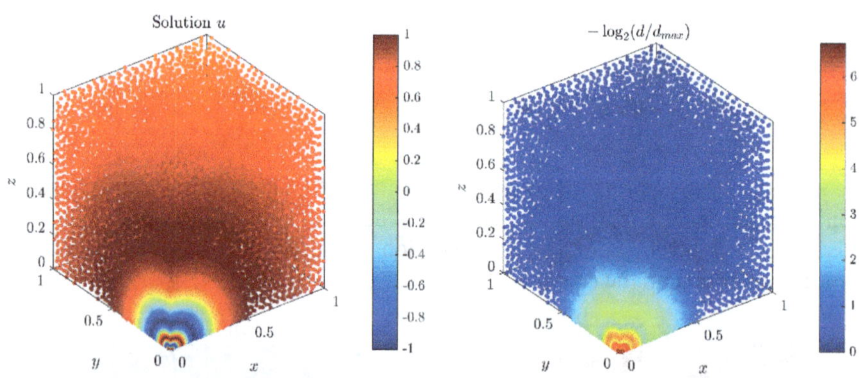

Figure 7: Solution u_i of the 3D Helmholtz equation (left) and the nodal density ρ_i (right).

5 CONCLUSIONS

We presented a fully automatic adaptive procedure for elliptic problems. RBF-FD method along with a node positioning algorithm and an ad hoc error indicator was used to iteratively refine initially uniform discretizations to obtain solution to Poisson's and Helmholtz equations in 2D and 3D, which would not have been feasible using only uniform discretizations. All computations were done using the Medusa library [14] for meshless solutions of PDEs. Future work will be focused on the development of better error indicators and more efficient refinement strategies.

ACKNOWLEDGEMENTS

The authors would like to acknowledge the financial support of the Research Foundation Flanders (FWO), The Luxembourg National Research Fund (FNR) and Slovenian Research Agency (ARRS) in the framework of the FWO Lead Agency project: G018916N Multi-analysis of fretting fatigue using physical and virtual experiments, the ARRS research core funding No. P2-0095 and the Young Researcher program PR-08346.

REFERENCES

[1] Bangerth, W. & Rannacher, R., *Adaptive Finite Element Methods for Differential Equations*, Birkhäuser, 2013.

[2] Oñate, E., Perazzo, F. & Miquel, J., A finite point method for elasticity problems. *Computers & Structures*, **79**(22–25), pp. 2151–2163, 2001.

[3] Tolstykh, A.I. & Shirobokov, D.A., On using radial basis functions in a "finite difference mode" with applications to elasticity problems. *Computational Mechanics*, **33**(1), pp. 68–79, 2003.

[4] Zienkiewicz, O.C. & Zhu, J.Z., A simple error estimator and adaptive procedure for practical engineering analysis. *International journal for Numerical Methods in Engineering*, **24**(2), pp. 337–357, 1987.

[5] Angulo, A., Pozo, L.P. & Perazzo, F., A posteriori error estimator and an adaptive technique in meshless finite points method. *Engineering Analysis with Boundary Elements*, **33**(11), pp. 1322–1338, 2009.

[6] Afshar, M.H., Naisipour, M. & Amani, J., Node moving adaptive refinement strategy for planar elasticity problems using discrete least squares meshless method. *Finite Elements in Analysis and Design*, **47**(12), pp. 1315–1325, 2011.

[7] Davydov, O. & Oanh, D.T., Adaptive meshless centres and RBF stencils for Poisson equation. *Journal of Computational Physics*, **230**(2), pp. 287–304, 2011.

[8] Kosec, G. & Šarler, B., H-adaptive local radial basis function collocation meshless method. *Computers Materials and Continua*, **26**(3), p. 227, 2011.

[9] Mitchell, W.F., A collection of 2D elliptic problems for testing adaptive grid refinement algorithms. *Applied Mathematics and Computation*, **220**, pp. 350–364, 2013.

[10] Slak, J. & Kosec, G., Adaptive radial basis function-generated finite differences method for contact problems. *International Journal for Numerical Methods in Engineering*, 2019. DOI: 10.1002/nme.6067.

[11] Oanh, D.T., Davydov, O. & Phu, H.X., Adaptive RBF-FD method for elliptic problems with point singularities in 2D. *Applied Mathematics and Computation*, **313**, pp. 474–497, 2017.

[12] Bayona, V., Flyer, N., Fornberg, B. & Barnett, G.A., On the role of polynomials in RBF-FD approximations: II. Numerical solution of elliptic PDEs. *Journal of Computational Physics*, **332**, pp. 257–273, 2017.

[13] Slak, J. & Kosec, G., On generation of node distributions for meshless PDE discretizations. arXiv preprint arXiv:1812.03160, 2018.

[14] Medusa. Medusa: coordinate free implementation of meshless methods, http://e6.ijs.si/medusa/. Accessed on: 10 Jun. 2019.

MESHLESS COLLOCATION METHODS FOR SOLVING PDES ON SURFACES

MENG CHEN[1], KA CHUN CHEUNG[2] & LEEVAN LING[3]
[1]ASTRI Hong Kong Applied Science and Technology Research Institute Company Limited, Hong Kong
[2]NVIDIA AI Technology Center (NVAITC), NVIDIA, USA
[3]Department of Mathematics, Hong Kong Baptist University, Hong Kong

ABSTRACT
We present three recently proposed kernel-based collocation methods in unified notations as an easy reference for practitioners who need to solve PDEs on surfaces $\mathcal{S} \subset \mathbb{R}^d$. These PDEs closely resemble their Euclidean counterparts, except that the problem domains change from bulk regions with a flat geometry of some surfaces, on which curvatures play an important role in the physical processes. First, we present a formulation to solve surface PDEs in a narrow band domain containing the surface. This class of numerical methods is known as the embedding types. Next, we present another formulation that works solely on the surface, which is commonly referred to as the intrinsic approach. Convergent estimates and numerical examples for both formulations will be given. For the latter, we solve both the linear and nonlinear time-dependent parabolic equations on static and moving surfaces.
Keywords: kernel-based collocation methods, elliptic partial differential equations on manifolds, convergence estimate.

1 INTRODUCTION

Throughout the paper, we use $\mathcal{S} \subset \mathbb{R}^d$ to denote some sufficiently smooth, connected, and compact surface with bounded geometry with $\dim(\mathcal{S}) = d - 1$. All surface functions and operators are labelled with subscripts $\mathcal{S}$ for clarity. In particular, the surface gradient $\nabla_{\mathcal{S}}$ and the Laplace-Beltrami $\Delta_{\mathcal{S}}$ operators (a.k.a. the surface Laplacian) can be defined in terms of the standard Euclidean gradient ∇ and Laplacian Δ operators for $\mathbb{R}^d$ via projections:

$$\nabla_{\mathcal{S}} := (I - \mathbf{n}\mathbf{n}^T)\nabla \quad \text{and} \quad \Delta_{\mathcal{S}} := \nabla_{\mathcal{S}} \cdot \nabla_{\mathcal{S}}, \tag{1}$$

where I is the identity operator and $\mathbf{n} = \mathbf{n}(\boldsymbol{\xi})$ denote the unit outward normal vector at $\boldsymbol{\xi} \in \mathcal{S}$. For bounded Euclidian domains Ω, we work on the standard Hilbert spaces $H^m(\Omega)$ and $H^{m-1/2}(\Omega)$ with standard Sobolev norms for some integer $m \geq d/2$. Sobolev spaces on surfaces can be defined by $H^m(\mathcal{S}) := (I - \Delta_{\mathcal{S}})^{-m/2}L^2(\mathcal{S})$, see [1].

The types of partial differential equations (PDEs) that we concerned are time-independent strongly elliptic equations on surfaces $\mathcal{S} \subset \mathbb{R}^d$ in the form of

$$\mathcal{L}_{\mathcal{S}} u_{\mathcal{S}} := (-a_{\mathcal{S}}\Delta_{\mathcal{S}} + \mathbf{b}_{\mathcal{S}} \cdot \nabla_{\mathcal{S}} + c_{\mathcal{S}}) u_{\mathcal{S}} = f_{\mathcal{S}} \tag{2}$$

for some surface differential operator $\mathcal{L}_{\mathcal{S}} : H^m(\mathcal{S}) \to H^{m-2}(\mathcal{S})$ with $W^m_\infty(\mathcal{S})$-bounded coefficients $a_{\mathcal{S}}$, $c_{\mathcal{S}} : \mathcal{S} \to \mathbb{R}$, and $\mathbf{b}_{\mathcal{S}} : \mathcal{S} \to \mathbb{R}^d$. We assume the existence [2] of classical solutions $u^*_{\mathcal{S}}$ to (2) in Hilbert spaces $H^m(\mathcal{S})$.

2 EMBEDDING KANSA METHODS

Methods in this section are variants of Kansa methods [3], [4] that built upon some constant-along-normal property. They are generalization of the finite difference based closest point method [5] and its meshfree extension [6]. Firstly, we define the closest point mapping cp

$$\text{cp}(\mathbf{x}) = \arg\inf_{\boldsymbol{\xi}\in\mathcal{S}} \|\boldsymbol{\xi} - \mathbf{x}\|_{\ell^2(\mathbb{R}^d)},$$

WIT Transactions on Engineering Sciences, Vol 126, © 2019 WIT Press
www.witpress.com, ISSN 1743-3533 (on-line)
doi:10.2495/BE420141

which maps each near-surface point $\mathbf{x} \in \mathbb{R}^d$ onto the nearest surface $\text{cp}(\mathbf{x}) \in \mathcal{S}$. For any smooth surface $\mathcal{S}$ of class C^{m+1}, we know that [7] the cp map is well-defined in some narrow-band domain containing $\mathcal{S}$

$$\Omega = \Omega_\delta := \left\{ \mathbf{x} \in \mathbb{R}^d : \inf_{\boldsymbol{\xi} \in \mathcal{S}} \|\boldsymbol{\xi} - \mathbf{x}\|_{\ell^2(\mathbb{R}^d)} < \delta \right\} \tag{3}$$

for some constant δ depending only on $\mathcal{S}$ and m such that $u_\mathcal{S} \circ \text{cp} \in H^m(\Omega)$ for all $u_\mathcal{S} \in H^m(\mathcal{S})$.

An embedded PDE associated with (2) is in the form of

$$\mathcal{L} := -a\Delta + \mathbf{b} \cdot \nabla + c \quad \text{in } \Omega, \tag{4}$$

with all coefficients in (4) being the cp-extensions of those in (2), i.e. $a = a_\mathcal{S} \circ \text{cp}$ and etc. for all $\mathbf{x} \in \Omega$. The following theorem gives the relationship of operators on $\mathcal{S}$ and Ω, and the solutions of (2) and (4).

Theorem 1. *[8, Thm. 2.1] Let $\mathcal{S} \subset \mathbb{R}^d$ be a codimension one C^3–smooth, connected and compact surface with well-defined normal $\mathbf{n} = \mathbf{n}(\mathbf{x})$ for all $\mathbf{x} \in \mathcal{S}$. Let $u \in C^2(\Omega) \cap H^{2+1/2}(\Omega)$ be an extension of $u_\mathcal{S} \in H^2(\mathcal{S})$ with $u_{|\mathcal{S}} = u_\mathcal{S}$. Then,*

$$\nabla_\mathcal{S} u := \nabla u - \mathbf{n}\partial_\mathbf{n} u \quad \text{and} \quad \Delta_\mathcal{S} u := \Delta u - H_\mathcal{S}\partial_\mathbf{n} u - \partial_\mathbf{n}^{(2)} u \quad \text{on } \mathcal{S}, \tag{5}$$

where $\partial_\mathbf{n} u := \mathbf{n}^T \nabla u$, $\partial_\mathbf{n}^{(2)} u := \mathbf{n}^T J(\nabla u)\mathbf{n}$ and $H_\mathcal{S}(\boldsymbol{\xi}) = \text{trace}(J(\mathbf{n})(I - \mathbf{n}\mathbf{n}^T))$, which is d times the mean curvature of $\mathcal{S}$ at $\mathbf{x}$, defined using the Jacobian operator J in Euclidean space. In particular, for any second order differential operator in the form of (2), if u satisfies the embedding conditions

$$\partial_\mathbf{n} u = 0 \quad \text{and} \quad \partial_\mathbf{n}^{(2)} u = 0 \quad \text{on } \mathcal{S}, \tag{6}$$

then $\mathcal{L}_\mathcal{S} u_\mathcal{S} = \mathcal{L}u$ on $\mathcal{S}$.

Note that (4) is simply an elliptic PDE posed in the narrow-band domain Ω. Once we solve (4) for u with constant-along-normal property, restricting it to the surface yields the surface solution $u_\mathcal{S} = u_{|\mathcal{S}}$.

2.1 Numerical algorithms

For the sake of convergence proof, we use kernels $\Phi_m : \mathbb{R}^d \times \mathbb{R}^d \to \mathbb{R}$ that are translation-invariant, symmetric positive definite with smoothness m, and satisfy

$$c(1 + \|\omega\|_2^2)^{-m} \leq \widehat{\Phi_m}(\omega) \leq C(1 + \|\omega\|_2^2)^{-m} \quad \text{for all } \omega \in \mathbb{R}^d, \tag{7}$$

for some constants $0 < c \leq C$. Commonly used kernels of this sort includes the standard Whittle-Matérn-Sobolev kernels, that are defined via the Bessel functions of the second kind in the form of

$$\Phi_m(\mathbf{x}) := \|\mathbf{x}\|_2^{m-\frac{d}{2}} \mathcal{K}_{m-\frac{d}{2}}(\|\mathbf{x}\|_2) \quad \text{for all } x \in \mathbb{R}^d, \tag{8}$$

and the class of compactly supported piecewise polynomial Wendland functions [9].

Let $Z = \{z_1, \ldots, z_{n_Z}\} \subset \Omega_\delta$ denote a discrete set of trial centers in the narrow-band domain with radius δ in (3) and we define the corresponding finite-dimensional trial space by

$$\mathcal{U}_{Z,\Omega,\Phi_m} := \text{span}\{\Phi_m(\cdot - z_j) : z_j \in Z\},$$

that is a subspace of the native space $\mathcal{N}_{\Omega,\Phi_m}$ of Φ_m [10], [11]. On the test side, we pick any discrete set $X = \{x_1, \dots, x_{n_X}\} \subset \mathcal{S}$ of sufficiently dense collocation points (with respect to Z) on the surface.

Under this surface-version setup for overdetermined Kansa methods [12], a convergent numerical solution [8, Thm. 4.1] is defined by

$$\arg\inf_{u\in\mathcal{U}_{Z,\Omega_\delta,\Phi_m}} \Big(\underbrace{\|\mathcal{L}u - f\|^2_{\ell^2(X)}}_{\text{PDE collocations}} + \underbrace{\|\partial_{\mathbf{n}}u\|^2_{\ell^2(X)} + \|\partial^{(2)}_{\mathbf{n}}u\|^2_{\ell^2(X)}}_{\text{Embedding conditions}} + h^\epsilon_X \underbrace{\|u - u_{|\mathcal{S}} \circ \mathrm{cp}\|^2_{\ell^2(Z)}}_{\text{Stability conditions}} \Big), \tag{9}$$

for some $m > 5/2 + d/2$, $\epsilon > 0$, and with the mesh norm of X defined by the shortest distance function $d_\mathcal{S} : \mathcal{S} \times \mathcal{S} \to \mathbb{R}$ on $\mathcal{S}$ as

$$h_X := \sup_{\xi\in\mathcal{S}} \min_{x\in X} d_\mathcal{S}(x, \xi).$$

Note that the Euclidean differential operator $\mathcal{L}$ in (9) is evaluated only at $X \subset \mathcal{S}$. It is unnecessary to cp-extend any coefficients of $\mathcal{L}_\mathcal{S}$ away from $\mathcal{S}$. The $(3n_X + n_Z) \times n_Z$ overdetermined linear system in (9) can be expressed in matrix form as

$$\begin{pmatrix} \mathcal{L}\Phi_m(X, Z) \\ \partial_{\mathbf{n}}\Phi_m(X, Z) \\ \partial^{(2)}_{\mathbf{n}}\Phi_m(X, Z) \\ h^\epsilon_X \left(\Phi_m(Z \setminus \mathcal{S}, Z) - \Phi_m(\mathrm{cp}(Z \setminus \mathcal{S}), Z)\right) \end{pmatrix} \boldsymbol{\lambda} = \begin{pmatrix} f(X) \\ \mathbf{0} \\ \mathbf{0} \\ \mathbf{0} \end{pmatrix}, \tag{10}$$

in which all differential operators act on the first argument of Φ_m. Solving (10) in the least-squares sense yields the unknown coefficient $\boldsymbol{\lambda} = \{\lambda_i\}_{i=1}^{n_Z}$. The Kansa solution can then be evaluated everywhere on $\mathcal{S}$ by

$$U = \sum_{\zeta_i\in Z} \lambda_i \Phi_m(\,\cdot\,, \zeta_i). \tag{11}$$

An error estimate for this solution is

$$\begin{aligned} &\|U - u^*_\mathcal{S}\|_{H^k(\mathcal{S})} \\ &\le C(h_Z^{m-k-1/2} + \delta^{1/2} h_X^{2-k} h_Z^{m-2-d/2} + \delta h_X^{d/2-k+3/2+\epsilon} h_Z^{m-d}) \|u^*_\mathcal{S}\|_{H^m(\mathcal{S})} \end{aligned} \tag{12}$$

for any $k \ge 2$ and $m > k + d/2 + 3/2$. By examining the right-hand side of this estimate, we see that using $\delta \approx h_Z$ is a reasonable choice for the radius of the narrow-band domain Ω.

Numerical evidence in [8] suggests that a reduced system by dropping the stability conditions in (9) has better numerically robustness when the set of collocation points X is not dense enough to fulfil the theoretical requirement, i.e., when h_X is larger then the theory allows. That is, we can solve the reduced $3n_X \times n_Z$ overdetermined linear system

$$\begin{pmatrix} \mathcal{L}\Phi_m(X, Z) \\ \partial_{\mathbf{n}}\Phi_m(X, Z) \\ \partial^{(2)}_{\mathbf{n}}\Phi_m(X, Z) \end{pmatrix} \boldsymbol{\lambda} = \begin{pmatrix} f(X) \\ \mathbf{0} \\ \mathbf{0} \end{pmatrix}, \tag{13}$$

for $\boldsymbol{\lambda}$ instead. In cases when the convergent theory applies, accuracy of (10) and (13) are nearly identical. For this reason, our numerical demonstration in the latter section will focus on the reduced system (13) instead.

3 INTRINSIC KANSA METHODS

Let the notation Z be redefined and reused. In this section, we are going to use a trial space defined solely on the surface $\mathcal{S}$ (instead of in some narrow-banded domain). Thus, we need a set of trial centers $Z = \{z_1, \dots, z_{n_Z}\} \subset \mathcal{S}$ to be defined on the surface. We still require the kernel to satisfy the decay condition (7) with smoothness order m.

Under the assumption in [13], [14] that $m > d/2$, we can obtain surface kernels

$$\Psi_{m-1/2}(\cdot\,,\,\cdot) := \Phi_m(\cdot\,,\,\cdot)_{|\mathcal{S}\times\mathcal{S}} : \mathcal{S}\times\mathcal{S} \to \mathbb{R}$$

that reproduce $H^{m-1/2}(\mathcal{S})$ simply by restricting the global kernels Φ_m on $\mathcal{S}$. The surface kernel here severs theoretical purpose and, in actual implementation, we are using the global kernel Φ_m of our choice.

Within the corresponding trial space, which is the span of translations of $\Psi_{m-1/2}$, i.e.,

$$\mathcal{U}_{Z,\mathcal{S},\Psi_{m-1/2}} := \text{span}\{\Phi_m(\cdot - \boldsymbol{\xi}) \mid \boldsymbol{\xi} \in Z\}, \tag{14}$$

we define an intrinsic least-squares solution by

$$U = \underset{u\in\mathcal{U}_{Z,\mathcal{S},\Psi_{m-1/2}}}{\arg\inf} \|\mathcal{L}_{\mathcal{S}}u(\boldsymbol{\xi}) - f(\boldsymbol{\xi})\|^2_{\ell^2(X)} \tag{15}$$

for some sufficiently dense set of collocation points $X \subset \mathcal{S}$. In [15], we show that the numerical solution U of (15) satisfies the estimate

$$\|U - u^*_{\mathcal{S}}\|_{H^k(\mathcal{S})} \le C h_Z^{m-d/2}\left(h_Z^{-k} + h_X^{-k+2} h_Z^{-2}\right)\|u^*_{\mathcal{S}}\|_{H^{m-1/2}(\mathcal{S})}, \tag{16}$$

for $k \ge 2$ and $m > k + d/2 + 1$ for any $u^*_{\mathcal{S}} \in H^{m-1/2}(\mathcal{S})$. Effectively, the intrinsic approach lower the smoothness requirement on the exact solution $u^*_{\mathcal{S}}$ by a half.

Note the presence of surface differential operator $\mathcal{L}_{\mathcal{S}}$ in (15), on which some extra analytical or numerical treatments are required.

3.1 Numerical algorithms

Some readers may be unfamiliar with surface differential operators. We begin with a simple example to demonstrate how one can *collocate* $\mathcal{L}_{\mathcal{S}}$ analytically.

3.1.1 Example

Let $\mathcal{S} \in \mathbb{R}^3$ be the unit sphere. We consider the differential operator $\mathcal{L}_{\mathcal{S}} = \Delta_{\mathcal{S}} - \mathbf{1}\cdot\nabla_{\mathcal{S}}$. Firstly, we compute the orthogonal projection operator

$$\mathcal{P} = I - \mathbf{n}\mathbf{n}^T = \begin{bmatrix} -x^2+1 & -xy & -xz \\ -xy & -y^2+1 & -yz \\ -xz & -yz & -z^2+1 \end{bmatrix}.$$

Let $\mathcal{P}_k$ denotes the kth column of $\mathcal{P}$. Using the fact that

$$\nabla_{\mathcal{S}} u = \mathcal{P}\nabla u = \begin{bmatrix} \mathcal{P}_1 \cdot \nabla \\ \vdots \\ \mathcal{P}_d \cdot \nabla \end{bmatrix} u,$$

and

$$\Delta_{\mathcal{S}} u = (\mathcal{P}\nabla)\cdot(\mathcal{P}\nabla)u = \text{trace}\left(\mathcal{P}\cdot J^T(\nabla_{\mathcal{S}} u)\right),$$

we can express the surface operator entirely in Cartesian coordinates after some symbolic manipulations as

$$\begin{aligned}\mathcal{L}_{\mathcal{S}} \;=\; & \Big((x-1)^2 + x(y+z) - 2\Big)\frac{\partial}{\partial x} + \Big((y-1)^2 + y(x+z) - 2\Big)\frac{\partial}{\partial y} \\ & + \Big((z-1)^2 + z(x+y) - 2\Big)\frac{\partial}{\partial z} \\ & + \Big(-x^2+1\Big)\frac{\partial^2}{\partial x^2} + \Big(-y^2+1\Big)\frac{\partial^2}{\partial y^2} + \Big(-z^2+1\Big)\frac{\partial^2}{\partial z^2} \\ & -2xy\frac{\partial^2}{\partial x\partial y} - 2xz\frac{\partial^2}{\partial x\partial z} - 2yz\frac{\partial^2}{\partial y\partial z},\end{aligned}$$

which no longer has any implicit dependency on $\mathcal{S}$ and can be used for collocation. We can solve (15) by solving the following $n_X \times n_Z$ overdetermined system

$$\mathcal{L}_{\mathcal{S}}\Phi_m(X,Z)\ \boldsymbol{\lambda} = f(X) \tag{17}$$

for $\boldsymbol{\lambda}$ in the least-squares sense. The numerical solution is given as in (11).

The above analytic approach works as long as we have analytic formula for the normal vector $\mathbf{n}$ of the surface, which is the case when $\mathcal{S}$ is given by some parametric formula or level set method. In cases when this is not so, i.e., when $\mathcal{S}$ is defined by a point cloud, we can opt for a pseudospectral approximation [16] and here is how.

We assume $\mathbf{n}(z)$, for all $z \in Z$, is known either a priori or by some approximations [17], [18]. Without explicitly knowing the function values of any surface function $u_{\mathcal{S}}$ at Z, we can still express the interpolant $I_Z u$ of u in the trial space $\mathcal{U}_{Z,\mathcal{S},\Psi_{m-1/2}}$ as

$$I_Z u_{\mathcal{S}} = [\Psi_{m-1/2}(\cdot, Z)][\Psi_{m-1/2}(Z,Z)]^{-1} u_{\mathcal{S}}(Z).$$

Since $Z \subset \mathcal{S}$, we have $\Psi_{m-1/2}(Z,Z) = \Phi_m(Z,Z)$ and no special treatment is required on the global kernel. The idea of pseudospectral is to approximate derivatives of $u_{\mathcal{S}}$ by that of $I_Z u_{\mathcal{S}}$. For surface gradient, we have

$$\nabla_{\mathcal{S}} u_{\mathcal{S}} \approx \widetilde{\nabla}_{\mathcal{S}} u_{\mathcal{S}} := \nabla_{\mathcal{S}}(I_Z u_{\mathcal{S}}) = \begin{bmatrix}\mathcal{P}_1\cdot\nabla \\ \vdots \\ \mathcal{P}_d\cdot\nabla\end{bmatrix}(I_Z u_{\mathcal{S}}).$$

One can, see [16] for details, write the kth component of $\nabla_{\mathcal{S}}(I_Z u_{\mathcal{S}})$ in the form of

$$[\widetilde{\nabla}_{\mathcal{S}} u_{\mathcal{S}}]_k = \underbrace{\Big(\mathcal{P}_k^T[\nabla\Phi_m(\cdot\,,Z)][\Phi_m(Z,Z)]^{-1}\Big)}_{=:G_k(\cdot\,,Z)} u_{\mathcal{S}}(Z). \tag{18}$$

We need all these $1\times n_Z$ vector function G_k to apply another round of pseudospectral for the surface Laplacian operator, namely

$$\Delta_{\mathcal{S}} u_{\mathcal{S}} \approx \widetilde{\nabla}_{\mathcal{S}}\cdot\nabla_{\mathcal{S}}(I_Z u_{\mathcal{S}})(Z),$$

in which the gradient of $u_{\mathcal{S}}$ is approximated, then evaluated, at Z. This specific choice allows simplification for a simple formula

$$\widetilde{\Delta}_{\mathcal{S}} u_{\mathcal{S}} = \sum_{k=1}^{d} G_k(\cdot, Z) G_k(Z, Z) u_{\mathcal{S}}(Z).$$

At this point, we can collocate at some set of collocation points $X \subset \mathcal{S}$ to yield an $n_X \times n_Z$ overdetermined matrix system

$$\left[a \sum_{k=1}^{d} G_k(X, Z) G_k(Z, Z) + \mathbf{b}^T G(X, Z) + c\Phi_m(X, Z)\Phi_m^{-1}(Z, Z) \right] u_{\mathcal{S}}(Z) = f(X) \tag{19}$$

for unknown $u(Z)$. By the relationship $\boldsymbol{\lambda} = \Phi_m^{-1}(Z, Z) u_{\mathcal{S}}(Z)$, which appears in all terms in the form of $G_k(\,\cdot\,, Z) u_{\mathcal{S}}(Z)$, we can cast the system in terms of unknown $\boldsymbol{\lambda}$ easily.

4 NUMERICAL EXAMPLES

We provide some numerical demonstrations using the (global) Whittle-Matérn-Sobolev kernels of order m as in (8). The first example will provide readers a sense of convergence of the methods in this paper. The next two aim to show how the methods can be applied in other applications, namely, solving surface pattern formation problems and convection-diffusion equations on moving surfaces. We denote our methods as follows:

Method 1: Embedding Kansa method as in (13),

Method 2: Approximated collocation method as in (19),

Method 3: Direct collocation analytically as in (17).

Sets of nodes X and Z are both quasi-uniform with some oversampling ratio h_Z/h_X based on fill distances. Having a specific number of points quasi-uniformly distributed on a surface is not an easy task that worths to be solved exactly. In our computations, we simply enforce $h_Z = \gamma h_X$ so that the oversampling ratio γ is roughly equal to some preassigned value.

4.1 Convergence rates and accuracy

We solve the surface elliptic PDE (2) with $a_{\mathcal{S}} = c_{\mathcal{S}} = 1$ and $\mathbf{b}_{\mathcal{S}} = \mathbf{0}$. Let the exact solution be

$$u_{\mathcal{S}}^* = (x + y + z) \exp\left(-x(x-1) - y(y-1) + z\right)$$

defined on a constant distance product (CPD) surface

$$\sqrt{(x-1)^2 + y^2 + z^2}\sqrt{(x+1)^2 + y^2 + z^2} \cdot \sqrt{x^2 + (y-1)^2 + z^2}\sqrt{x^2 + (y+1)^2 + z^2} = 1.1.$$

We apply all methods with an oversampling ratio of $\gamma \approx 1.4$. Fig. 1 shows the error distributions on the surface obtained by Method 3 with $m = 4$. In Fig. 2, the $L^2(\mathcal{S})$ and $H^2(\mathcal{S})$ error profiles with the kernels of order $m = k + d/2 + 1/2 = 4$ and $m = k + d/2 + 3/2 = 5$ are plotted respectively. Note that the former m is below what error estimates (12) and (16) call for. All subfigures were made with the same scale for easy comparison.

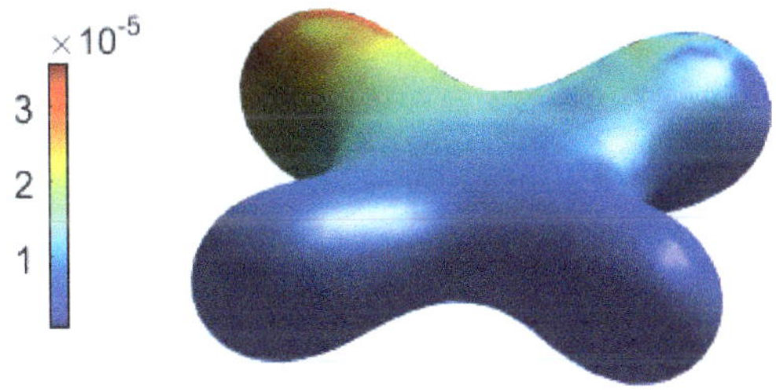

Figure 1: Error functions, obtained by Method 2 with the oversampling collocations: $h_Z \approx 1.4h_X \approx 0.02$ and kernels of smoothness order $m = 4$.

Clearly, kernels with smoothness order below the theoretical requirement for the error estimates, i.e., $m = 4$, remain practical. In terms of accuracy, errors of Method 3 the best among the three. With our smooth solution, larger value of m is also beneficial to all methods in terms of accuracy. The $H^2(\mathcal{S})$-error convergence rates of Method 1 are close to the predicted ones $(m - 2.5)$ in theory. Though the theoretical rate $(m - 3.5)$ for Method 3 is lower compared with Method 1, it achieves higher orders of convergence in this example. Convergence rate of Method 2 sits between those of both approaches, which justifies our choice of approximation methods to the surface differential operators.

4.2 Pattern formation on surfaces

In this section, we solve a system of coupled reaction-diffusion equations for pattern formations on surfaces. In general, a specific pattern can be obtained by solving the following equations:

$$\begin{cases} \dfrac{\partial u}{\partial t} &= \delta_u \Delta_{\mathcal{S}} u + f_u(u,v), \\ \dfrac{\partial v}{\partial t} &= \delta_v \Delta_{\mathcal{S}} v + f_v(u,v), \end{cases} \tag{20}$$

where u and v are the activator and inhibitor of the system. In particular, we select

$$\begin{cases} f_u(u,v) &= \alpha u(1 - \tau_1 v^2) + v(1 - \tau_2 u), \\ f_v(u,v) &= \beta v \left(1 + \dfrac{\alpha \tau_1}{\beta} uv\right) + u(\gamma + \tau_2 v), \end{cases} \tag{21}$$

as the reaction terms in (20). In this numerical experiment, we generate two different patterns on a Torus Knot. The non-linear problem (20) is solved by Method 1, which replaces the surface Laplacian operator with traditional Laplacian with two additional constraints. After converting the surface PDE to its embedding space, we solve the embedded PDE by a second-order semi-implicit backward differentiation formula (SBDF2) until the solution reaches steady state. In Fig. 3, we show the steady-state solutions of (20) with two different sets of parameters $(\alpha, \beta, \tau_1, \tau_2)$; one for a spot pattern and the other for some stripe pattern. It is convincing that our method is a robust alternative for pattern formation problems. The simulation was implemented by MATLAB parallel computing toolbox. The computation time using a consumer-graded NVIDIA GeForce GTX 1080 took around 204 seconds to attain its

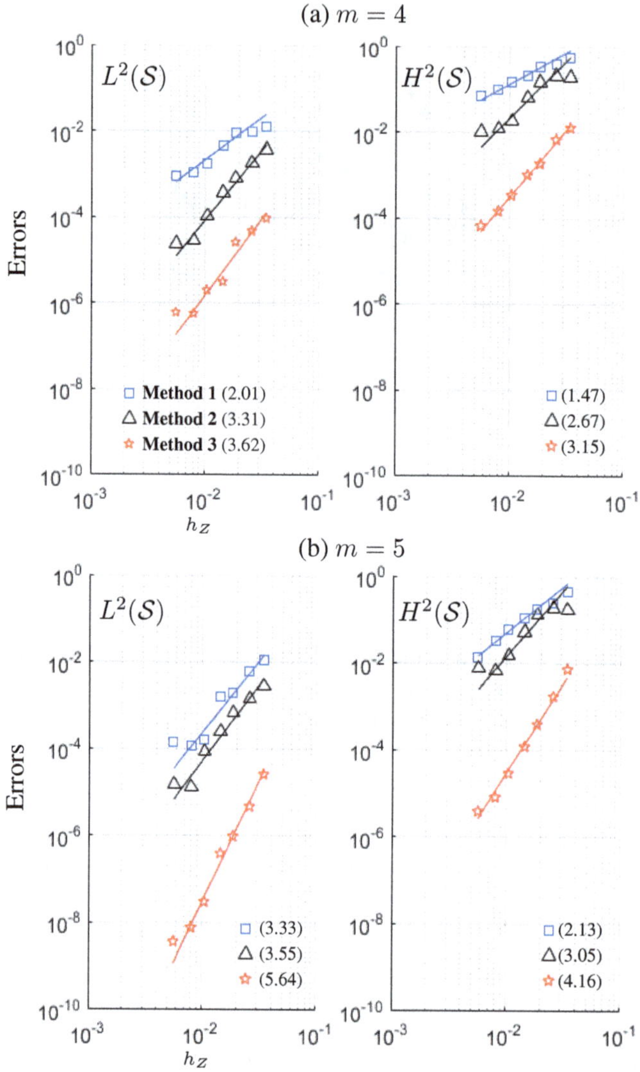

Figure 2: For the kernels of (a) $m = 4$ and (b) $m = 5$, the $L^2(\mathcal{S})$ and $H^2(\mathcal{S})$ error profiles with convergence rates listed in legends by using our three methods, based on the same seven pairs of X and Z ($h_Z \approx 1.4h_X$).

steady state solution while it took 4228 seconds on CPU. Such 20X speed-up is done simply by adding few lines of codes to convert the system matrix to GPU Array.

4.3 Merging surfaces by curvature motion

Let the evolving surface with time $t \in [0, T]$ be denoted as $\mathcal{S}(t)$. In this test, initial surfaces are two spheres of radius 0.72 on the upper left side and 1.8 on the lower right side, shown in Fig. 4. First, we let they approach to each other at velocities $\mathbf{v} = \frac{\sqrt{2}}{2}(1, 0, -1)$ and $-\mathbf{v}$ respectively. Then they collide at $t = 0.01$ to generate a new closed and smooth surface, see

Figure 3: Steady state solutions of the reaction diffusion equation obtained by Method 1.

Fig. 5(a). The merged surface further evolved under geometric motion $\mathbf{v} = -\kappa\mathbf{n}$ by the mean curvature $\kappa = \nabla_{\mathcal{S}} \cdot \mathbf{n}$, see [19]. Along the evolution, we compute point clouds by the method in [20].

Now, on $\mathcal{S}(t)$, we want to solve convection-diffusion equations

$$\partial_t u + \mathbf{v} \cdot \nabla u + (\nabla_{\mathcal{S}} \cdot \mathbf{v})u - \Delta_{\mathcal{S}} u = 0 \quad \text{on } \mathcal{S}(t) \times [0, T], \tag{22}$$

subject to some Dirichlet initial conditions. In (22), vector $\mathbf{v}$ denotes the velocity of surface motion.

We first discretize (22) by the Crank-Nicolson (CN) method in time. Then, we apply Method 2 in space to solve the discretized problem and complete the time stepping. Note that Method 3 is not appropriate for this problem due to the lack of analytical information. Implementing Method 1 also involves certain difficulties, e.g., the high curvature occurring right after the merge poses a constraint and we must use some *very narrow*-band domains.

At each time step before merging ($t < 0.01$), we respectively solve the problem (22) on both spheres, via our oversampled Method 2 to the time semi-discretized problem. We employ kernels of low order of smoothness $m = 2.5$ and the time step size is set to be $\triangle t = 0.001$. At the moment of merging at $t = 0.01$, the numerical solutions at merged points are obtained by

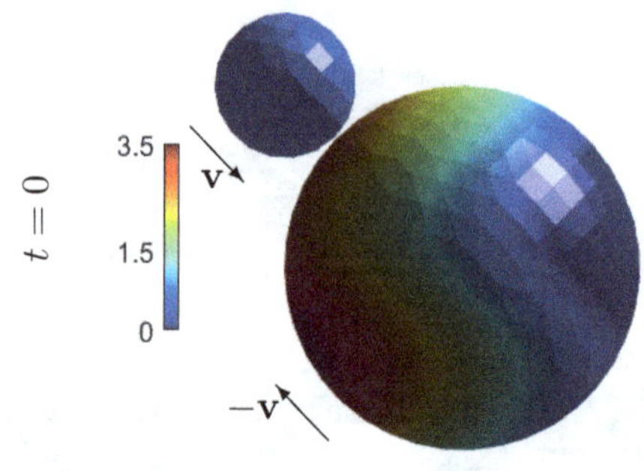

Figure 4: The initial conditions (color) with $n_X = 567$, $n_Z = 365$ totally on two merging different spheres at constant velocities $\mathbf{v} = \frac{\sqrt{2}}{2}(1, 0, -1)$ and $-\mathbf{v}$ (arrow).

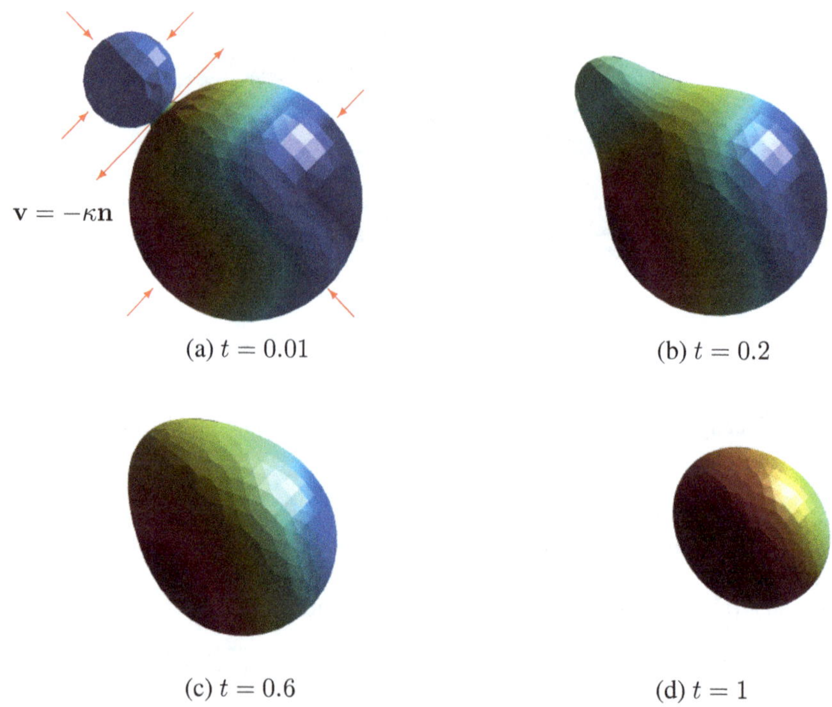

Figure 5: The corresponding snapshots of numerical solutions (color) with evolving velocity (arrow) at several time points, obtained by Method 3 with CN, using $n_X = 666$, $n_Z = 427$, $m = 2.5$ and $\triangle t = 10^{-3}$ with the same scalars of color and coordinates as in Fig. 4, on a mean-curvature evolving surface.

interpolation in Fig. 5(a). After then, we use approximated normal vectors to simulate mean-curvature motion. Numerical results of (22) obtained by Method 3 at three times $t = 0.2, 0.6$ and 0.1 are color mapped in Fig. 5(b)–(d) respectively.

5 CONCLUSION

This paper reviews three alternatives for solving surface PDEs. We focus on surfaces with codimension one for simplicity; all methods in this paper can be easily generalized to higher codimensions. In our presented example, we see some higher than predicted convergence rates, which is not uncommon to see in Kansa-type methods for surface PDEs. Generally speaking, convergence rate will gradually slow down to the predicted one as the surface curvature increases. Although all presented methods are global (because we have rigorous convergence theorems), we want to point out to our readers that there are many local meshfree methods that solve surface PDEs. Readers are recommended to include those in consideration before pinpointing their method of choice.

REFERENCES

[1] Strichartz, R.S., Analysis of the Laplacian on the complete Riemannian manifold. *J. Funct. Anal.*, **52**, pp. 48–79, 1983.
[2] Wloka, J., *Partial Differential Equations*. Cambridge University Press, 1987.
[3] Kansa, E.J., Application of Hardy's multiquadric interpolation to hydrodynamics. *Proceedings of the 1986 Simulation Conference*, vol. 4, pp. 111–117, 1986.
[4] Kansa, E.J., Multiquadrics—a scattered data approximation scheme with applications to computational fluid-dynamics. I. Surface approximations and partial derivative estimates. *Comput. Math. Appl.*, **19**(8–9), pp. 127–145, 1990.
[5] Ruuth, S.J. & Merriman, B., A simple embedding method for solving partial differential equations on surfaces. *J. Comput. Phys.*, **227**(3), pp. 1943–1961, 2008.
[6] Piret, C., The orthogonal gradients method: A radial basis functions method for solving partial differential equations on arbitrary surfaces. *J. Comput. Phys.*, **231**(14), pp. 4662–4675, 2012.
[7] Chen, Y. & Macdonald, C.B., The closest point method and multigrid solvers for elliptic equations on surfaces. *SIAM J. Sci. Comput.*, **37**(1), pp. A134–A155, 2015.
[8] Cheung, K.C. & Ling, L., A kernel-based embedding method and convergence analysis for surfaces PDEs. *SIAM J. Sci. Comput.*, **40**(1), pp. A266–A287, 2018.
[9] Wendland, H., Error estimates for interpolation by compactly supported radial basis functions of minimal degree. *J. Approx. Theory*, **93**(2), pp. 258–272, 1998.
[10] Buhmann, M.D., *Radial basis functions: Theory and implementations*, vol. 12 of *Cambridge Monographs on Applied and Computational Mathematics*. Cambridge University Press: Cambridge, 2003.
[11] Wendland, H., *Scattered data approximation*, vol. 17 of *Cambridge Monographs on Applied and Computational Mathematics*. Cambridge University Press: Cambridge, 2005.
[12] Cheung, K.C., Ling, L. & Schaback, R., H^2–convergence of least-squares kernel collocation methods. *SIAM J. Numer. Anal.*, **56**(1), pp. 614–633, 2018.
[13] Fuselier, E.J. & Wright, G.B., Scattered data interpolation on embedded submanifolds with restricted positive definite kernels: Sobolev error estimates. *SIAM J. Numer. Anal.*, **50**(3), pp. 1753–1776, 2012.
[14] Narcowich, F.J., Sun, X. & Ward, J.D., Approximation power of RBFs and their associated SBFs: A connection. *Adv. Comput. Math.*, **27**(1), pp. 107–124, 2007.
[15] Chen, M. & Ling, L., Intrinsic meshless collocation methods for PDEs on manifolds. *SIAM J. Num. Anal.*, to be published.
[16] Fuselier, E.J. & Wright, G.B., A high-order kernel method for diffusion and reaction-diffusion equations on surfaces. *J. Sci. Comput.*, **56**, pp. 535–565, 2013.

[17] Fornberg, B., Driscoll, T.A., Wright, G. & Charles, R., Observations on the behavior of radial basis function approximations near boundaries. *Comput. Math. Appl.*, **43**(3–5), pp. 473–490, 2002.
[18] Marchandise, E., Piret, C. & Remacle, J.F., CAD and mesh repair with radial basis functions. *J. Comput. Phys.*, **231**(5), pp. 2376–2387, 2012.
[19] Deckelnick, K., Dziuk, G. & Elliott, C.M., Computation of geometric partial differential equations and mean curvature flow. *Acta Numer.*, **14**, pp. 139–232, 2005.
[20] Bronsard, L. & Wetton, B.T.R., A numerical method for tracking curve networks moving with curvature motion. *J. Comput. Phys.*, **120**(1), pp. 66–87, 1995.

EXACT TIME DEPENDENT CALCULATIONS FOR FLUIDS AND SOLIDS

EDWARD J. KANSA
Convergent Solutions, USA

ABSTRACT

Meshfree transport calculations have been traditionally performed in the Lagrangian or particle velocity mode connected to an underlying distorting mesh. That is remapped (interpolated) onto a regular mesh. This primitive procedure was necessary with extremely primitive electronic computers that has extremely restrictive memory storage and execution speed. However, the meshless radial basis function approach allows points to move; without diffusive re-interpolation; one can choose to allow the motion at each interior point to convert the partial differential equation into an exact differential, thereby eliminating truncation errors. Given the proper tools to utilize extended arithmetic precision on computers, electronic computers can asymptotically approach the power of ideal academic mathematics and solve numerically problems approaching theoretical methods.

Keywords: hyperbolic partial differential equations, radial basis functions, exact solutions of time expansion coefficients, exponential convergence, meshless methods.

1 INTRODUCTION

Hyperbolic partial differential equations (PDEs) are modified using well-established local mesh-based methods by the introduction of upwind differencing to prevent destabilizing truncation errors. The typical excuse for justifying the modification of the physics is that "in the limit", the artificial unphysical numerical diffusion vanishes. However, on a finite computer, this limit is never reached. This paper demonstrates a numerical meshfree scheme valid for gas, liquid, and solid mechanics that allows for transport without unphysical numerical diffusion by utilizing modern computational methods.

Consider a set of inviscid gas dynamic hyperbolic equations:

$$\partial\rho/\partial t + \nabla\bullet m = 0, \tag{1}$$

$$\partial m/\partial t + \nabla\bullet(mm/\rho + pI) = 0, \tag{2}$$

$$\partial E/\partial t + \nabla\bullet((m/\rho)*(p+E)) = 0, \tag{3}$$

where m =ρu, ρ= gas density, u is the velocity vector, m, is the momentum density vector, p as the ideal gas pressure, e is the internal energy, mm is a 2nd order momentum dyadic, I is 2nd order dyadic, E = (ρe +p), $\gamma = c_p/c_v$.

Although gas dynamic equations are primarily longitudinal, i.e. the compressions and rarefactions are in the direction of the rarefactions, in higher dimensions, there is a small amplitude angular momentum wave that propagates in the normal direction of the compression wave.

For elastic solid dynamics, Hooke's law applied to small displacements, u. An ideal solid elastic body can be subject to both compressive and shear stress given by a tensor, σ.

The conservation of mass is:

$$\partial\rho/\partial t + \nabla\bullet(\rho u) = 0. \tag{4}$$

The conservation of momentum is now in the form of Hooke's law:

$$\partial^2(\rho\mathbf{u})/\partial t^2 + \nabla^2\boldsymbol{\sigma} = 0. \tag{5}$$

WIT Transactions on Engineering Sciences, Vol 126, © 2019 WIT Press
www.witpress.com, ISSN 1743-3533 (on-line)
doi:10.2495/BE420151

Two dimensional elastic solutions allow for longitudinal waves and transverse waves. Longitudinal waves produce compressions and rarefactions in the direction of wave propagation, whereas transverse waves produce compressions and rarefactions normal to the wave propagation direction.

2 NUMERICAL SIMULATION OF HYPERBOLIC WAVES

It must be emphasized here that the mathematics executed on electronic computers is not the same as the mathematics practiced in academia because academic mathematics is non-material existing in the ideal Platonic world in which all rational and irrational numbers have infinite precision, and all transcendental functions are complete to infinite order whereas in computer mathematics, rational and irrational numbers are designated by a finite number of bits comprising a computer word and all transcendental functions are expanded as truncated series expansions or rational fractions. In other words, Plato's description of shadows flickering on a cave wall is most appropriate for mathematics on electronic computers.

Most applied computational engineers, physicists, and mathematicians still use the techniques that were absolutely necessary in early days of electronic computers that hade very limited memory and were extremely slow. Because of severe memory restrictions, it was assumed that the unknown solutions to partial differential and integral equations can be constructed as simple piece-wise constant, linear, or quadratic polynomials.

Kansa [1] showed that the usual derivative expressions of finite differences are really related to a linear or quadric interpolation fit of two or three dependent variables at two or three independent locations. In practice, derivative approximations are limited to quadratic polynomials because derivative approximations are plagued by polynomial snaking.

The next step upward are radial basis functions that depends only upon the geodesic between pairs of points in any geometry. A radial basis function, (RBF), $\phi(x;y)$, is a function of the geodesic distance, d_g, in Rieman geometry from a data center, y, to an evaluation center, x, where x, $y \in \Re^n$. RBFs can be: 1.Polynomials of d_g having compact support over a limited range, 2. A polynomial with global support, or 3. A transcendental function, continuously differentiable, (C^∞), with global support. The C^∞ RBFs with global support are known theoretically to converge exponentially fast. Madych and Nelson [2], [3] showed that for interpolation, C^∞RBFs converge as:

$$\theta^{(\langle c\rangle/\langle h\rangle)}, \quad \theta< 1, \qquad (6)$$

where <c> is the average shape parameter, <h> is the average geodesic distance. Note that in most applications of C^∞ RBFs use the Euclidean radial distance between pairs of points. It was shown in Kansa [4] that partial derivative approximations with C^∞ RBFs are very accurate negating the need for the weak formulation of finite elements. Because of the accurate approximation of partial derivative, Kansa [4] applied the C^∞RBFs to the solution of hyperbolic, parabolic, and elliptic PDEs.

While the C^∞ RBFs have impressive theoretical properties such as exponential convergence rates, in practice, these RBFs can become so ill-conditioned that the solution is worthless. This phenomenon was explored in Kansa and Holoborodko [5] as being caused by the finite size of the computer words tied to the machine epsilon, ε. Computer chip manufacturers have extremely small demand for computer chips with extended precision, hence chips have 64 or 128 bits. The most prevalent way to obtain extended precision at present is by means of software methods (see www.Advanpix.com).

Because of the possibility of either severely ill-conditioned or even numerically singular linear equations using single or double precision computers, many researchers were afraid to use C∞ RBFs even though the vast amount of domain decomposition literature was available to counter-act ill-conditioning as well as extended precision software.

3 ELIMINATING DESTABILIZING TRUNCATION ERRORS

Lagrangian methods have been used extensively to control finite difference, element, and volume truncation errors by allowing mesh points to move at each local convective velocity, u_j. After the mesh points are advected and the local mesh becomes severely distorted, even to the point of having unphysical negative volumes, the mesh is interpolated onto new locations that are not severely distorted. The operation of "remapping" introduces unphysical numerical diffusion, so this process is used sparingly. This is yet another example in which physics is sacrificed for obsolete numerics.

Kansa et al. [6], Kansa [7] and Kansa et al. [8] used meshless C^∞ RBFs for inviscid Euler equations. In the fixed earth or Eulerian frame, the conservation equations have the form:

$$\partial U/\partial t + \nabla \bullet F = f, \quad \text{over } \Omega \backslash \partial\Omega, \tag{7}$$

$$U \text{ specified on } \partial\Omega, \tag{8}$$

where F is the appropriate flux vector, and f is the forcing function.

However, if each data center velocity, η, is chosen to make each PDE into an exact differential,

$$\partial U/\partial t + \eta \bullet \nabla U = 0, \quad \text{over } \Omega \backslash \partial\Omega, \tag{9}$$

$$U \text{ specified on } \partial\Omega, \tag{10}$$

where each moving velocity is chosen so that

$$\{F \;\square\!+ \eta \bullet \nabla U\} = 0, \tag{11}$$

$$\text{along each } dx_j/dt = \boldsymbol{\eta}. \tag{12}$$

Although data centers and evaluation points have the freedom to move, this motion must be constrained to prevent points from leaving the domain and any interpolation onto new centers must be constrained to conserve total mass, momentum components, and total energy. Strict conservation is enforced volumetrically by integrating the basis over the evaluation centers.

In the moving frame over Ω\∂Ω, it is assumed that space and time are separable. Then the time dependent expansion coefficients over the interior forms a set of time dependent ordinary different equations that have exact solutions.

An excellent numerical test of the theoretical discussion present above is the ideal gas self-similar rarefaction fan. As long as the rarefaction fan is freely propagating without any interactions with solid walls, shocks, contact surfaces, or other rarefaction fans, the sample problem is obtained from Torro's book [9], as well as the physics. This rarefaction fan will have the same shape and will be invariant in magnitude, but not position. A rarefaction wave, in primitive variables, is presented in Tables 1 and 2 at a start time and an ending time that has propagated 1e4 time steps.

The two parts of the table are identical, as they should be if there are negligibly small truncation errors contaminating the solution. This attests to the "exact" nature of the time marching scheme at least to the 43rd decimal place.

Table 1: Profile of left rarefaction wave at every fifth data center using the multiquadric radial basis function and extended precision package (50 digits) from www.Advanpix.com.

Time = 0.12425000					
Gas density (ρ)	Gas velocity (U)	Gas pressure (P)	Left characteristic (u-c)	Gas density (R)	Gas velocity (x)
1.000000	0.000000	1.000000	-1.183216	5.916080	1.615159
0.830353	0.215926	0.770846	-0.924105	5.916080	1.699435
0.684553	0.185243	0.588261	-0.665994	5.916080	1.783711
0.559993	0.647779	0.444075	=0.405992	5.916080	1.867987
0.047759	0.971668	0.284816	-0.017215	5.916080	1.997848

All the eigenvalues are equal to 1.00000000.

Table 2: Profile of left rarefaction wave at every fifth data center using the multiquadric radial basis function and extended precision package (50 digits) from www.Advanpix.com and final time 2.3750000.

Time = 2.3750000					
ρ	U	P	u-c	R^-	x
1.000000	0.000000	1.000000	-1.183216	5.916080	0.810138
0.830353	0.215926	0.770846	-0.924105	5.916080	-0.194748
0.684553	0.185243	0.588261	-0.665994	5.916080	1.783711
0.559993	0.647779	0.444075	=0.405992	5.916080	1.867987
0.047759	0.971668	0.284816	-0.017215	5.916080	1.959115

All the eigenvalues are equal to 1.00000000.

The variable R^- is the left Riemann invariant, for all corresponding values of u and c.

$$R^- = u + 2c/(\gamma-1). \tag{13}$$

4 DISCUSSION

The mathematics practiced at universities is not the same as the mathematics exercised on electronic computers because all numbers on computers have finite prescribed precision and all functions, especially transcendental functions, are truncated expansions. When extended precision is activated on electronic computers, not only are the word sizes increased, but more software terms are added to library functions, especially for transcendental functions. Persons are familiar with infinite Fourier series expansions for various functions. Although sines and cosines are transcendental functions, these are primarily applied to one-dimensional expansion problems. In addition, such expansions are used for one-dimensional functions without the benefit of any explicit wavelet properties.

The entire class of transcendental functions can be made into C^∞ RBFs by using either the Euclidean or geodesic distances instead of a point in $\Re^n$. C^∞ RBF is much more flexible in application than a transcendental function. Theoretically, in the Platonic ideal world, C^∞ RBFs in interpolation and approximation context, converge exponentially fast as $\theta^{(\langle c\rangle/\langle h\rangle)}$, $\theta<1$. This exponential convergence rate can be accelerated by a finer discretization or by increasing the average shape parameter, c. Using a finer discretization increases the total number of points, whereas increasing the shape parameter does not, but has the effect of increasing the condition number. Whenever the condition number exceeds the inverse of the computer machine epsilon, the round-off errors can render any solution meaningless. However, by decreasing the machine epsilon by increasing the number of active digits by either hardware or software methods enables computers to approach asymptotically the ideal computer.

For time dependent problems, the dependent variable, U(x,t) is written as an expansion of spatial C^∞ RBFs and products of unknown time dependent expansion coefficients, $\alpha(t)$, allowing each interior point to move at velocity at which the transformed PDE is transformed into an exact differential. The result is that the ODE in the expansion coefficients in the moving frame have exact solutions, and any errors are due to the truncated functions stored in memory. For strictly hyperbolic PDEs, the Riemann solvers, see Toro [9] are recommended for interactions of shock and contact surface discontinuities that produce new waves. For the continuous rarefaction waves, it must be remembered that the function is continuous at the endpoints of the rarefaction, but the derivatives are discontinuous in order to obtain the correct physics. It was demonstrated that a non-interacting rarefaction wave is self-similar, only the data centers expanded; the density, gas velocity, pressure, the characteristic velocity, and the Riemann invariant should be invariant in time, at least to very high order round-off errors, since no computer can compute exactly.

Using more powerful spatial interpolation and approximation schemes without the constraints of local mesh stencils allows researchers to exploit improved software methods that asymptotically approach ideal Platonic academic mathematics.

REFERENCES

[1] Kansa, E.J., Multiquadrics – A scattered data approximation scheme with applications to computational fluid dynamics: I. Surface approximations and partial derivative estimates. *Comput. Math. Appl.*, **19**(6–8), pp. 127–145, 1990.

[2] Madych, W.R. & Nelson, S.A., Multivariate interpolation and conditionally positive definite functions, II. *Math. Comput.*, **54**, pp. 211–230, 1990.

[3] Madych, W.R. & Nelson, S.A., Bounds on multivariate polynomials and exponential error estimates for multiquadric interpolation. *J. Approx. Theory*, **70**, pp. 94–114, 1992.

[4] Kansa, E.J., Multiquadrics – A scattered data approximation scheme with applications to computational fluid dynamics: II. Solutions to parabolic, hyperbolic, and elliptic partial differential equations. *Comput. Math. Appl.*, **19**(6–8), pp. 147–161, 1990.

[5] Kansa, E.J. & Holoborodko, P., On the ill-conditioned nature of C∝-RBF strong collocation. *Eng. Anal. Bound. Elem.*, **78**, pp. 26–30, 2017.

[6] Kansa, E.J., Power, H. Fasshauer, G.E. & Ling, L., A volumetric integral radial basis function method for time-dependent partial differential equations: I. Formulation. *Eng. Anal. Bound. Elem.*, **28**, pp. 1191–1206, 2004.

[7] Kansa, E.J., Exact explicit time integration of hyperbolic partial differential equations with radial basis functions. *Eng. Anal. Bound. Elem.*, **31**, pp. 577–585, 2007.

[8] Kansa, E.J., Aldredge, R.C. & Ling, L., Numerical simulation of two-dimensional combustion using mesh free methods. *Eng, Anal. Bound. Elem.*, **33**, pp. 940–950, 2009.

[9] Toro, E.F., *Riemann Solvers and Numerical Methods for Fluid Mechanics*, 2nd ed., Springer: Berlin, 1999.

FAST SINGULAR BOUNDARY METHOD: MATHEMATICAL BACKGROUND AND APPLICATION IN WAVE PROPAGATION ANALYSIS

ZHUO-JIA FU[1,2], JUN-PU LI[1] & QIANG XI[1]
[1]Center for Numerical Simulation Software in Engineering and Sciences, Hohai University, China
[2]Institute of Continuum Mechanics, Leibniz University Hannover, Germany

ABSTRACT
Despite active research of many decades, numerical solution of the large-scale exterior wave problems remains a great challenge. The finite element method (FEM) needs to be coupled with some effective special treatments for handling unbounded domains, which are often tricky and largely based on trial–error experiences. The boundary element method (BEM) appears attractive to exterior problems, because the fundamental solutions satisfy the governing equation and Sommerfeld radiation condition at infinity. However, the BEM encounters computationally expensive singular numerical integration of fundamental solutions. To avoid this troublesome issue, the method of fundamental solutions (MFS) distributes the source nodes on a fictitious boundary outside the physical domain and is meshless, integration-free, and highly accurate. However, the placement of fictitious boundary is still an open issue for complex-shaped boundary and multiply connected domain problems. Recently, an alternative meshless boundary collocation approach, singular boundary method (SBM), has been proposed to solve various wave propagation problems. The key issue of the SBM is to determine the accurate source intensity factors instead of the singularities of fundamental solutions at origin. Several techniques have been proposed and investigated. The SBM is mathematically simple, easy-to-program, meshless, and applies the concept of source intensity factors to eliminating the singularity of the fundamental solutions. The method avoids singular numerical integrals in the BEM and circumvents the troublesome placement of the fictitious boundary in the MFS. This paper first presents the mathematical background of the SBM, and then applies it to wave propagation analysis.
Keywords: fundamental solutions, singular boundary method, boundary element method, wave propagation.

1 INTRODUCTION

Wave propagation has extensive and wide-range applications in engineering and sciences, especially in geological exploration, photovoltaic devices and metamaterials and ultrasonic non-destructive testing (NDT). The finite element method (FEM) and boundary element method (BEM) are two dominant numerical methods in the field of computational wave propagation. Since wave propagation analysis frequently refers to infinite domain problem, the domain-discretization FEM requires some additional efficient treatments to carefully truncate the infinite domain to finite domain, which are often tricky and largely based on trial–error experiences. It is natural to employ the BEM [1] in this kind of application without domain discretization since the semi-analytical fundamental solutions are used as basis functions, which satisfy the governing equation and the Sommerfeld radiation condition at infinity. Therefore, the BEM is an attractive alternative to the FEM especially for wave propagation under infinite domain. However, the BEM encounters sophisticated mathematical and computationally expensive numerical integration over the singularities.

To avoid this inconvenience in the BEM, the method of fundamental solutions (MFS) [2] places the source nodes on a fictitious boundary outside the physical domain, and implements a linear combination of fundamental solutions to approximate the numerical solution. In comparison with the BEM, the MFS provides more accurate solutions with the meshless and

WIT Transactions on Engineering Sciences, Vol 126, © 2019 WIT Press
www.witpress.com, ISSN 1743-3533 (on-line)
doi:10.2495/BE420161

integration-free properties. However, the placement of source nodes on the fictitious boundary is vital for its numerical accuracy and stability, which is still an open issue especially for complex-shaped or multi-connected domains.

During the last two decades, great attentions have been paid to develop an accurate and stable semi-analytical boundary discretization scheme to eliminate the singularities of fundamental solutions in the BEM and also avoid the troublesome placement of fictitious boundary in the MFS. Generally speaking, there are two ways to achieve this goal. One is to derive nonsingular radial basis function general solutions instead of the singular fundamental solutions as basis function, for example, boundary knot method, boundary collocation method and modified collocation Trefftz method etc. The other one is to introduce the regularized technique to overcome the singularities of fundamental solutions, for example, regularized meshless method, modified method of fundamental solutions, singular boundary method (SBM) and boundary distributed source method etc.

This study focuses on the singular boundary method, which has been first proposed by Prof. Chen in 2009 [3]. Later, Fu and Chen [4] applied the SBM to wave propagation analysis. It inherits the merits from the BEM and the MFS and introduces the concept of the "origin intensity factors" or "source intensity factors" to eliminate the singularity in the fundamental solutions and instead of the numerical computation of the singular integrals in the BEM. There are several approaches [5]–[7] proposed to determine the origin intensity factors of both the fundamental solutions and their derivatives in the SBM application. The efficiency and accuracy of these approaches in the SBM have been verified for Laplace equations, Helmholtz equations, acoustic waves, elastic waves, and water waves with complicated-geometry domains.

This paper presents the fast singular boundary method for large-scale wave propagation application. The key issue is to overcome the ill-conditioning dense matrix generated from the SBM. So far, several fast dense matrix solvers have been employed, such as the Fast Fourier Transform (FFT) [7], the fast multipole methods (FMM) [8] dual-level scheme [9] and the multi-level scheme [10] etc.

A brief outline of the paper is as follows: in Section 2, a detailed description of the proposed fast SBM model is presented. Section 3 provides several numerical examples to demonstrate the effectiveness of the proposed method. Finally, in Section 4 some conclusions are drawn from the present analysis.

2 FAST SINGULAR BOUNDARY METHOD FOR WAVE PROPAGATION

Consider the time-harmonic wave propagating in a homogeneous medium D exterior to an object bounded with the closed curve Γ, which can be described by Helmholtz equation:

$$\nabla^2 u(x) + k^2 u(x) = 0, \qquad x \in D, \tag{1}$$

subjected to the boundary conditions

$$u(x) = \overline{u} \qquad x \in \Gamma_D, \tag{2}$$

$$q(x) = \frac{\partial u(x)}{\partial \mathbf{n}} = \overline{q} \qquad x \in \Gamma_N, \tag{3}$$

where u is the complex-valued amplitude of the radiated and/or scattered waves (velocity potential or wave pressure):

$$u=\begin{cases} u_R = u_T, & \text{if radiation,} \\ u_S = u_T - u_I, & \text{if scattering,} \\ u_{R+S} = u_T - u_I, & \text{if both,} \end{cases}$$

in which the subscripts *T, R, S, I* denote the total, radiated, scattered and incident waves, respectively. $k=\omega/c$ is the wavenumber, ω the angular frequency, c the wave speed in homogeneous medium *D*, and $\mathbf{n}$ the outward normal unit vector on the closed curve $\Gamma=\Gamma_D+\Gamma_N$, in which Γ_D and Γ_N represent the boundaries with the essential boundary (Dirichlet) conditions and the natural boundary (Neumann) conditions, respectively. For the exterior wave propagation problems, all scattered and radiated waves should be outgoing. This constraint condition is also known as the well-known Sommerfeld radiation condition:

$$\lim_{r\to\infty} r^{\frac{1}{2}(\dim-1)}\left(\frac{\partial u}{\partial r}-iku\right)=0, \tag{4}$$

where *dim* is the dimension of the acoustic problems, and $i=\sqrt{-1}$.

2.1 Singular boundary method

Similar to the MFS, the SBM approximate solutions *u(x)* and *q(x)* of exterior wave propagation problem (eqns (1) and (2)) can be represented by a linear combination of the singular fundamental solutions or their partial derivation terms as follows:

$$u(x_m)=\begin{cases} \sum_{j=1}^{N}\alpha_j G(x_m,s_j), & x_m\in D\backslash\Gamma, \\ \sum_{\substack{j=1\\ j\neq m}}^{N}\alpha_j G(x_m,s_j)+\alpha_m U_S^{jj}, & x_m\in\Gamma, \end{cases} \tag{5}$$

$$q(x_m)=\frac{\partial u(x_m)}{\partial \mathbf{n}_x}=\begin{cases} \sum_{j=1}^{N}\alpha_j \frac{\partial G(x_m,s_j)}{\partial \mathbf{n}_x}, & x_m\in D\backslash\Gamma, \\ \sum_{\substack{j=1\\ j\neq m}}^{N}\alpha_j \frac{\partial G(x_m,s_j)}{\partial \mathbf{n}_x}+\alpha_m Q_S^{jj}, & x_m\in\Gamma, \end{cases} \tag{6}$$

where *N* is the number of source nodes s_j, α_j the *j*th unknown coefficient, $\mathbf{n}_x$ the outward normal unit vector on the collocation nodes x_m, the fundamental solutions $G(x_m,s_j)=\begin{cases} iH_0(kr_{mj})/4, & \dim=2 \\ e^{ikr_{mj}}/(4\pi r_{mj}), & \dim=3 \end{cases}$, $H_n^{(1)}$ denotes the *n*th order Hankel function of the first kind, the Euclidean distance $r_{mj}=\|x_m-s_j\|_2$. If the collocation nodes and source nodes coincide, i.e., $x_m=s_j$, the well-known singularities in the fundamental solutions are encountered. The concept of the origin intensity factors U_S^{jj} and Q_S^{jj} is introduced into the SBM for avoiding these singularities. Thanks to the same order of the singularities in the Laplace and Helmholtz fundamental solutions [6], the origin intensity factors U_S^{jj} and Q_S^{jj} can be represented by the following asymptotic expressions:

$$U_S^{jj} = U_{S0}^{jj} + B, \quad r_{mj} \to 0, \tag{7}$$

$$Q_S^{jj} = Q_{S0}^{jj}, \quad r_{mj} \to 0, \tag{8}$$

where U_{S0}^{jj} and Q_{S0}^{jj} are the origin intensity factors of the corresponding Laplace equation [6] $B = \begin{cases} -\left(\ln(k/2) + \gamma - i\pi/2\right)/(2\pi), & \dim = 2 \\ ik/(4\pi), & \dim = 3 \end{cases}$ and $\gamma = 0.57721566490153286\cdots$ is the Euler constant. In this paper, the subtracting and adding-back technique is implemented to derive the formulation of the origin intensity factors Q_{S0}^{jj} as follows:

$$Q_{S0}^{jj} = \frac{1}{L_m} - \sum_{\substack{j=1 \\ j \neq m}}^{N} \frac{L_j}{L_m} \frac{\partial G_0(x_m, s_j)}{\partial \mathbf{n}_s}, \tag{9}$$

where the fundamental solutions of Laplace equation are given by $G_0 = \begin{cases} -\ln(r_{mj})/(2\pi), & \dim = 2 \\ 1/(4\pi r_{mj}), & \dim = 3 \end{cases}$, $\mathbf{n}_s$ is the outward unit normal vector on the source nodes s_j, L_j is half-length of the curve $\widehat{s_{j-1}s_{j+1}}$ between the source nodes s_{j-1} and s_{j+1} for 2D problems or the infinitesimal area of the source nodes s_j for 3D problems as shown in Fig. 1.

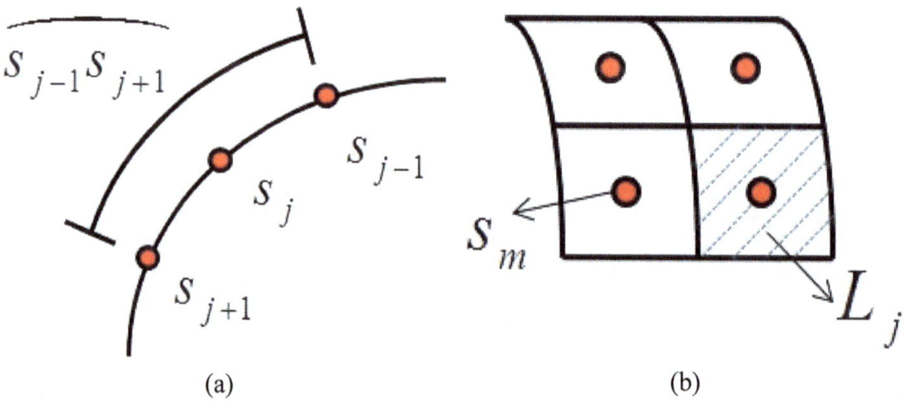

Figure 1: Schematic configuration. (a) Yhe source points s_j and the curve $\widehat{s_{j-1}s_{j+1}}$ for 2D problems; and (b) The source points s_j and the corresponding infinitesimal area L_j for 3D problems.

For the determination of the origin intensity factors U_{S0}^{jj}, the following empirical formulation is used in 2D problems [7]:

$$U_{S0}^{jj} = -\frac{1}{2\pi}\ln\left(\frac{L_j}{2\pi}\right), \tag{10}$$

and the following formulation is used in 3D problems [11]:

$$U_S^{jj} = -\frac{1}{kL_m}\left\{\sum_{\substack{j=1\\ j\neq m}}^{N}\left[G(x_m,s_j)q(s)-\frac{\partial G(x_m,s_j)}{\partial n_s}u(s)\right]L_j - C(x_m)u(x_m)\right\}, \tag{11}$$

where the associated general solution $u(s)$ and its partial derivation terms $q(s)$ can be written as follows:

$$u(s) = \sum_{d=1}^{3}\sin\left(k(s_{jd}-x_{md})\right)\cdot n(x_{md}), \tag{12}$$

$$q(s) = k\sum_{d=1}^{3}\cos\left(k(s_{jd}-x_{md})\right)\cdot n(x_{md})\cdot n(s_{jd}), \tag{13}$$

in which $n(x_{md}), n(s_{jd})$ denotes the Cartesian component of $\mathbf{n}_x$ and $\mathbf{n}_s$, respectively.

By adapting either eqn (5) or (6) at all the collocation nodes, the SBM formulation for exterior wave propagation problems (eqns (1) and (2)) can be discretized to the following linear system of equations:

$$\mathbf{A}\boldsymbol{\alpha} = \mathbf{b}, \tag{14}$$

where $\mathbf{A}$ is the SBM resultant matrix, $\boldsymbol{\alpha}$ and $\mathbf{b}$ denote the unknown coefficient vector and the known boundary condition vector, respectively. After determining the vector $\boldsymbol{\alpha}$, the complex-valued amplitude of velocity potential or wave pressure at any discretization nodes inside the domain or on the surface boundary can be evaluated via eqns (5) and (6).

2.2 Fast Fourier Transform Singular boundary method

For periodic obstacles, the Fast Fourier Transform can be used to accelerate the SBM matrix computation. Consider two typical periodic soft/rigid obstacles as shown in Fig. 2.

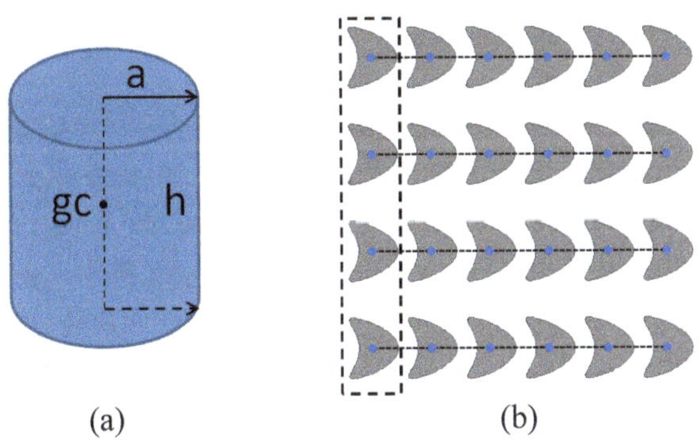

Figure 2: Schematic configuration of two typical periodic structures. (a) Axisymmetric cylinder structure; (b) Four periodic arrays of scatterers.

The surface of the kite-shaped obstacle in Figs 2(b) can be represented as

$$\Gamma=\left\{(x_1,x_2):x_1=\frac{\cos\varphi+0.65\cos 2\varphi-0.65}{6},x_2=0.25\sin\varphi,0\le\varphi\le 2\pi\right\}.$$

Taking a column of four kite-type scatterers as one obstacle, both the resultant matrices $\mathbf{A}$ constructed by the SBM for the typical periodic structures in Fig. 2 are block Toeplitz matrix $\mathbf{T}$ with the form

$$\mathbf{A}=\mathbf{T}=\begin{pmatrix} \mathbf{T}_0 & \mathbf{T}_1 & \cdots & \mathbf{T}_{M-1} \\ \mathbf{T}_{-1} & \mathbf{T}_0 & \ddots & \mathbf{T}_{M-2} \\ \cdots & \ddots & \ddots & \vdots \\ \mathbf{T}_{1-m} & \mathbf{T}_{2-m} & \cdots & \mathbf{T}_0 \end{pmatrix}, \tag{15}$$

where each $\mathbf{T}_i$ denotes an $N\times N$ matrix. The unique block number is $2M-1$, where M is the number of the scatterers in a row of the periodic structures. As we know, the block Toeplitz matrix can be embedded into a larger block circulant matrix $\mathbf{C}'$

$$\mathbf{C}'=\begin{pmatrix} \mathbf{T}_0 & \mathbf{T}_1 & \cdots & \mathbf{T}_{-2} & \mathbf{T}_{-1} \\ \mathbf{T}_{-1} & \mathbf{T}_0 & \mathbf{T}_1 & \ddots & \mathbf{T}_{-2} \\ \vdots & \mathbf{T}_{-1} & \ddots & \ddots & \vdots \\ \mathbf{T}_2 & \vdots & \ddots & \ddots & \mathbf{T}_1 \\ \mathbf{T}_1 & \mathbf{T}_2 & \cdots & \mathbf{T}_{-1} & \mathbf{T}_0 \end{pmatrix}. \tag{16}$$

According to the property of block circulant matrix, only the first block row $\mathbf{R}'=[\mathbf{T}_0 \ \ \mathbf{T}_1 \ \ \cdots \ \ \mathbf{T}_{M-1} \ \ \mathbf{T}_{1-M} \ \ \mathbf{T}_{2-M} \ \ \cdots \ \ \mathbf{T}_{-1}]$ of $\mathbf{C}'$ needs to be computed and stored. Obviously, the computational storage can be reduced to $(2M-1)/M^2$ storage of the full SBM resultant matrix $\mathbf{T}$. Eqn (14) can be rewritten as follows:

$$\mathbf{C}'\boldsymbol{\alpha}'=\mathbf{b}'\Leftrightarrow\begin{pmatrix} \mathbf{C}'_{11} & \mathbf{C}'_{12} \\ \mathbf{C}'_{21} & \mathbf{C}'_{22} \end{pmatrix}\begin{pmatrix} \boldsymbol{\alpha} \\ \mathbf{0} \end{pmatrix}=\begin{pmatrix} \mathbf{b} \\ \mathbf{d} \end{pmatrix}, \tag{17}$$

where $\mathbf{C}'_{11}$=$\mathbf{T}$ and $\mathbf{d}$=$\mathbf{C}'_{21}\boldsymbol{\alpha}$. It should be mentioned that the linear system of equations (eqn (17)) cannot be solved directly due to the unknown vectors $\boldsymbol{\alpha}$ and $\mathbf{d}$ appeared in both the left-hand and right-hand side vectors $\boldsymbol{\alpha}'$ and $\mathbf{b}'$. Therefore, the following simple iterative algorithm is implemented to determine unknown vectors $\boldsymbol{\alpha}$ via eqn (17).

i) Substitute the known elements of $\mathbf{b}'$ into $\mathbf{b}'_k$, namely, replace $\mathbf{b}_k$ by $\mathbf{b}$ and apply the FFT to compute the following block circulant matrix-vector multiplication:

$$\mathbf{C}'^{-1}\mathbf{b}'_k=\boldsymbol{\alpha}'_k\Leftrightarrow\mathbf{C}'^{-1}\begin{pmatrix} \mathbf{b}_k \\ \mathbf{d}_k \end{pmatrix}=\begin{pmatrix} \boldsymbol{\alpha}_k \\ \mathbf{e}_k \end{pmatrix}. \tag{18}$$

ii) Substitute the known elements of $\boldsymbol{\alpha}'$ into $\boldsymbol{\alpha}'_k$, namely, replace $\mathbf{e}_k$ by zero vector and apply the FFT to compute the following block circulant matrix-vector multiplication:

$$\mathbf{C}'\boldsymbol{\alpha}'_k = \mathbf{b}'_{k+1} \Leftrightarrow \mathbf{C}'\begin{pmatrix}\boldsymbol{\alpha}_k \\ \mathbf{0}\end{pmatrix} = \begin{pmatrix}\mathbf{b}_{k+1} \\ \mathbf{d}_{k+1}\end{pmatrix}. \tag{19}$$

iii) Stop the iteration process if the convergence is achieved with

$$\left\| \begin{pmatrix}\mathbf{b}_{k+1} - \mathbf{b}_k \\ \mathbf{e}_k\end{pmatrix} \right\|_\infty < \tau_1, \tag{20}$$

where τ_1 denotes the convergence tolerance. Otherwise restart the loop.

2.3 Fast multipole singular boundary method

The fast multipole method (FMM) has been proposed by Greengard and Rokhlin. It consists of computing nearby solutions and far-fields solutions and can be employed to dramatically accelerate the solution procedure of the SBM system of equations (eqn (14)), in which matrix **A** is a dense and non-symmetric matrix. The main idea of the fast multipole SBM is to employ iterative solvers (such as GMRES) to solve the matrix system of equations and then employ the FMM to accelerate the matrix-vector multiplication in each iteration step, without forming the matrix **A** explicitly in general.

In the fast multipole SBM, one can rewrite fundamental solutions as follows:

$$G\left(x_m, s_j\right) = \sum_i G_i^{(1)}\left(x_m, z\right) G_i^{(2)}\left(s_j, z\right). \tag{21}$$

Then the node-to-node computations in the conventional SBM are replaced by cell-to-cell interactions using a hierarchical tree structure of cells containing groups of nodes. This is possible by introducing the multipole and local expansions of the kernels and employing certain translations. The roadmap of the FMMSBM can be written as follows:

Step 1 Generate a hierarchical tree partitioning for computational domain;

Step 2 Upward pass. Accumulate multipole expansions for far field interactions through a backward visiting of the tree;

Step 3 Downward pass. Translate multipole moments to local expansions, and construct local expansions through a forward visiting of the tree;

Step 4 Evaluation of all the interactions. Evaluate far-field interactions to particles by local expansions, and evaluate directly near-field interactions.

The details of the FMMSBM formulation and implementation can be found in the literatures [8]. However, for large-scale wave propagation analysis, the FMMSBM still produces a very ill-conditioning resultant matrix, which slows down the convergence rate of matrix-iterative solvers even leads to incorrect results.

2.4 Dual-level singular boundary method

To further improve the computable ability of the FMMSBM, the dual level preconditioning technique is introduced to transform the fully populated resultant matrix to a local-supported sparse matrix on the fine level, which may overcome the highly ill-conditioning and excessive storage requirement in the FMMSBM solution of large-scale wave propagation. The procedure of the DLFMMSBM (Fig. 3) includes the following steps:

Step 1 (on the coarse level) Evaluation of initial approximate solution on the fine level;
Step 2 (on the fine level) Evaluation of initial accurate solution on the fine level;
Step 3 (on the fine level) Assessment of initial average relative error on the fine level;
Step 4 (on the fine level) Evaluation of kth accurate residual potential on coarse level;
Step 5 (on the coarse level) Evaluation of kth residual potential solution on the coarse level;
Step 6 (on the fine level) Evaluation of (k+1)th accurate solution on fine level;
Step 7 (on the fine level) Assessment of *(k +1)*th average relative error on the fine level.

The details of the DLFMMSBM formulation and implementation can be found in the literatures [9].

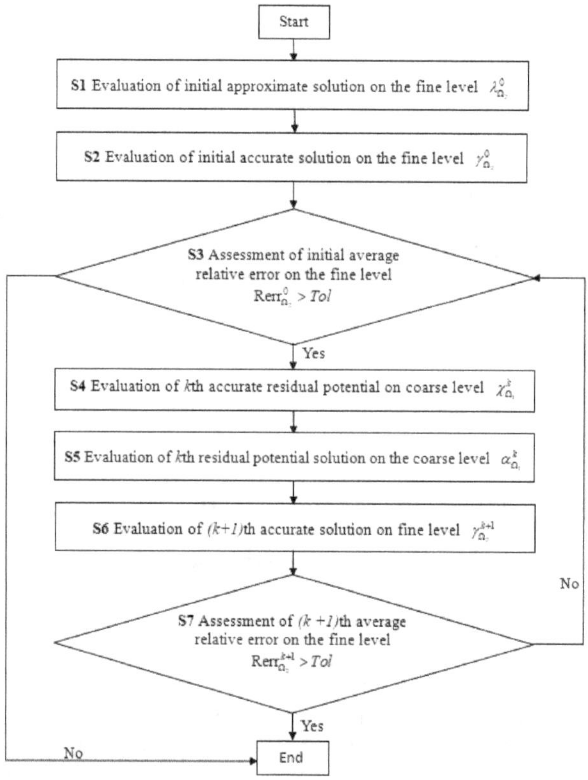

Figure 3: Algorithm diagram of the DLFMMSBM.

3 NUMERICAL RESULTS AND DISCUSSIONS

In this section, the efficiency and accuracy of the FFTSBM are first tested on wave propagation by 3D cylinder structures and four periodic arrays of kite-shaped scatterers. Subsequently, the numerical comparison between DLFMMSBM and FMMSBM is given for wave radiation from a submarine model. The numerical accuracy is measured by the maximum error *Merr(u)* defined as

$$Merr(u) = \max_{1 \le i \le NT} \left| u(i) - \overline{u}(i) \right|, \tag{22}$$

where $\overline{u}(i)$ and $u(i)$ are the analytical and numerical solutions at x_i, respectively, and NT is the total number of the test points in the interested domain. The test points $\{\mathbf{y}_j\}$ are placed on a similar geometric surface to the scatterer with its center at the geometric center gc of the scatterer, and the off-distance parameter od is defines as

$$od = \frac{\mathbf{y}_j - gc}{\mathbf{y}_j^0 - gc}, \tag{23}$$

where $\{\mathbf{y}_j^0\}$ are the points with a uniform angular distribution on the surface of the scatterer. Unless otherwise specified, NT=200, od=2, $\tau_1=10^{-4}$ and $Tol=10^{-4}$ are used in all the following numerical examples on a personal computer with an Intel Core i7 Processor with 16GB RAM.

Example 1: 3D wave radiation by a circular cylinder with soft or rigid boundaries.

First consider the wave propagation by a circular cylinder as shown in Fig. 2(a), where the pressure (u) and the normal velocity (q) on the cylinder surface are induced by a point source of the spherical dilatation wave with the unit intensity located at the geometric center. The analytical solution is presented by

$$u_R(r_{gc}, \theta_{gc}, x_3) = e^{ikr_{gc}} / r_{gc}, \tag{24}$$

where (r_{gc}, θ_{gc}) denotes the polar coordinates with the geometric center gc as the origin of the plane $x_3=0$. Fig. 4 plots the convergence rate of the FFTSBM with the CPU time versus boundary node number for 3D wave propagation problem with $ka = 3/16$ and $a = 3h/16$.

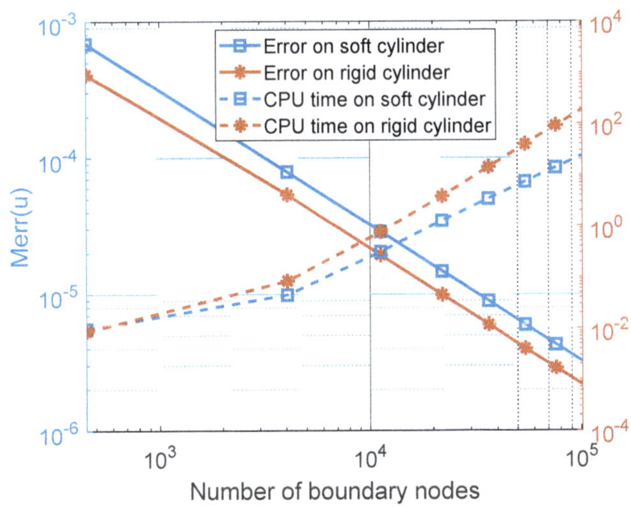

Figure 4: Convergence rate of the FFTSBM with the CPU time versus the boundary node number.

WIT Transactions on Engineering Sciences, Vol 126, © 2019 WIT Press
www.witpress.com, ISSN 1743-3533 (on-line)

It can be observed from Fig. 4 that the FFTSBM has a rapid convergence rate with the curve-slope 0.85, and it only requires about 3 minutes for periodic structure with 1 million boundary nodes.

Example 2: 2D wave scattering by 4×6 infinite kite-shaped impermeable cylinders.

Next consider the wave interaction with four periodic arrays of infinite kite-shaped impermeable scatterers as shown in Fig. 2(b), which can be simplified to 2D scattering problem of an incident plane wave $e^{ik(x_1\cos\theta+x_2\sin\theta)}$ scattered by 4×6 rigid kite-shaped obstacles. Fig. 5 presents the wave pressure amplitude $|u|$ around this four-periodic-array structure at the incidence wave angle $\theta=\pi/4$ and the dimensionless wavenumber $ka=\pi/2$ by using 50 boundary nodes on each kite-shaped obstacle surface.

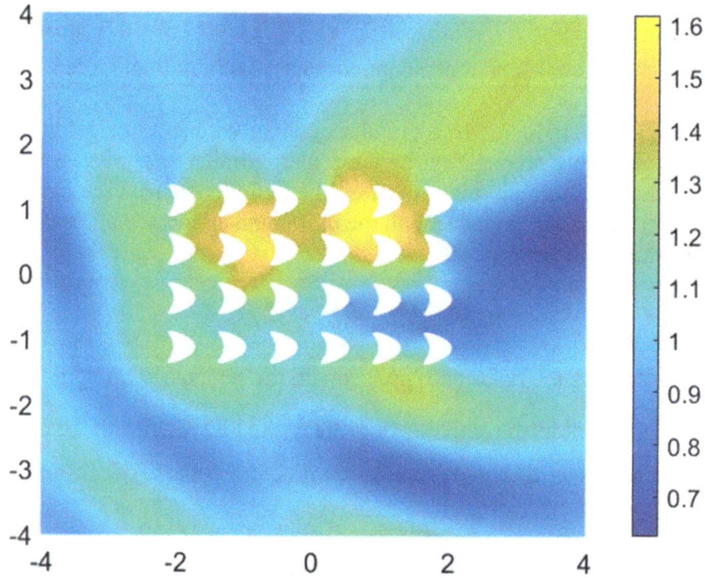

Figure 5: Wave pressure amplitude $|u|$ for the 4×6 kite-shaped cylinder array at the incidence wave angle $\theta=\pi/4$ and the dimensionless wavenumber $ka=\pi/2$ by using FFTSBM.

Example 3: 3D wave radiation by a submarine model (Fig. 6)

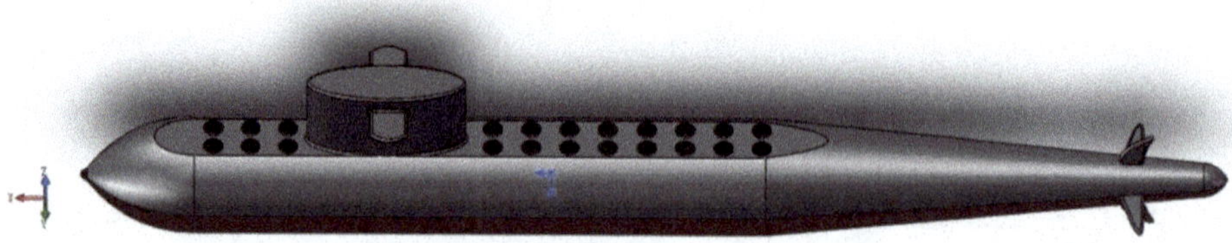

Figure 6: A submarine model.

We consider the wave propagation by a submarine model with a size of 120m×12m×12m as shown in Fig. 6, where acoustic wave speed is set as 1500m/s. The analytical solution is given by

$$u_R(x_1,x_2,x_3)=x_3e^{ikr}\left(1+\frac{i}{kr}\right)\Big/r^2, \tag{25}$$

The test points are uniformly distributed on the surface of sphere with radius 100m. Table 1 lists the numerical comparison between FMMSBM and DLFMMSBM in Example 3 at two certain wave frequencies (*f*=87 and 217 Hz). It can be found from Table 1 that the FMM is introduced to allow the SBM to solve the large-scale wave propagation by complicated-geometry structure with more than 1 million boundary nodes. Moreover, the DLFMMSBM provides more accurate solutions with less CPU time and fewer boundary node requirement than those of the FMMSBM.

Table 1: Numerical comparison between FMMSBM and DLFMMSBM in Example 3 at two certain wave frequencies (*f*=87 and 217 Hz).

	DLFMM+SBM		FMM+SBM [8]	
kd (*d=120m*)	44.4	110.9	44.4	110.9
f (*Hz*)	87	217	87	217
Coarse-level nodes	5053	8019	/	/
Fine-level nodes	65,058	100,096	63,488	1,030,876
Rerr (*Tol*=1e-4)	1.93E–4	5.04E– 4	1.13E-3	2.30E-2
CPU time (*s*)	6.76E+1	2.09E+2	7.57E+2	9.24E+3

4 CONCLUSIONS

This paper presents the singular boundary method (SBM) in conjunction with the fast dense matrix solvers for large-scale wave propagation analysis. The Fast Fourier Transform (FFT) is introduced to the proposed SBM for solving periodic structure with 1 million boundary nodes about 3 minutes. The fast multipole method allows the SBM to solve the large-scale wave propagation by complicated-geometry structure with more than 1 million boundary nodes about 150 minutes. Furthermore, the dual-level preconditioning scheme enables the FMMSBM only to place 2 – 3 coarse-level nodes to obtain the acceptable numerical results. This node requirement almost approaches the minimum requirement specified by the Shannon's sampling theorem. It is worth noting that the proposed DLFMMSBM for large-scale wave propagation analysis with complicated-geometry structure at high wavenumber is still under intense study.

ACKNOWLEDGEMENTS

The work described in this paper was supported by the National Science Fund of China (Grant No. 11772119, 11572111), Alexander von Humboldt Research Fellowship (ID: 1195938) and Qing Lan Project.

REFERENCES

[1] Ciskowski, R.D. & Brebbia, C.A. (eds), *Boundary Element Methods in Acoustics*, Springer-Verlag: Berlin and New York, 1991.

[2] Fairweather, G., Karageorghis, A. & Martin, P.A., The method of fundamental solutions for scattering and radiation problems. *Engineering Analysis with Boundary Elements*, **27**(7), pp. 759–769, 2003.

[3] Chen, W., Singular boundary method: a novel, simple, meshfree, boundary collocation numerical method. *Acta Mechanica Solida Sinica (Chinese)*, **30**, pp. 592–599, 2009.

[4] Fu, Z.J. & Chen, W., A novel boundary meshless method for radiation and scattering problems. *Advances in Boundary Element Techniques XI*, pp. 83–90, 2010.

[5] Fu, Z.J., Chen, W., Chen, J.T. & Qu, W.Z., Singular boundary method: three regularization approaches and exterior wave applications. *Computer Modeling in Engineering & Sciences*, pp. 1–26, 2014.

[6] Fu, Z.J., Chen, W. & Gu, Y., Burton–Miller-type singular boundary method for acoustic radiation and scattering. *Journal of Sound and Vibration*, **333**(16), pp. 3776–3793, 2014.

[7] Fu, Z., Chen, W., Wen, P. & Zhang, C., Singular boundary method for wave propagation analysis in periodic structures. *Journal of Sound and Vibration*, **425**, pp. 170–188, 2018.

[8] Qu, W., Chen, W. & Zheng, C., Diagonal form fast multipole singular boundary method applied to the solution of high-frequency acoustic radiation and scattering. *International Journal for Numerical Methods in Engineering*, **111**(9), pp. 803–815, 2017.

[9] Li, J., Chen, W. & Fu, Z., A modified dual-level algorithm for large-scale three-dimensional Laplace and Helmholtz equation. *Computational Mechanics*, **62**(4), pp. 893–907, 2018.

[10] Li, J., Chen, W., Qin, Q.H. & Fu, Z., A modified multilevel algorithm for large-scale scientific and engineering computing. *Computers & Mathematics with Applications*, **77**(8), pp. 2061–2076, 2019.

[11] Li, J., Chen, W., Fu, Z., & Qin, Q.H., A regularized approach evaluating the near-boundary and boundary solutions for three-dimensional Helmholtz equation with wideband wavenumbers. *Applied Mathematics Letters*, **91**, pp. 55–60, 2019.

SINGULAR BOUNDARY METHOD IN A FREE VIBRATION ANALYSIS OF COMPOUND LIQUID-FILLED SHELLS

VASYL I. GNITKO[1], KYRYL G. DEGTYARIOV[1], ARTEM O. KARAIEV[2*] & ELENA A. STRELNIKOVA[1,2†]
[1]A.N. Podgorny Institute of Mechanical Engineering Problems of the Ukrainian Academy of Sciences, Ukraine
[2]V.N. Karazin Kharkiv National University, Ukraine

ABSTRACT

A new numerical approach is proposed to address problems of free liquid vibrations in axisymmetric compound rigid shells using a singular boundary method. The liquid is supposed to be perfect and incompressible, and its flow is irrotational so the liquid velocity can be presented as a potential gradient. The approximation of a small fluid surface elevation is used, and the free surface function is presented as a sum of the fluid-filled shell height and the small elevation function. A series expansion of the potential function about stationary states is used. To find the stationary states, an eigenvalue problem is formulated. Eigenvalues and eigenvectors are obtained using the singular boundary method with the origin intensity factor as the singular integral over the singular boundary element. The numerical results for free vibration analysis of cylindrical shells, obtained by singular boundary method and direct boundary element methods, are compared. Different compound rigid shells are considered in numerical simulations of free liquid vibrations.

Keywords: free vibrations, axisymmetric problems, potential theory, singular boundary method, compound shells, surface tension, original intensity factor.

1 INTRODUCTION

The boundary-type methods are widely used in different research areas, such as potential theory, elasticity and electromagnetic theories, heat transfer, fracture mechanics, fluid dynamics, biomedicine [1], [2]. Compared with domain-type techniques, such as finite elements methods, boundary element methods (BEM) have certain advantages as stability of calculation processes, high accuracy, fast convergence, and boundary only discretization, i.e., reducing the dimension of problems by one and so essentially shortening computer time and storage. However, the mathematical tools used in boundary element formulations often seem to be unfamiliar and difficult to customers, particularly at numerical simulations based on evaluating singular and hypersingular integrals. In recent years, the alternative boundary-type techniques are developed such as mesh-free and integration-free approaches [3]–[5]. Both of them use the method of fundamental solution (MFS) with prescribed source point locations as basic concept. The MFS is one of the collocation-based boundary-type, mesh-free method. In order to avoid the singularity of fundamental solutions, the source points are located on fictitious boundaries inside or outside the calculation domain relatively to internal or external boundary value problems. This method is mathematically simple, because it does not require thorough gridding and avoids calculation of complicated singular integrals. But the main problem of MFS is in proper choosing the set of source points. The new concept of origin intensity factors (OIFs) to isolate the singularity of fundamental solutions is introduced in singular boundary method (SBM) [3], [6]. In this method the source points are placed on the boundaries in coincidence with collocation points. The OIF concept is first proposed in Chen et al. [3], where an inverse interpolation technique is developed with using the set of sample points

* *ORCID: https://orcid.org/0000-0003-3176-8496*
† *ORCID: https://orcid.org/0000-0003-0707-7214*

WIT Transactions on Engineering Sciences, Vol 126, © 2019 WIT Press
www.witpress.com, ISSN 1743-3533 (on-line)
doi:10.2495/BE420171

placed inside the domain. In recent research the SBM is successfully applied to potential [7], heat conduction [8], acoustic stokes flow [9], and biharmonic [10] problems.

In this paper, the problem of free liquid vibrations is formulated and solved by using the direct and indirect traditional BEM, and with singular boundary method. Here mixed boundary value problems are considered.

2 PROBLEM STATEMENT OF LIQUID VIBRATIONS IN RIGID SHELLS OF REVOLUTION

2.1 Boundary value problem

The problem of free harmonic axisymmetric liquid vibrations in rigid compound shells of revolution is considered. We designate a moistened shell surface by S_1, and a free surface by S_0. Suppose the Cartesian coordinate system $0xyz$ is connected with the shell, the liquid free surface S_0 coincides with the plane $z = H$ at the state of rest, Fig. 1.

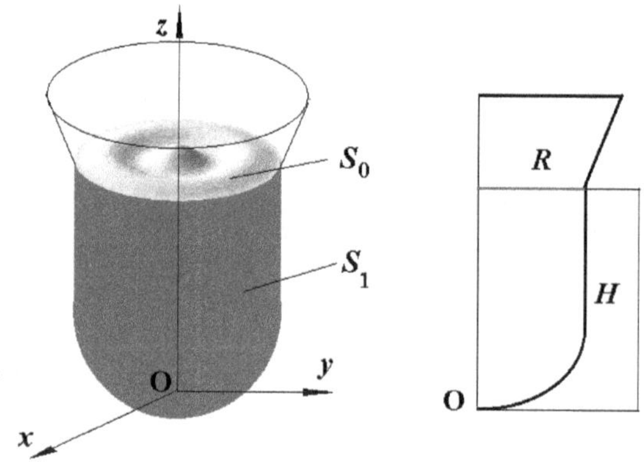

Figure 1: Liquid-filled shell of revolution and its draft.

Fluid velocity v inside inelastic shells of revolutions satisfies Euler's system of equations for an ideal fluid [11]:

$$\frac{\partial \rho}{\partial t} + \operatorname{div} \rho \mathbf{v} = 0, \quad \rho\left(\frac{\partial \mathbf{v}}{\partial t} + (\mathbf{v}, \nabla)\mathbf{v}\right) = -\nabla P + \mathbf{F}. \tag{1}$$

Here, F is an external force acting on the dynamical system. Let g be the gravity acceleration, ρ_l is for the liquid density, and r is a radius-vector of the point inside the liquid domain. Suppose that the liquid is incompressible, so $\rho_l = \text{const}$, and the fluid motion is irrotational so there exist the velocity potential $\mathbf{v} = \nabla\varphi$. Assume that the fluid is located in the gravity field, and the only external force acting on the system is the gravity force $\mathbf{F} = \rho_l \mathbf{g}$.

With these assumptions, Euler's system (1) is transformed to [11]:

$$\Delta\varphi = 0, \quad \nabla\left(\frac{\partial \varphi}{\partial t} + \frac{1}{2}|\nabla\varphi|^2 + \frac{1}{\rho_l}P - (\mathbf{g}, \mathbf{r})\right) = 0. \tag{2}$$

Last equality in eqn (2) leads to the Bernoulli equation:

$$\frac{\partial\varphi}{\partial t}+\frac{1}{2}|\nabla\varphi|^2+\frac{1}{\rho}P-(\mathbf{g},\mathbf{r})=C(t). \tag{3}$$

The constant $C(t)$ on the right-hand side of eqn (3) can be absorbed in the potential. We can set this constant as the derivative of another function $B(t)$:

$$C(t)=\frac{dB(t)}{dt},\quad \varphi(\mathbf{r},t)\to\varphi(\mathbf{r},t)-B(t). \tag{4}$$

The gravity acceleration is directed against Oz axis, so the Bernoulli equation takes the following form:

$$\frac{\partial\varphi}{\partial t}+\frac{1}{2}|\nabla\varphi|^2+\frac{1}{\rho_l}P+gz=0. \tag{5}$$

The Laplace equation for the potential φ is the governing equation of the fluid motion. When solution of this equation is received, then the fluid velocity v can be obtained as the gradient of the velocity potential φ, and the pressure p can be defined from Bernoulli's equation (5).

So it is necessary to formulate the boundary conditions for the Laplace equation. The shell walls are inelastic, so the velocity v has only tangential component, and [12]:

$$\left.\frac{\partial\varphi}{\partial\mathbf{n}}\right|_{S_1}=0, \tag{6}$$

where n is an external unit normal to the shell wetted surface.

There are some difficulties with the free surface boundary conditions. Firstly, introduce an unknown function ζ that describes the z-coordinate of the free surface. Any point on the free surface is defined by its radius-vector r that in the axisymmetric case has two components, $\mathbf{r}=\mathbf{r}(\rho(t),\zeta(\rho,t))$.

To define Euler's velocity in this point, we need to take the following total time derivative:

$$\mathbf{v}(\mathbf{r},t)=\frac{d\mathbf{r}}{dt}=\left(\frac{d\rho}{dt},\frac{d\zeta}{dt}\right). \tag{7}$$

This procedure gives two following identities for the velocity on the free surface:

$$v_\rho\big|_{S_0}=\frac{d\rho}{dt}=\left.\frac{\partial\varphi}{\partial\rho}\right|_{S_0},\ v_z\big|_{S_0}=\frac{d\zeta}{dt}=\frac{\partial\zeta}{\partial t}+\frac{\partial\zeta}{\partial\rho}\frac{d\rho}{dt}=\frac{\partial\zeta}{\partial t}+v_\rho\frac{\partial\zeta}{\partial\rho}=\frac{\partial\zeta}{\partial t}+\left.\frac{\partial\varphi}{\partial\rho}\right|_{S_0}\frac{\partial\zeta}{\partial\rho}=\left.\frac{\partial\varphi}{\partial z}\right|_{S_0}. \tag{8}$$

So, the time derivative of the free-surface function $\zeta(\rho,t)$ satisfies the next identity:

$$\frac{\partial\zeta}{\partial t}=\left.\frac{\partial\varphi}{\partial z}\right|_{S_0}-\left.\frac{\partial\varphi}{\partial\rho}\right|_{S_0}\frac{\partial\zeta}{\partial\rho}. \tag{9}$$

Now we define a unit outward normal to the free surface as below:

$$\mathbf{n}\big|_{S_0} = \left(-\frac{\partial \zeta}{\partial \rho}, 1\right) \Big/ \sqrt{1+\zeta_\rho'^{\,2}}. \tag{10}$$

From here we get relationship between Euler's and Lagrange's coordinates on the free surface:

$$\frac{\partial \zeta}{\partial t} == \frac{1}{\sqrt{1+\zeta_\rho'^{\,2}}}\left(n_\rho \frac{\partial \varphi}{\partial \rho}\bigg|_{S_0} + n_z \frac{\partial \varphi}{\partial z}\bigg|_{S_0} \right) = \frac{1}{\sqrt{1+\zeta_\rho'^{\,2}}} \frac{\partial \varphi}{\partial \mathbf{n}}\bigg|_{S_0}. \tag{11}$$

Combining these conditions to get an equation with the potential function only, we need to take a time derivative from Bernoulli eqn (5) on the free surface:

$$\frac{\partial^2 \varphi}{\partial t^2}\bigg|_{S_0} + \frac{1}{2}\frac{\partial |\nabla \varphi|^2}{\partial t}\bigg|_{S_0} + \frac{1}{\rho_l}\frac{\partial P}{\partial t}\bigg|_{S_0} = 0. \tag{12}$$

As the pressure on the free surface is equal to atmospheric one, its time derivative can be expressed in the following manner:

$$\frac{\partial P}{\partial t}\bigg|_{S_0} + \frac{\partial P}{\partial z}\bigg|_{S_0} \frac{\partial \zeta}{\partial t} = \frac{\partial P_0}{\partial t} = 0 \Rightarrow \frac{\partial P}{\partial t}\bigg|_{S_0} = -\frac{\partial P}{\partial z}\bigg|_{S_0} \frac{\partial \zeta}{\partial t} = -\frac{\partial P}{\partial z}\bigg|_{S_0} \frac{1}{\sqrt{1+\zeta_\rho'^{\,2}}} \frac{\partial \varphi}{\partial \mathbf{n}}\bigg|_{S_0}. \tag{13}$$

The second component of the pressure gradient is received from second Euler's equation using eqn (5) as:

$$-\frac{1}{\rho_l}\frac{\partial P}{\partial z} = \frac{\partial v_z}{\partial t} + \frac{1}{2}\left(\frac{\partial v_\rho^2}{\partial z} + \frac{\partial v_z^2}{\partial z} \right) + g = \frac{\partial^2 \varphi}{\partial t \partial z} + \frac{1}{2}\left(\frac{\partial}{\partial z}\left(\frac{\partial \varphi}{\partial \rho}\right)^2 + \frac{\partial}{\partial z}\left(\frac{\partial \varphi}{\partial z}\right)^2 \right) + g. \tag{14}$$

Thus, we receive the next relation:

$$\frac{\partial^2 \varphi}{\partial t^2}\bigg|_{S_0} + \frac{1}{2}\frac{\partial |\nabla \varphi|^2}{\partial t}\bigg|_{S_0} + \left(\frac{\partial^2 \varphi}{\partial t \partial z} + \frac{1}{2}\left(\frac{\partial}{\partial z}\left(\frac{\partial \varphi}{\partial \rho}\right)^2 + \frac{\partial}{\partial z}\left(\frac{\partial \varphi}{\partial z}\right)^2 \right) + g \right)\Bigg|_{S_0} \cdot \frac{1}{\sqrt{1+\zeta_\rho'^{\,2}}} \frac{\partial \varphi}{\partial \mathbf{n}}\bigg|_{S_0} = 0. \tag{15}$$

Considering the small vibrations one can neglect the velocity's squares. In assumption that the pressure gradient along z-axis depends only on the gravitational field, the velocity alterations haven't large influence on pressure. So, the linear boundary condition on the free surface takes the following form:

$$\frac{\partial^2 \varphi}{\partial t^2}\bigg|_{S_0} + g\frac{\partial \varphi}{\partial \mathbf{n}}\bigg|_{S_0} = 0. \tag{16}$$

We have got the complete boundary and initial conditions for the dynamical system, and finally the boundary value problem for the Laplace equation:

$$\Delta \varphi = 0; \quad \frac{\partial^2 \varphi}{\partial t^2}\bigg|_{S_0} + g\frac{\partial \varphi}{\partial \mathbf{n}}\bigg|_{S_0} = 0; \quad \frac{\partial \varphi}{\partial \mathbf{n}}\bigg|_{S_1} = 0; \quad \frac{\partial \zeta}{\partial t} = \frac{\partial \varphi}{\partial \mathbf{n}}\bigg|_{S_0}. \tag{17}$$

The following solvability condition needs to be added [12]:

$$\int_{S_0} \frac{\partial \varphi}{\partial \mathbf{n}} dS_0 = 0. \tag{18}$$

It would be noted that the formulation of the boundary value problem in eqn (17) is received as the result of linearization of free surface boundary conditions and can be considered as a first approximation of nonlinear sloshing in the shells.

2.2 Stationary states

To find stationary states of system (17), we separate the velocity potential into spatial and temporal functions [13]:

$$\varphi(\mathbf{r},t) = \psi(\mathbf{r})e^{i\omega t}, \quad i^2 = -1. \tag{19}$$

Using eqn (19), the following boundary value problem for the function ψ is received:

$$\Delta\psi = 0, \quad \left.\frac{\partial \psi}{\partial \mathbf{n}}\right|_{S_0} = \frac{\omega^2}{g}\psi\Big|_{S_0}, \quad \left.\frac{\partial \psi}{\partial \mathbf{n}}\right|_{S_1} = 0, \quad \iint_{S_0} \frac{\partial \psi}{\partial \mathbf{n}} dS_0 = 0. \tag{20}$$

These relations lead to an eigenvalue problem. Its eigenvalues ω_k are natural frequencies of sloshing, and its eigenvectors $\psi_k(\mathbf{r})$ are sloshing natural modes. Each of these eigenvectors is the solution of system (20), and every velocity potential can be expanded in terms of potentials of the stationary states as below [13]:

$$\varphi(\mathbf{r},t) = \sum_{k=1}^{\infty} \alpha_k \varphi_k(\mathbf{r},t) = \sum_{k=1}^{\infty} \alpha_k \psi_k(\mathbf{r}) e^{i\omega_k t}. \tag{21}$$

Expression for the surface function $\zeta(\rho,t)$ is obtained from the Bernoulli equation on the free surface, namely we have:

$$\zeta(\rho,t) = -\frac{1}{g}\left.\frac{\partial \varphi}{\partial t}\right|_{S_0} = -\frac{1}{g}\left.\frac{\partial \varphi}{\partial t}\right|_{z=\zeta(\rho,t)} = -\frac{i}{g}\sum_{k=1}^{\infty} \alpha_k \omega_k \varphi_k(\rho, \zeta(\rho,t), t). \tag{22}$$

To find unknown coefficients in (22) we need to know the initial shape of the surface function:

$$\zeta(\rho,t_0) = \zeta_0(\rho) = -\frac{i}{g}\sum_{k-1}^{\infty} \alpha_k \omega_k \varphi_k(\rho, \zeta_0(\rho), t_0). \tag{23}$$

Using the approximation of small fluid surface elevation, one can present the free surface function as the sum of the fluid-filled shell height and a small elevation function $\varepsilon(\rho,t)$:

$$\zeta(\rho,t) = H + \varepsilon(\rho,t). \tag{24}$$

We can expand $\varphi_k(\rho,z,t)$ into Taylor series near the unperturbed free surface state considering constant values of ρ and *t*:

$$\varphi_k(\rho,z,t)=\varphi_k(\rho,H,t)+\left.\frac{\partial\varphi_k}{\partial z}\right|_{z=H}(z-H)+\frac{1}{2}\left.\frac{\partial^2\varphi_k}{\partial z^2}\right|_{z=H}(z-H)^2+\cdots. \quad (25)$$

It is possible to retain the only linear members in expansion (25) in the point $z=\zeta(\rho,t)$ due to the small values of $\varepsilon^2(\rho,t)$ and higher terms:

$$\varphi_k(\rho,\zeta(\rho,t),t)=\varphi_k(\rho,H,t)+\left.\frac{\partial\varphi_k}{\partial z}\right|_{z=H}\varepsilon(\rho,t). \quad (26)$$

Using eqn (26), we received the elevation function $\varepsilon(\rho,t)$ from eqn (21) as follows:

$$\zeta(\rho,t)=H-\frac{gH+i\sum_{k=1}^{\infty}\alpha_k\omega_k\varphi_k(\rho,H,t)}{g+i\sum_{k=1}^{\infty}\alpha_k\omega_k\left.\frac{\partial\varphi_k}{\partial z}\right|_{z=H}}. \quad (27)$$

So, we come to the spectral boundary problem on the natural sloshing modes and frequencies. Thus, the problem of liquid vibrations in gravitational field is reduced to finding the natural frequencies and basis eigenvectors.

2.3 Eigen frequencies and eigenvectors problem

To find our own frequencies and eigenvectors, we use a singular boundary method, but not in its classical formulation. Naturally, singular boundary methods are developed from the source density formulation. In this formulation, the unknown function and its normal derivative are presented in the next form [1]:

$$\psi(\xi)=\oint_S\sigma(\mathbf{r})u^*(\xi,\mathbf{r})dS(\mathbf{r}),$$
$$\frac{\partial\psi(\xi)}{\partial\mathbf{n}(\xi)}=2\pi\sigma(\xi)+\oint_S\sigma(\mathbf{r})q^*(\xi,\mathbf{r})dS(\mathbf{r}). \quad (28)$$

Here, $\sigma(\mathbf{r})$ is a surface density of the unknown function $\psi(\xi)$:

$$u^*(\xi,\mathbf{r})=\frac{1}{|\xi-\mathbf{r}|},\quad q^*(\xi,\mathbf{r})=\frac{\partial}{\partial\mathbf{n}(\xi)}\left[\frac{1}{|\xi-\mathbf{r}|}\right]. \quad (29)$$

However, using representation (28) creates difficulties in solving the problem under consideration in corner points, because it is impossible to define correctly the normal vector in corner points in classical sense, and in vicinities of these points the potential normal derivative tends to satisfy both conditions (6) and (16). This means that the normal derivative in corner points cannot satisfy Lipschitz continuity [14] so the surface density is not continuous in these points. If we use above formulation, we need to present integrals in eqn (28) as partial sums, but in the vicinities of the corner points the partial sums tend to disconverge.

In the current research, authors use third Green's identity as governing equation [1]:

$$2\pi\psi(\xi)+\oint_S \psi(\mathbf{r})q^*(\xi,\mathbf{r})dS(\mathbf{r})=\oint_S \frac{\partial\psi(\mathbf{r})}{\partial\mathbf{n}(\mathbf{r})}u^*(\xi,\mathbf{r})dS(\mathbf{r}). \tag{30}$$

We divide boundary into N boundary elements; at that m elements belong to the free surface and others $N-m$ belong to the shell walls. At each boundary element we select the collocation point. In doing so, we represent the surface integrals in eqn (30) as sums of integrals over surfaces $S_i, i=\overline{1,N}$ of each boundary element. For each collocation point the following equation is valid:

$$2\pi\psi(\mathbf{r}_k)+\sum_{i=1}^{N}\int_{S_i}\psi(\mathbf{r})q^*(\mathbf{r}_k,\mathbf{r})dS_i=\sum_{i=1}^{N}\int_{S_i}\frac{\partial\psi(\mathbf{r})}{\partial\mathbf{n}(\mathbf{r})}u^*(\mathbf{r}_k,\mathbf{r})dS_i\,. \tag{31}$$

We apply Cauchy's average theorem to eqn (31) and receive:

$$2\pi\psi(\mathbf{r}_k)+\sum_{i=1}^{N}q^*(\mathbf{r}_k,\mathbf{r}_i)\int_{S_i}\psi(\mathbf{r})dS_i=\sum_{i=1}^{N}u^*(\mathbf{r}_k,\mathbf{r}_i)\int_{S_i}\frac{\partial\psi(\mathbf{r})}{\partial\mathbf{n}(\mathbf{r})}dS_i\,. \tag{32}$$

Using boundary condition (16) one can obtain:

$$\begin{aligned}&2\pi\psi(\mathbf{r}_k)+\sum_{i=1}^{m}q^*(\mathbf{r}_k,\mathbf{r}_i)\int_{S_i}\psi(\mathbf{r})dS_i+\sum_{i=m+1}^{N}q^*(\mathbf{r}_k,\mathbf{r}_i)\int_{S_i}\psi(\mathbf{r})dS_i\\&=\frac{\omega^2}{g}\sum_{i=1}^{m}u^*(\mathbf{r}_k,\mathbf{r}_i)\int_{S_i}\psi(\mathbf{r})dS_i.\end{aligned} \tag{33}$$

Let us introduce the following denominations:

$$\beta_{0i}=\int_{S_i}\psi(\mathbf{r})dS_i \text{ if } S_i\in S_0,\ \beta_{1i}=\int_{S_i}\psi(\mathbf{r})dS_i \text{ if } S_i\in S_1. \tag{34}$$

Then we can express β_{1i} through β_{0i} as:

$$\left(\frac{2\pi\delta_{ik}}{s_k}+Q_{ik}^{*11}\right)\beta_{1k}=\left(\frac{\omega^2}{g}U_{ik}^{*10}-Q_{ik}^{*10}\right)\beta_{0k}, \tag{35}$$

where $s_k=\int_{S_k}dS_k$, $Q_{ik}^{*\alpha\beta}=q^*(\mathbf{r}_i,\mathbf{r}_k)$ and $U_{ik}^{*\alpha\beta}=u^*(\mathbf{r}_i,\mathbf{r}_k)$, $\alpha,\beta=0$ for points on the free surface, and $\alpha,\beta=1$ for side wall points, namely:

$$\begin{aligned}&\alpha=0,\beta=0\Rightarrow\mathbf{r}_i\in S_0,\mathbf{r}_k\in S_0,\alpha=0,\beta=1\Rightarrow\mathbf{r}_i\in S_0,\mathbf{r}_k\in S_1,\\&\alpha=1,\beta=0\Rightarrow\mathbf{r}_i\in S_1,\mathbf{r}_k\in S_0,\alpha=1,\beta=1\Rightarrow\mathbf{r}_i\in S_1,\mathbf{r}_k\in S_1.\end{aligned} \tag{36}$$

Thus, the system to determine the eigenvalues and eigenvectors using singular boundary method is received as following:

$$\begin{gathered}\left(\frac{2\pi\delta_{ik}}{s_k}+Q_{ik}^{*00}-Q_{il}^{*01}\left(\frac{2\pi\delta_{lp}}{s_p}+Q_{lp}^{*11}\right)^{-1}Q_{pk}^{*10}\right)\beta_{0k}\\=\frac{\omega^2}{g}\left(U_{ik}^{*00}-Q_{il}^{*01}\left(\frac{2\pi\delta_{lp}}{s_p}+Q_{lp}^{*11}\right)^{-1}U_{pk}^{*10}\right)\beta_{0k}\end{gathered}\tag{37}$$

In equations beforehand, Einstein's summation convention is applied.

2.4 Origin intensity factors

It is ascertained that the diagonal elements of matrixes $Q_{ik}^{*\alpha\beta}$ and $U_{ik}^{*\alpha\beta}$ cannot be defined in the singular points as Green's function and its normal derivative. To overcome this difficulty, we use the following presentation of diagonal elements as average values of Green's function and its normal derivative:

$$U_{kk}^{*\alpha\beta}=\frac{\int_{S_k}u^*(\mathbf{r}_k,\mathbf{r})dS_k}{s_k},\quad Q_{kk}^{*\alpha\beta}=\frac{\int_{S_k}q^*(\mathbf{r}_k,\mathbf{r})dS_k}{s_k}.\tag{38}$$

These expressions are called origin intensity factors in singular boundary methods.

In the axisymmetric case, we can calculate surface integrals as curve integrals of Green's function integrated by the angle variable [15]. Green's function and its normal derivative, integrated by the angle variable, are expressed in terms of complete elliptic integrals of the first and second kind [15], [16] using the following formulae [17]

$$\iint_{S_k}\frac{\partial}{\partial\mathbf{n}}\left(\frac{1}{|\mathbf{r}_k-\mathbf{r}|}\right)dS_k=\int_{\Gamma_k}\Theta(\mathbf{r}_k,\mathbf{r})\rho(z)d\Gamma_k,\quad \iint_{S_k}\frac{1}{|\mathbf{r}_k-\mathbf{r}|}dS_k=\int_{\Gamma_k}\Phi(\mathbf{r}_k,\mathbf{r})\rho(z)d\Gamma_k,$$

$$\Theta(\mathbf{r}_k,\mathbf{r})=\frac{4}{\sqrt{a+b}}\left\{\frac{1}{2r}\left[\frac{\rho^2-\rho_k^2+(z_k-z)^2}{a-b}E(m)-K(m)\right]n_r+\frac{z_k-z}{a-b}E(m)n_z\right\},$$

$$\Phi(\mathbf{r}_k,\mathbf{r})=\frac{4}{\sqrt{a+b}}K(m),\quad \mathrm{E}(k)=\int_0^{\pi/2}\sqrt{1-k^2\sin^2\theta}\,d\theta,\quad K(m)=\int_0^{\pi/2}\frac{d\theta}{\sqrt{1-m^2\sin^2\theta}},$$

$$a=\rho^2+\rho_k^2+(z-z_k)^2,\quad b=2\rho\rho_k,\quad m^2=\frac{2b}{a+b}.\tag{39}$$

The complete elliptic integral of the second kind $K(m)$ can be presented in the next form near singular points:

$$K(m)=\frac{2}{\pi}\ln\frac{1}{m'}+\frac{4}{\pi}\int_0^1\frac{1}{\sqrt{(1-x^2)(1-m'^2x^2)}}\ln\frac{1}{x}dx,\quad m'^2=1-m^2.\tag{40}$$

So, evaluating the original intensity factors is reduced to calculation of integrals with logarithmic singularity. To treat with this singularity, we use quadrature formulae with quadrature nodes in the roots of orthogonal polynomials with the logarithmic weight [1] and apply the technique described in Gnitko et al. [17].

3 NUMERICAL RESULTS

The problems of sloshing in rigid shells of revolution are solved using boundary element method and singular boundary method for comparison. Cylindrical, cylindrical-spherical, and cylindrical-conic shells are under consideration, Fig. 2.

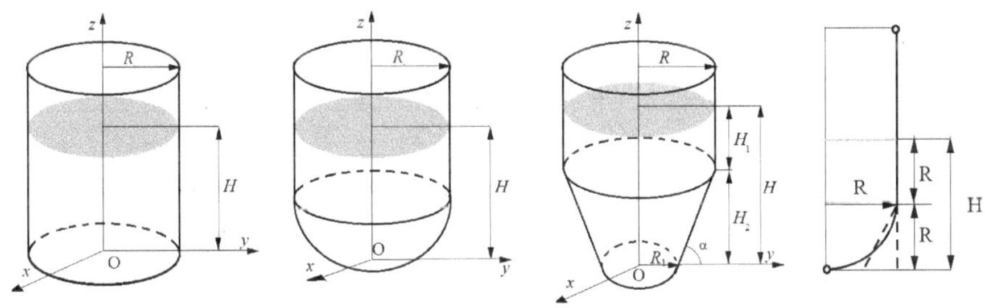

Figure 2: Liquid-filled shells of revolution.

For all shells we suppose that $R=1$ and $H=2$. The height of cylindrical parts for cylindrical-conic and cylindrical-spherical shells is equal to *R*. For the cylindrical-conic shell, we consider $R_1=0.5R$.

Axisymmetric vibrations of liquid are studied. For cylindrical shell, comparison is done between numerical BEM and SBM results, and the following analytical solution [18]:

$$\frac{\omega_k^2}{g}=\frac{\mu_k}{R}\tanh\left(\frac{\mu_k H}{R}\right),\quad \psi_k=J_0\left(\mu_k\frac{\rho}{R}\right)\cosh\left(\frac{\mu_k z}{R}\right). \tag{41}$$

Here values μ_k are the roots of the equations

$$\frac{dJ_0(x)}{dx}=0, \tag{42}$$

where $J_0(x)$ is the Bessel function of the first kind.

Results of comparison with analytical solution (41) for two methods are shown in Table 1 for different numbers *m* of boundary elements on the free surface. Here, the errors (maximal difference between analytical and numerical solutions) for modes $k=\overline{0,4}$ are presented.

In the BEM simulation, the one-dimensional axisymmetrical formulation is in use [15], [16], and the SBM is described beforehand. The equal numbers of elements on the free surface (*m*) and on wetted parts (3*m*) of the shell are applied both in the BEM and SBM.

Analyzing results of Table 1, one can conclude that both methods give errors less than 1%. The results obtained by the boundary element method are more accurate, but the singular boundary method has essential advantages in CPU time calculations.

The modes of vibrations for cylindrical, cylindrical-conic, cylindrical-spherical shells are shown in Figs 3, 4, and 5, respectively. Solid lines present modes of vibrations obtained using the boundary element method, and dashed lines present the singular boundary method's solutions.

It should be noted that all vibration modes demonstrate the Bessel-like behaviour. For all shells of revolution considered in this study, the liquid frequencies of axisymmetric vibrations are very close for each mode.

Table 1: Comparison between analytical and numerical solutions of BEM and SBM.

m		0th mode error	1st mode error	2nd mode error	3rd mode error	4th mode error	CPU time
20	BEM	4.2054e-07	0.0191	0.0094	0.0115	0.0228	0.14
	SBM	0.0033	0.0907	0.0277	0.0250	0.0417	0.043
75	BEM	2.2917e-07	3.9247e-04	0.0010	0.0020	0.0054	1.952
	SBM	9.1500e-04	0.0058	0.0089	0.0122	0.0247	0.474
150	BEM	1.7993e-07	1.3658e-04	6.45e-04	0.0017	0.0018	8
	SBM	4.5902e-04	0.0038	0.0097	0.0186	0.0151	2
400	BEM	1.4327e-07	2.4538e-05	7.98e-05	1.80e-04	1.50e-04	82
	SBM	1.7552e-04	0.0018	0.0032	0.0049	0.0032	29

4 CONCLUSION

The variant of singular boundary method is proposed with a new calculation scheme for the origin intensity factors. The presentation of diagonal elements as average values of Green's function and its normal derivative is used, and the technique is elaborated for numerical estimation of the diagonal elements based on quadrature formulae with nodes in the roots of orthogonal polynomials with the logarithmic weight. This study demonstrates the applicability of the SBM to problems of liquid sloshing in rigid shells of revolution with arbitrary meridians. The emphasis is done to possibility of using SBM for fluid domains with corner points. Numerical experiments proved that the results obtained with proposed SBM are in good agreement with BEM for axisymmetric problems of liquid vibrations. The singular boundary method has the essential advantage in CPU time calculations, does not require thorough gridding and avoids calculation of complicated singular integrals. The proposed scheme of SBM reveals new possibilities for the analysis and modeling the linear and nonlinear process of the liquid movement in different reservoirs.

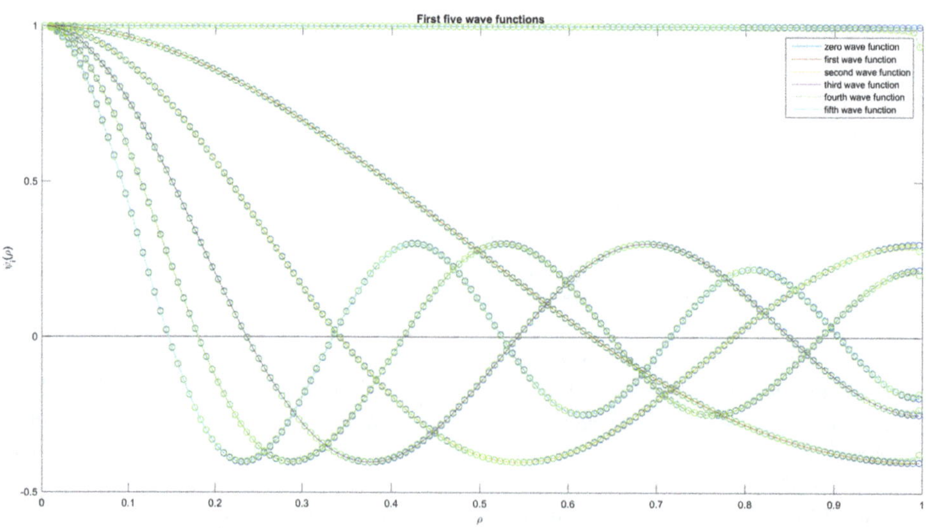

Figure 3: First five eigenmodes for cylindrical shell using BEM and SBM.

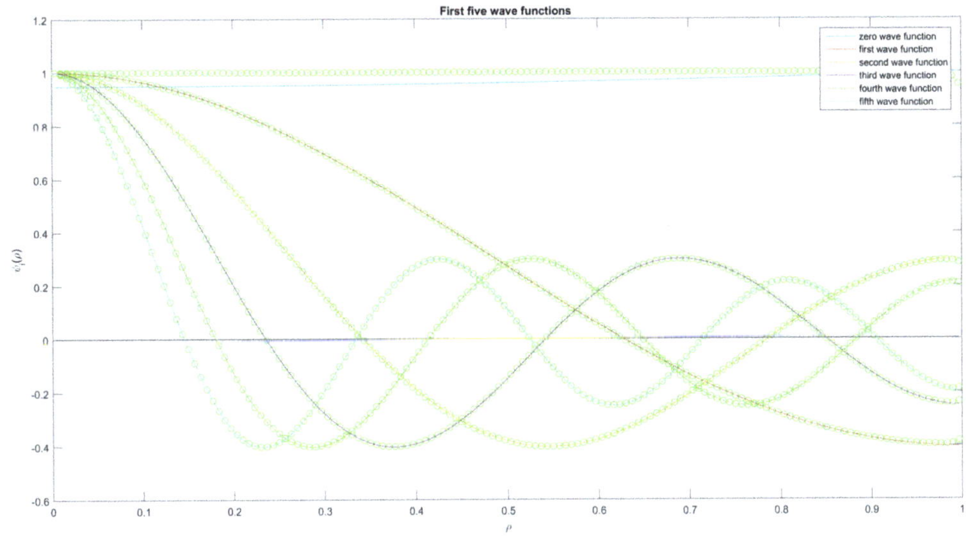

Figure 4: First five eigenmodes for cylindrical-spherical shell using BEM and SBM.

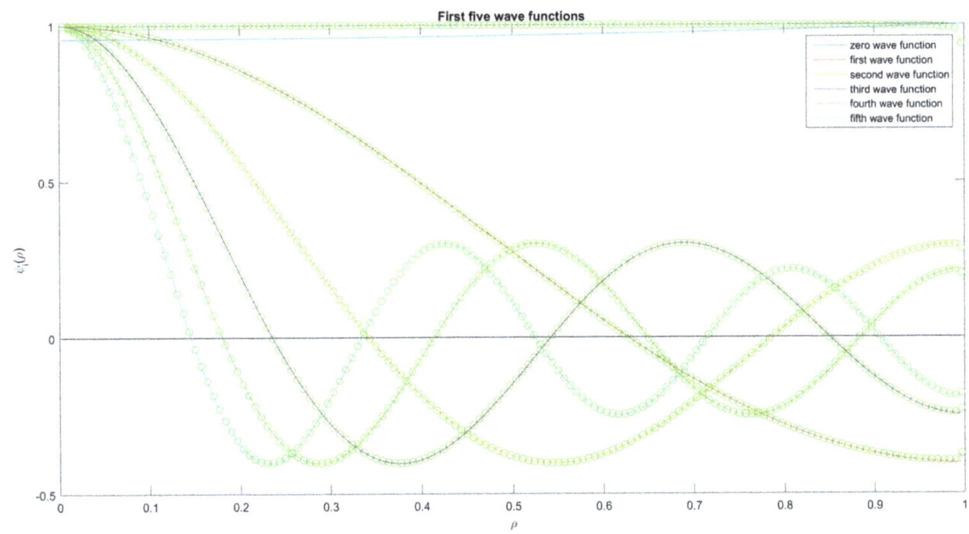

Figure 5: First five eigenmodes for cylindrical-conic shell using BEM and SBM.

ACKNOWLEDGEMENTS

Financial support for the project from STCU under grant No. 6247 is gratefully acknowledged. The authors would also like to thank our foreign collaborator, Professor Alexander Cheng, University of Mississippi, USA, for his constant support and interest to our research.

REFERENCES

[1] Brebbia, C.A., Telles, J.C.F. & Wrobel, L.C., *Boundary Element Techniques: Theory and Applications in Engineering*. Springer-Verlag, 1984.

[2] Katsikadelis, J.T., The *Boundary Element Method for Engineers and Scientists*, 2nd ed., Academic Press, Elsevier: Oxford, 2016.

[3] Chen, W., Zhang, J.Y. & Fu, Z.J., Singular boundary method for modified Helmholtz equations. *Eng. Anal. Bound. Elem.*, **44**, pp. 112–119, 2014.

[4] Chen, W. & Wang, F.Z., A method of fundamental solutions without fictitious boundary. *Engineering Analysis with Boundary Elements*, **34**, pp. 530–532, 2010.

[5] Kleefeld, A. & Pieronekz, L., The method of fundamental solutions for computing acoustic interior transmission eigenvalues. *Inverse Problems*, **34**(3), 2018.

[6] Sun, L., Chen, W. & Cheng, A.H.-D., Evaluating the origin intensity factor in the singular boundary method for three-dimensional dirichlet problems. *Advances in Applied Mathematics and Mechanics*, **9**(6), pp. 1289–1311, 2017.

[7] Gu, Y. & Chen, W., Infinite domain potential problems by a new formulation of singular boundary method. *Applied Mathematical Modelling*, **37**(4), pp. 1638–1651, 2013.

[8] Gu, Y., Chen, W., Zhang, C. & He, X., A meshless singular boundary method for three-dimensional inverse heat conduction problems in general anisotropic media. *International Journal of Heat and Mass Transfer*, **84**, pp. 91–102, 2015.

[9] Qu, W., Chen, W., Fu, Z.-J. & Gu, Y., Fast multipole singular boundary method for Stokes flow problems. *Mathematics and Computers in Simulation*, **146**, pp. 57–69, 2017.

[10] Yang, C. & Li, X., Meshless singular boundary methods for biharmonic problems. *Engineering Analysis with Boundary Elements,* **56**, pp. 39–48, 2015.

[11] Landau, L.D. & Lifshitz, E.M., *Fluid Dynamics*, Pergamon Press, 1987.

[12] Gnitko, V., Degtyariov, K., Naumenko, V. & Strelnikova, E., BEM and FEM analysis of the fluid-structure Interaction in tanks with baffles. *Int. Journal of Computational Methods and Experimental Measurements*, **5**(3), pp. 317–328, 2017.

[13] Landau, L.D. & Lifshitz, E.M., *Quantum Mechanics,* Pergamon Press, 1965.

[14] O'Searcoid, M., *Metric Spaces,* Springer, 2006.

[15] Gnitko, V., Degtyariov, K., Karaiev, A. & Strelnikova, E., Multi-domain boundary element method for axisymmetric problems in potential theory and linear isotropic elasticity. *WIT Transactions on Engineering Sciences*, vol. 122, WIT Press: Southampton and Boston, pp. 13–25, 2019.

[16] Ravnik, J., Strelnikova, E., Gnitko, V., Degtyarev, K. & Ogorodnyk, U., BEM and FEM analysis of fluid-structure interaction in a double tank. *Engineering Analysis with Boundary Elements*, **67**, pp. 13–25, 2016.

[17] Gnitko, V., Degtyarev, K., Naumenko, V. & Strelnikova, E., Coupled BEM and FEM Analysis of fluid-structure interaction in dual compartment tanks. *International Journal of Computational Methods and Experimental Measurements*, **6**(6), pp. 976–988, 2018.

[18] Ibrahim, R.A., *Liquid Sloshing Dynamics*, Cambridge University Press: Cambridge, 2005.

GLOBAL SINGULAR BOUNDARY METHOD FOR SOLVING 2D NAVIER–STOKES EQUATIONS

JURAJ MUŽÍK & ROMAN BULKO
Department of Geotechnics, Faculty of Civil Engineering, University of Zilina, Slovakia

ABSTRACT

This article presents a numerical algorithm based on the singular boundary method (SBM) for the incompressible steady Navier–Stokes equations formulated using primitive variables. The SBM with the Stokeslet fundamental solution and dual reciprocity (DR) principle has been chosen to solve the nonlinear flow equations. The particular solution of the non-homogeneous Stokes equations is constructed as a linear combination of implicitly local radial basis function. The simple direct iterative scheme was used to handle nonlinearities of Navier–Stokes equations with a variation of the non-homogeneous term of the particular solution. The non-homogeneous term is formed using the nonlinear convective terms of the momentum equations, evaluated using values from previous iterations. It is found that SBM with a localized DR principle gives reasonable results for numerical problems of lid-driven cavity flow up to Re = 3200, and the backward-facing step at Re = 800.

Keywords: singular boundary method, dual reciprocity, Navier–Stokes equation, Stokeslet.

1 INTRODUCTION

The solution of Navier–Stokes (NS) equations is one of the basic tasks of computational hydromechanics. Together with the continuity equation, it is a system of nonlinear equations. This system has already been clarified by a number of numerical methods, starting with the finite difference method through finite element method to modern meshless methods. The meshless methods consist of the smooth particle hydrodynamics (SPH) [1], the least square collocation meshless method [2]–[4], the meshless local Petrov–Galerkin (MLPG) method [5], [6], the local boundary integral element method (LBIEM) [7]–[9], and the radial basis integral equation method (RBIEM) [10]–[12]. All of these methods have advantages as well as shortcomings. One category of numerical methods used to solve the NS equations is a group of the formulas based on the boundary integral, such as e.g. boundary element method (BEM). The methods based on the boundary integral theory are represented by the BEM [1], [2], the method of fundamental solutions (MFS) [3], [4], and the singular boundary method. In the case of BEM, the singularities of the fundamental solution of the solved problem are handled by proper integration method, the MFS overcomes the singularity using a fictitious boundary, but the optimum location of this boundary remains the open problem especially for complex-shaped domains. To bypass the fictitious boundary construction, the singular boundary method (SBM) formulation adopts a concept of the origin intensity factors (OIFs). Several techniques have been developed to determine the source intensity factors, namely, inverse interpolation technique (IIT), subtracting and adding-back regularization and empirical formulas [8], [9]. In this article, the regularized boundary integral equation is used to derive a formula for Dirichlet origin intensity factors.

For the problems governed by the NS equation, these methods have to use a connection with complementary methods to be able to effectively handle nonlinear transport members of this system. The final solution is represented as a combination of the singular boundary integral and particular solutions. In this work, we present the possibility how to solve the steady fluid flow governed by the primitive variable form of NS equations using SBM with dual reciprocity (DR) scheme used to tackle the in-homogeneities brought by nonlinear transport terms of NS equations. Section 2 of our paper introduces NS equations in primitive

WIT Transactions on Engineering Sciences, Vol 126, © 2019 WIT Press
www.witpress.com, ISSN 1743-3533 (on-line)
doi:10.2495/BE420181

variable form. Section 3 contains a brief description of the SBM and the connection with the DRM. Section 4 is dedicated to the two numerical examples and the last section, Section 5, contains the conclusions.

2 GOVERNING EQUATIONS

Incompressible steady flow is governed by Navier–Stokes equations in the closed domain Ω with the boundary Γ, which can be written in their primitive variables as

$$\frac{\partial u_i}{\partial x_i} = 0, \tag{1}$$

$$\mu \frac{\partial^2 u_i}{\partial x_j \partial x_j} - \frac{\partial p}{\partial x_i} = \rho \frac{\partial}{\partial x_j}(u_j u_i) + f_i, \tag{2}$$

where u_i is the velocity vector component in the direction i, p is the pressure, v is the kinematic viscosity, ρ is the density of a liquid, and f_i represents body forces component in the direction i. The flow should satisfy boundary conditions in the form

$$u_i(x_D) = f_i(x_D) x_D \in \Gamma_D, \tag{3}$$

$$t_i(x_N) = \sigma_{ij} n_j(x_N) x_N \in \Gamma, \tag{4}$$

where u_i are the velocities and t_i are the tractions on the boundary and n_j represents the outward normal to the boundary. The stress tensor is defined as

$$\sigma_{ij} = -p\delta_{ij} + \mu(\frac{\partial u_i}{\partial x_j} + \frac{\partial u_j}{\partial x_i}). \tag{5}$$

The solution of eqns (1) and (2) can be defined as eqn (6):

$$u_i = u_i^{\mathrm{H}} + u_i^{\mathrm{P}},\ p = p^{\mathrm{H}} + p^{\mathrm{P}}, \tag{6}$$

where u^{H} are the solutions of the homogeneous problem that captures the Stokes equations together with the boundary conditions (eqns (3) and (4)) and $u^{\mathrm{P}}{}_{\mathrm{i}}$ and p^{P} are the particular solutions of the non-homogeneous eqns (1) and (2). The homogeneous problem can be determined using the SBM and the particular solution is formulated using the method of the particular solution (MAPS).

3 SINGULAR BOUNDARY METHOD

The SBM is based on the same principle as the MFS however the SBM utilizes the same set of collocation and source points placed on the boundary Γ. The Green's function G_{ij} is the fundamental solution of the Stokes equation known as Stokeslet:

$$G_{ij}(x,y) = -\frac{1}{4\pi}\left(ln\left(\frac{1}{|x-y|}\right)\delta_{ij} + \frac{(x_i-y_i)(x_j-y_j)}{|x-y|^2}\right). \tag{7}$$

The traction fundamental solution K_{ij} is

$$K_{ij}(x,y) = -\frac{1}{\pi}\frac{(x_k-y_k)(x_i-y_i)(x_j-y_j)}{|x-y|^4} n_j(y), \tag{8}$$

and corresponding pressure solution p_j is

$$p_j(x,y) = -\frac{1}{2\pi}\frac{(x_j-y_j)}{|x-y|^2}. \tag{9}$$

The SBM defines the solution of the homogeneous problem using the Green's functions as

$$u_i^H(x^m) = \sum_{k=1,k\neq m}^{n} \alpha_k G_{ij}(x^m, y^k) + \alpha_m U_{ij}(x^m, y^m), \tag{10}$$

$$t_i^H(x^m) = \sum_{k=1,k\neq m}^{n} \alpha_k K_{ij}(x^m, y^k) + \alpha_m T_{ij}(x^m, y^m), \tag{11}$$

where α_k are the unknown coefficients, x^m and y^k are the collocation and source points and U_{ij}, T_{ij} are the origin intensity factors, respectively. Origin intensity factors (OIFs) are a key problem for the SBM. In order to prevent problems concerning the OIF of the Stokeslet for the 2D domains, the regularized boundary integral formulation described in Zhang et al. [13] and Sun et al. [14] has been used.

The OIFs associated with the Laplace fundamental solution for the Neumann type boundary condition can be calculated using subtracting and adding-back technique [7], [13]:

$$Q_{ij}(x^m, y^m) = -\frac{1}{L_m}\sum_{k=1,k\neq m}^{n} L_k \frac{\partial G_{ij}(x^m, y^k)}{\partial n(y^k)}, \tag{12}$$

where L_j is the length of the appropriate part of the boundary around point $\mathbf{x}$ (see Fig. 1). In the case of the Dirichlet boundary condition, one of the numerous possibilities of how to evaluate OIF is eqn (13), based on the get regularized boundary integral:

$$U_{ij}(x^m, y^m) = -\frac{1}{L_m}\sum_{k=1,k\neq m}^{n} A_k[n_i(x^m)\cdot n_j(y^k)G_{ij}(x^m, y^k) - n_i(x^m)(y^k - x^m)\frac{\partial G_{ij}(x^m, y^k)}{\partial n(y^k)}]. \tag{13}$$

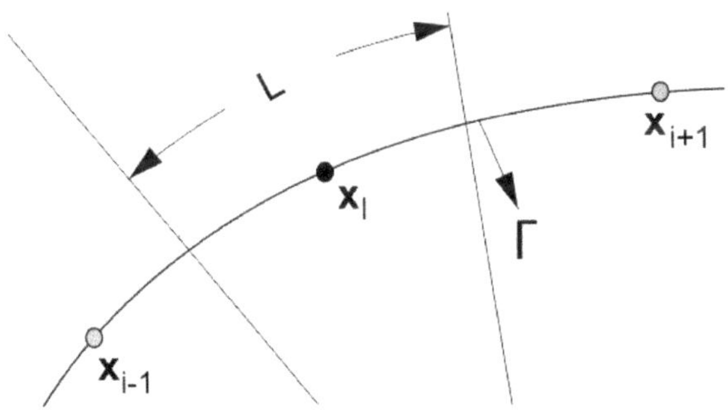

Figure 1: Part of boundary L around point $\mathbf{x}_i$.

3.1 Dual reciprocity method: particular solution

The particular solutions u^P, and p^P of eqn (6) is approximated using a linear superposition of the radial basis functions [15]:

$$u_i^P = \sum_{j=1}^{M} \beta_{jk}\bar{u}_{ik}(r_j), p^P = \sum_{j=1}^{M} \beta_{jk}\bar{p}k\,(r_j), \tag{14}$$

where M is the number of the internal points, β_j are the unknown coefficients, and $\bar{u}_{ik}$ and $\bar{p}_k$ are the radial basis functions which can be defined as the particular solutions of the following non-homogeneous system of eqns (15) and (16):

$$\mu \frac{\partial^2 \bar{u}_{ik}}{\partial x_j \partial x_j} - \frac{\partial \bar{p}_k}{\partial x_i} = \varphi(r)\delta_{ik}$$

$$\frac{\partial \bar{u}_{ik}}{\partial x_i} = 0, \tag{15}$$

where $\varphi(r)$ is the implicitly compactly supported RBF function (MQ) defined as

$$\varphi(r) = \left(1 - \frac{r}{\alpha}\right), \tag{16}$$

where r is Euclidean distance from the function origin and the α is scaling parameter used to represent the radius of circular support domain, the values outside the support domain are set as zeros. One of the main difficulties is the problem of how to find a closed-form expression of the particular solutions $\bar{u}_{ik}$ and $\bar{p}_k$. To solve the problem, we employ the way more detailed one described in more detail in Florez and Power [15], Bustamante et al. [16] and Happel and Brenner [17]. As the first step, an auxiliary potential Φ is defined and is used to express the velocity particular solution as follows:

$$\bar{u}_{ik} = \frac{\partial^2 \Phi}{\partial x_j \partial x_j}\delta_{ik} - \frac{\partial^2 \Phi}{\partial x_i \partial x_k}. \tag{17}$$

The velocity particular solution captures implicitly the continuity equation, which can be checked by supplying the above velocity field into the continuity equation. Substituting eqn (17) into the momentum eqn (15) allows us to write

$$\mu\left(\frac{\partial^4 \Phi}{\partial x_j \partial x_j \partial x_k \partial x_k}\delta_{ik} - \frac{\partial^4 \Phi}{\partial x_j \partial x_j \partial x_i \partial x_k}\right) - \frac{\partial \bar{p}_k}{\partial x_i} = \varphi\delta_{ik}. \tag{18}$$

The particular solution we explore is represented by the RBF approximation of the auxiliary potential used to express velocity, pressure and stress fields. The particular solution can be found splitting the momentum eqn (18) into two parts and satisfying these two equations. Then the following system is obtained:

$$\mu \frac{\partial^4 \Phi}{\partial x_j \partial x_j \partial x_k \partial x_k} = \varphi(r), \tag{19}$$

$$\mu \frac{\partial^4 \Phi}{\partial x_j \partial x_j \partial x_i \partial x_k} + \frac{\partial \bar{p}_k}{\partial x_i} = 0. \tag{20}$$

The second, eqn (20), can be used to obtain pressure particular solution

$$\bar{p}_k = \mu \frac{\partial^3 \Phi}{\partial x_j \partial x_j \partial x_k}. \tag{21}$$

The first, eqn (19), is used to compute the auxiliary potential Φ by the direct integration

$$\Phi(r) = \frac{r^4}{14400}\left(225 + \frac{r}{\alpha}\left(25\frac{r}{\alpha} - 128\right)\right), r \leq \alpha$$

$$\Phi(r) = \frac{\alpha^2}{48}\ln\frac{r}{\alpha}\left(r^2 + \frac{\alpha^2}{5}\right) + \frac{\alpha^2 r^2}{576} + \frac{97\alpha^4}{14400}, r > \alpha. \tag{22}$$

Substituting the auxiliary potential Φ to eqns (17) and (21), the pressure and velocity field of a particular solution is obtained.

3.2 System of equations

Substituting eqns (14) and (15) into eqns (1) and (2), we have got the following M equations:

$$\sum_{j=1}^{M} \beta_{jk}\, \varphi(r_{mj}) = \rho u_k \frac{\partial u_i}{\partial x_k}, \quad m = 1,2,\ldots,M. \tag{23}$$

To complete the whole system, we must add N boundary conditions in all boundary nodes (see Fig. 2). These additional N equations can be formulated using boundary operator B as

$$\sum_{j=1}^{M} \beta_{jk}\, B\left(\bar{u}_{ik}(r_{mj})\right) + \sum_{j=1,m\neq j}^{N} \alpha_{jm} B\left(G(r_{mj})\right) + \alpha_{mm} B(U_m) = b_0(x_m,t),$$
$$m = 1,2,\ldots,N. \tag{24}$$

The system of linear equations can be written in the following form:

$$\begin{bmatrix} \mathbf{K} & \mathbf{L} \\ \mathbf{0} & \mathbf{N} \end{bmatrix} \begin{Bmatrix} \alpha \\ \beta \end{Bmatrix} = \begin{Bmatrix} \mathbf{b}_0 \\ \mathbf{S} \end{Bmatrix}, \tag{25}$$

where

$$K_{ij} = B\left(G(r_{ij})\right), \quad L_{ij} = B\left(\bar{u}_{ik}(r_{mj})\right), \tag{26}$$

and

$$N_{ij} = \varphi(r_{mj}), \quad S_i = \rho u_k \frac{\partial u_i}{\partial x_k}. \tag{27}$$

The computational algorithm deals with the nonlinearity caused by transport terms adopting Pickard iterative scheme together with the variation of right-hand side evaluated using transport terms S_i (eqn (27)). The part of the equation system eqn (25) used to construct a particular solution will be sparse and the system will be computationally more efficient. The other effect is that the local nature of kernel functions used to create particular solution enables the solution of problems with higher Reynolds number.

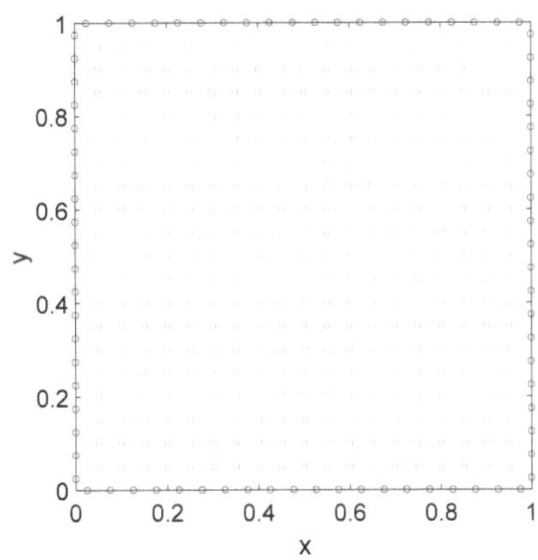

Figure 2: The boundary and internal nodes distribution covering the global computational domain used for SBMDR method.

4 SIMULATION OF THE LID-DRIVEN CAVITY PROBLEM

Lid-driven cavity flow is being used as a standard test case for the validation of numerical solutions of incompressible Navier–Stokes flow. The upper wall of the cavity motions with a velocity $u = 1$, and no-slip impermeable boundary conditions are considered along the other three walls. The geometry and velocity boundary conditions are demonstrated in Fig. 3. The dilemma of this problem remains in the presence of the singularities of the pressure and velocity at the two top corners of the cavity. Therefore, the amount of the computational nodes utilized decreases towards the corners of the cavity (see Fig. 3) to catch the high pressure and velocity function gradients.

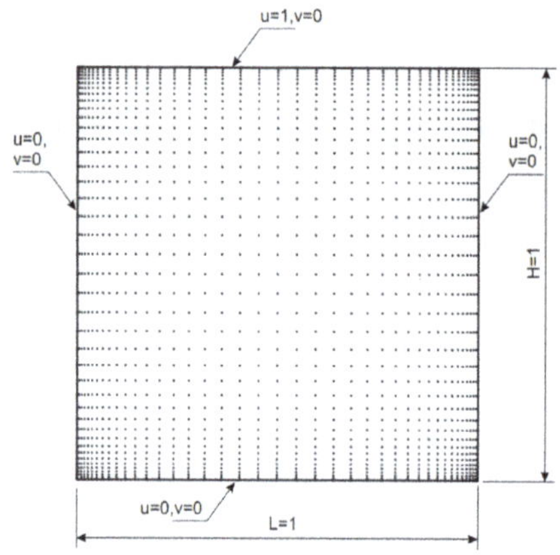

Figure 3: The lid-driven cavity problem geometry, boundary conditions, and distribution of internal nodes.

The steady solution using the SBMDR is achieved when the discrepancy between two consecutive time steps (in the definition of the infinity norm) is lower than specified, $\varepsilon_{tol} = 1\times10^{-6}$ is applied for presented numerical computations. The results of the SBMDR are confronted with the solution represented in AbdelMigid et al. [18] for the Reynolds number Re = 3200; the vertical velocity components along a horizontal line and the horizontal velocity components along a vertical line, straight to the center of the cavity, are plotted in Fig. 4. Fig. 5 shows streamlines and pressure contours for Reynolds number of Re = 3200.

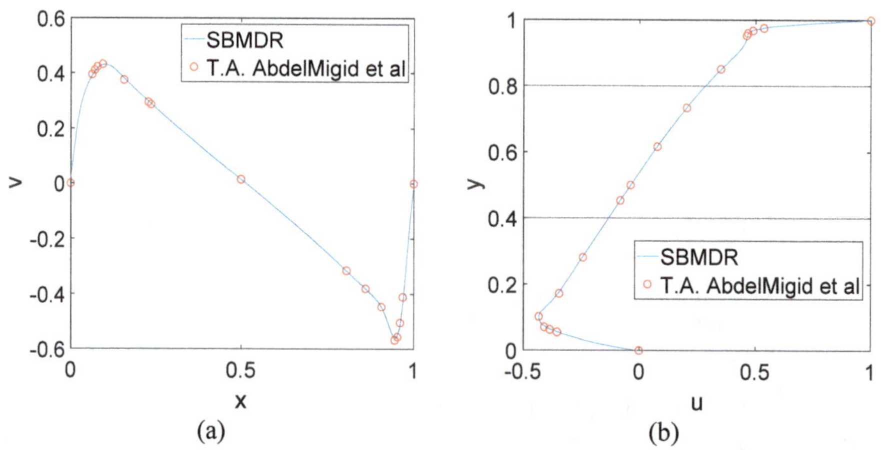

Figure 4: Lid-driven cavity, Re = 3200. (a) Vertical velocity components along horizontal line $y = 0.5$; and (b) Horizontal velocity components along vertical line $x = 0.5$.

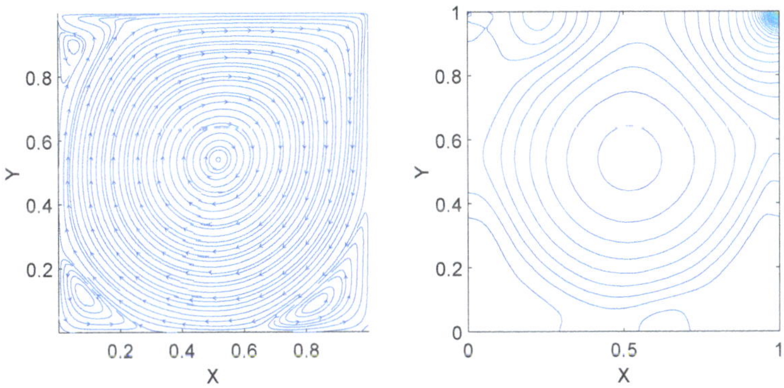

Figure 5: Streamlines and pressure contours for lid-driven cavity problem, Re = 3200.

5 SIMULATION OF THE BACKWARD-FACING STEP FLOW

The flow over a backward-facing step is extensively tested arrangement of the fluid flow in a channel. A couple of the numerical and experimental studies of the 2D steady incompressible flows have been disclosed (e.g. Erturk [19]). The inlet velocity is considered to be horizontal with a parabolic distribution, and maximum inflow velocity is $u_{max} = 1.5$, and the average inflow velocity is $u_a = 1$. The Reynolds number could be defined as

$$Re = \frac{u_a H}{\nu}, \tag{28}$$

where H is the height of the channel (see Fig. 6). The geometry configuration and boundary conditions are shown in Fig. 6.

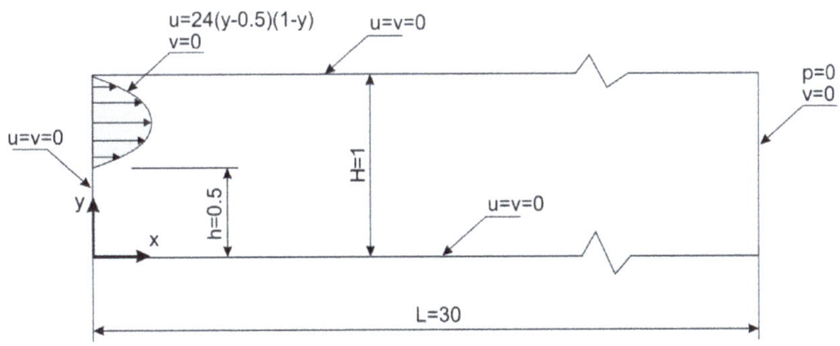

Figure 6: The geometry of the backward-facing step problem.

The entire domain is covered by a uniform distribution of computational nodes, 261 nodes in the x-direction and 31 in the y-direction. The primary values of all quantities are set to zero. The steady solution is again attained when the tolerance between two consecutive time steps is lower than a designated value.

Fig. 7 demonstrates streamlines and pressure contours for Re = 800, and Fig. 8 demonstrates the horizontal velocity components in vertical profiles. Comparing the results of the SBMDR for Re = 800 with results of Erturk [19] show the close agreement of the proposed formulation.

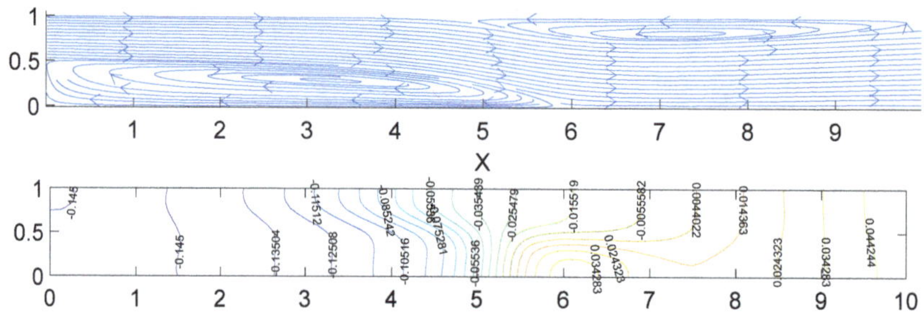

Figure 7: Stream function contours (streamlines) and pressure contours for the backward-facing step, Re = 800.

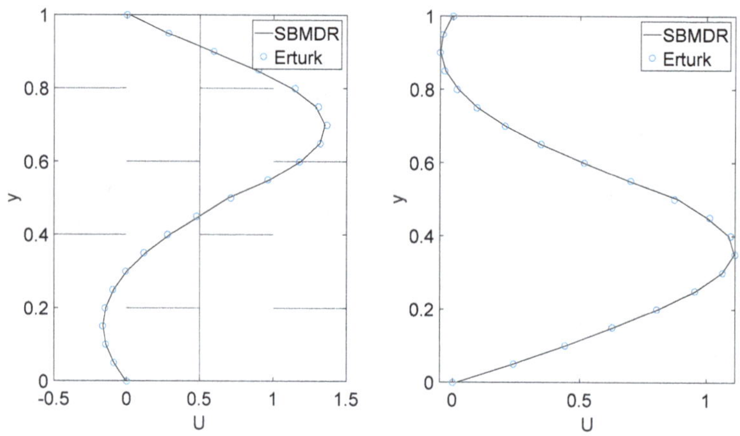

Figure 8: Backward-facing step, the horizontal velocities at x = 3 and x = 7 profiles for Re = 800.

6 CONCLUSIONS

This paper documents the experiment to apply the global SBM and DRM (SBMDR) for the solution the incompressible laminar flow outlined by a primitive variable form of Navier–Stokes equation. Two types of benchmarking problems, lid-driven cavity and backward-facing step, have been successfully tested. This type of boundary collocation method seems to be quite a useful alternative to the solutions of incompressible fluid flow using the boundary based numerical schemes. The dual reciprocity technique implemented using implicitly compactly supported RBFs generates sparse characteristic matrix and enables the method to obtain reasonable results even for high Reynolds number. The presented numerical code and numerical results need to be more analyzed in the sense of precision, convergence and computational demands.

ACKNOWLEDGEMENTS

This contribution is the result of the project funded by the Scientific Grant Agency of Slovak Republic (VEGA) No. 1-0716-17. Internal grant UNIZA "Vyhodnotenie statickej penetračnej skúšky použitím neurónových sietí".

REFERENCES

[1] Gingold, R.A. & Monaghan, J.J., Smoothed particle hydrodynamics: Theory and applications to non-spherical stars. *Monthly Notices of the Royal Astronomical Society*, **181**, pp. 375–389, 1977.

[2] Kansa, E.J., Multiquadrics – a scattered data approximation scheme with application to computational fluid dynamics. *Computers & Mathematics with Applications*, **19**, pp. 127–145, 1990.

[3] Sanyasiraju, Y. & Chandhini, G., Local radial basis function based grid-free scheme for unsteady incompressible viscous flows. *Journal of Computational Physics*, **227**, pp. 8922–8948, 2008.

[4] Shu, C., Ding, H. & Yeo, K.S., Local radial basis function-based differential quadrature method and its application to solve two-dimensional incompressible Navier–Stokes equations. *Computer Methods in Applied Mechanics and Engineering*, **192**, pp. 941–954, 2003.

[5] Atluri, S.N., Kim, H.-G. & Cho, J.Y., A critical assessment of the truly Meshless Local Petrov–Galerkin (MLPG), and Local Boundary Integral Equation (LBIE) methods. *Computational Mechanics*, **24**, 348–372, 1999.

[6] Lin, H. & Atluri, S.N., The Meshless Local Petrov–Galerkin (MLPG) method for solving incompressible Navier–Stokes equations. *Computer Modeling in Engineering & Sciences*, **2**, pp. 117–142, 2001.

[7] Zhu, T., Zhang, J. & Atluri, S., A meshless numerical method based on the local boundary integral equation (LBIE) to solve linear and nonlinear boundary value problems. *Engineering Analysis with Boundary Elements*, **23**, pp. 375–389, 1999.

[8] Sladek, V., Sladek, J., Atluri, S.N. & Van Keer, R., Numerical integration of singularities in meshless implementation of local boundary integral equations. *Computational Mechanics*, **25**, pp. 394–403, 2000.

[9] Kovářík, K. & Mužík, J., A meshless solution for two dimensional density-driven groundwater flow. *Engineering Analysis with Boundary Elements*, **37**, pp. 187–196, 2013.

[10] Bui, T.T. & Popov, V., Radial basis integral equation method for Navier–Stokes equations. *Boundary Elements and Other Mesh Reduction Methods*, vol. 49, WIT Press, pp. 141–150.

[11] Popov, V. & Thanh Bui, T., A meshless solution to two-dimensional convection-diffusion problems. *Engineering Analysis with Boundary Elements*, **34**, pp. 680–689, 2010.

[12] Kovářík, K., Mužík, J. & Mahmood, M.S., A meshless solution of two dimensional unsteady flow. *Engineering Analysis with Boundary Elements*, **36**, pp. 738–743, 2012.

[13] Zhang, Y.M., Sun, F.L., Young, D.L., Chen, W. & Gu, Y., Average source boundary node method for potential problems. *Engineering Analysis with Boundary Elements*, **70**, pp. 114–125, 2016.

[14] Sun, F.L., Zhang, Y.M., Young, D.L. & Chen, W., A new boundary meshfree method for potential problems. *Advances in Engineering Software*, **100**, pp. 32–42, 2016.

[15] Florez, W.F. & Power, H., Comparison between continuous and discontinuous boundary elements in the multidomain dual reciprocity method for the solution of the two-dimensional Navier–Stokes equations. *Engineering Analysis with Boundary Elements*, **25**, pp. 57–69, 2001.

[16] Bustamante, C.A., Florez, W.F. & Power, H., A global meshless collocation particular solution method for solving the two-dimensional NavierStokes system of equations. *Computers and Mathematics with Applications*, **65**, pp. 1939–1955, 2013.

[17] Happel, J. & Brenner, H., *Low Reynolds Number Hydrodynamics*, 2nd ed., Martinus Nijhoff Publishers: The Hague, 1983.
[18] AbdelMigid, T.A., Saqr, K.M., Kotb, M.A. & Aboelfarag, A.A., Revisiting the lid-driven cavity flow problem: Review and new steady state benchmarking results using GPU accelerated code. *Alexandria Engineering Journal*, **56**(1), pp. 123–135, 2017.
[19] Erturk, E., Numerical solutions of 2-D steady incompressible flow over a backward-facing step, Part I: high Reynolds number solutions. *Computers & Fluids*, **37**, pp. 633–55, 2008.

MESH-FREE ANALYSIS OF PLATE BENDING PROBLEMS BY MOVING FINITE ELEMENT APPROXIMATION

VLADIMIR SLADEK, JAN SLADEK & MIROSLAV REPKA
Institute of Construction and Architecture, Slovak Academy of Sciences, Slovakia

ABSTRACT

The governing equations for plate bending problems are given by the partial differential equations of the 4th order. Therefore C^1 continuous elements are required for approximation and discretization with using the standard finite element method (FEM). A novel discretization method is proposed and developed for numerical solution of plate bending boundary value problems. In contrast to the standard FEM, the analysed domain is not covered by a mesh of fixed non-overlapping finite elements, but only a net of nodes is used for discretization. Around each node, there is properly created a Lagrange finite element and the spatial variation of field variables is interpolated within this element in terms of nodal values and polynomial shape functions defined in the intrinsic coordinate space of the finite element. Thus, the Lagrange finite element associated with a node is moving within the analysed domain from node to node. Since there are no element interfaces known in standard FEM with a fixed finite element mesh, the difficulties with continuity of derivatives of field variables on such interfaces are avoided and higher order derivatives are available within the moving finite element. This makes the moving finite element (MFE) approximation to be applicable also to the development of the strong formulation of a boundary value problem with collocation of both the governing equations and boundary conditions at interior and boundary nodes, respectively. To decrease the order of the polynomial interpolation, the original set of the governing PDE is decomposed into a system of 2nd order PDEs by introducing a new field variable. Then, the boundary conditions are to be modified and the bi-quadratic Lagrange finite element is sufficient for approximation. Both the strong and the local weak formulations are derived and employed in the numerical test examples with focusing on verification of the reliability (accuracy and stability) and efficiency of the new method.

Keywords: Lagrange finite element, strong and weak formulation, convergence and accuracy, numerical stability, computational efficiency.

1 INTRODUCTION

The most frequently used methods in engineering computations, such as finite element method (FEM), finite volume method (FVM), boundary element method (BEM) are based on weak formulations of boundary value problems. Although, the strong formulations offer better computational efficiency because of elimination of integrations, the price which should be paid is the need to approximate higher order derivatives. The standard finite elements (FE) give only C^0 continuity, while the recently developed mesh-free approximations utilize the higher order continuous shape functions, but these are not expressed in terms of elementary functions and hence, such mesh-free approximations result into time consuming evaluation of shape functions. Utilization of advantages of the polynomial interpolation (like in FEM) and the element-free discretization of the analysed domain (like in mesh-free methods) has been proposed originally in works [1]–[3] and recently elaborated in [4]–[6]. In this paper, the moving finite element (MFE) approximation [4] is developed and implemented for the solution of plate bending boundary value problems, with studying numerical aspects of the proposed method.

WIT Transactions on Engineering Sciences, Vol 126, © 2019 WIT Press
www.witpress.com, ISSN 1743-3533 (on-line)
doi:10.2495/BE420191

2 FORMULATION OF BOUNDARY VALUE PROBLEMS FOR THIN ELASTIC PLATES

The bending of elastic plates can be described in a unique way [7] by including the assumptions of three plate bending theories, such as the Kirchhoff–Love theory (KLT), shear deformation plate bending theories of the 1st and 3rd order (FSDPT and TSDPT). Considering the thin elastic, homogeneous plates, the tensor of bending moments is expressed in terms of derivatives of dimensionless deflections $w^*(\mathbf{x}) = w(\mathbf{x})/h$ as

$$M^{*(w)}_{\alpha\beta} = (1-\nu)w^*_{,\alpha\beta} + \nu\delta_{\alpha\beta}\nabla^2 w^* , \quad M^{*(w)}_{\alpha\beta}(\mathbf{x}) = \frac{L^2}{Dh} M^{(w)}_{\alpha\beta}(\mathbf{x}), \tag{1}$$

where h, ν, D, L are the plate thickness, Poisson ratio, bending stiffness, a characteristic length in the mid-plane domain Ω, respectively, and the superscript * stands for notation of dimensionless quantities. The dimensionless in-plane Cartesian coordinates are defined as $x^*_\alpha = x_\alpha / L$, and the superscript will be omitted in Cartesian coordinates in what follows, because we shall deal only with the dimensionless coordinates. The governing equation of the plate subjected to transversal loading $t_3(\mathbf{x})$ is given by

$$M^{*(w)}_{\alpha\beta,\alpha\beta} = -t^*_3, \quad t^*_3 = \frac{L^4}{Dh} t_3, \tag{2}$$

and can be rewritten as

$$\nabla^2\nabla^2 w^* = t^*_3 . \tag{3}$$

In order to decrease the order of the derivatives of deflection field, it is convenient to introduce the additional field variable m^* by eqn (4), when the 4th order PDE (3) is replaced by the system of two 2nd order PDE

$$\nabla^2 w^* - m^* = 0 , \quad \nabla^2 m^* = t^*_3 . \tag{4}$$

The homogeneous boundary conditions can be written on the plate edge as

(i) clamped edge (CE):

$$w^*\Big|_{\partial\Omega} = 0 , \quad \partial w^* / \partial\mathbf{n}\Big|_{\partial\Omega} = 0 ; \tag{5a}$$

(ii) simply supported edge (SSE):

$$w^*\Big|_{\partial\Omega} = 0 , \quad n_\alpha n_\beta M^{*(w)}_{\alpha\beta}\Big|_{\partial\Omega} = 0; \tag{5b}$$

(iii) free edge (FrE):

$$n_\alpha n_\beta M^{*(w)}_{\alpha\beta}\Big|_{\partial\Omega} = 0 , \quad V^*\Big|_{\partial\Omega} = 0 , \tag{5c}$$

with

$$V^*(\mathbf{x}) := n_\alpha(\mathbf{x}) M^{(w)*}_{\alpha\beta,\beta}(\mathbf{x}) + \frac{\partial T^{(w)*}}{\partial \mathbf{t}}(\mathbf{x}) - \sum_c \left[\!\left[T^{(w)*}(\mathbf{x}^c) \right]\!\right] \delta(\mathbf{x}-\mathbf{x}^c)$$

$$= n_\alpha(\mathbf{x}) m^*_{,\alpha}(\mathbf{x}) + \frac{\partial T^{(w)*}}{\partial \mathbf{t}}(\mathbf{x}) - \sum_c \left[\!\left[T^{(w)*}(\mathbf{x}^c) \right]\!\right] \delta(\mathbf{x}-\mathbf{x}^c),$$

where $T^{(w)*} = \frac{L^2}{Dh} T^{(w)}$, $T^{(w)} := t_\alpha n_\beta M^{(w)}_{\alpha\beta}$ is the twisting moment, and the jump term is defined at corner points $\mathbf{x}^c$ on the boundary edge $\partial\Omega$ as

$$\left[\!\left[T^{(w)}(\mathbf{x}^c) \right]\!\right] := T^{(w)}(\mathbf{x}^c - 0) - T^{(w)}(\mathbf{x}^c + 0).$$

Recall that at corner points the $\partial w^* / \partial \mathbf{n}\big|_{\partial\Omega}$ is not defined uniquely because of the discontinuity of the normal vector. Therefore it is convenient to replace this boundary condition at $\mathbf{x}^c$ by the local weak form of the governing eqn (4_1)

$$\int_{\Omega^c} \left(\nabla^2 w^* - m^* \right) d\Omega = 0 \;\Rightarrow\; \int_{L^c} n_\alpha w^*_{,\alpha} d\Gamma - \int_{\Omega^c} m^* d\Omega = 0, \tag{6}$$

on the local subdomain around the point $\mathbf{x}^c$, where we have utilized that the boundary of the subdomain Ω^c is $\partial\Omega^c = \Gamma^c \cup L^c$ and $\partial w^* / \partial \mathbf{n}\big|_{\Gamma^c} = 0$.

In view of (1) the boundary condition

$$n_\alpha n_\beta M^{*(w)}_{\alpha\beta}\Big|_{\partial\Omega} = 0, \tag{7}$$

can be rewritten as $\left[\nabla^2 w^* - (1-\nu) t_\alpha t_\beta w^*_{,\alpha\beta} \right]\Big|_{\partial\Omega} = 0$. Furthermore,

$$t_\alpha t_\beta w^*_{,\alpha\beta} = \frac{\partial}{\partial \mathbf{t}}\left(t_\beta w^*_{,\beta} \right) - \frac{\partial t_\beta}{\partial \mathbf{t}} w^*_{,\beta} = \frac{\partial}{\partial \mathbf{t}}\left(\frac{\partial w^*}{\partial \mathbf{t}} \right) - \frac{\partial t_\beta}{\partial \mathbf{t}}\left(n_\beta n_\gamma + t_\beta t_\gamma \right) w^*_{,\gamma} = \frac{\partial}{\partial \mathbf{t}}\left(\frac{\partial w^*}{\partial \mathbf{t}} \right) - \kappa \frac{\partial w^*}{\partial \mathbf{n}}$$

with $\kappa = n_\beta \frac{\partial t_\beta}{\partial \mathbf{t}}$ being the curvature of the boundary edge. If $w^*\big|_{\partial\Omega} = 0$, then $\partial w^* / \partial \mathbf{t}\big|_{\partial\Omega} = 0$ as well as $\partial\left(\partial w^* / \partial \mathbf{t} \right) / \partial \mathbf{t}\big|_{\partial\Omega} = 0$ and the boundary condition (7) on the SSE becomes $\left[\nabla^2 w^* + \kappa \partial w^* / \partial \mathbf{n} \right]\Big|_{\partial\Omega} = 0$. Eventually, the boundary conditions on the SSE can be rewritten as

$$w^*\Big|_{\partial\Omega} = 0, \quad \left[m^* + \kappa \partial w^* / \partial \mathbf{n} \right]\Big|_{\partial\Omega} = 0. \tag{8}$$

The advantage of this expression of the boundary conditions is the greater convenience for numerical treatment than the expression given by (5b) because of the lower order of the derivatives involved. The curvature is vanishing on straight edges as well as at corners, if the edges on both sides of the corner are straight lines. Then, the boundary condition (8_2) is simplified as $m^*\big|_{\partial\Omega}=0$ and both the boundary conditions on SSE are Dirichlet type. Nevertheless, a special treatment is required at corner points, in general. One possibility is to consider (8_2) in weak sense on the very short segments Γ^{cs} and Γ^{cf} of the length ε . Then,

$$\int_{\Gamma^{cs}\cup\Gamma^{cf}}\left[m^*+\kappa\,\partial w^*/\partial\mathbf{n}\right]d\Gamma\approx 2\varepsilon m^*(\mathbf{x}^c)+\varepsilon\left(\kappa(\mathbf{x}^{cs})\frac{\partial w^*}{\partial\mathbf{n}}(\mathbf{x}^{cs})+\kappa(\mathbf{x}^{cf})\frac{\partial w^*}{\partial\mathbf{n}}(\mathbf{x}^{cf})\right)=0,$$

and finally, the boundary conditions at a corner on the SSE can be rewritten as

$$w^*\Big|_{\partial\Omega}=0,\quad m^*(\mathbf{x}^c)+\frac{1}{2}\left(\kappa(\mathbf{x}^{cs})\frac{\partial w^*}{\partial\mathbf{n}}(\mathbf{x}^{cs})+\kappa(\mathbf{x}^{cf})\frac{\partial w^*}{\partial\mathbf{n}}(\mathbf{x}^{cf})\right)=0. \tag{9}$$

In the case of free-edge, it is necessary to rearrange the boundary condition $V^*\big|_{\partial\Omega}=0$. Owing to the Dirac-delta function as well as the discontinuity of the normal vector at a corner point, it is necessary to consider this boundary condition in weak sense around arbitrary point on the boundary edge, i.e. in weak sense on $\Gamma^{cs}\cup\Gamma^{cf}$ and/or $\Gamma^{bs}\cup\Gamma^{bf}$ considered around $\mathbf{x}^c$ and/or $\mathbf{x}^b$, respectively (see Fig. 1). For the sake of brevity, we shall use the superscript "b" for arbitrary point on the boundary edge. Then, the weak form of the considered boundary condition takes the form

$$\int_{\Gamma^{bs}\cup\Gamma^{bf}}\left[n_\alpha m^*_{,\alpha}+\frac{\partial T^{(w)*}}{\partial\mathbf{t}}-\sum_c\left[\!\left[T^{(w)*}(\mathbf{x}^c)\right]\!\right]\delta(\mathbf{x}-\mathbf{x}^c)\right]d\Gamma=\int_{\Gamma^{bs}\cup\Gamma^{bf}}n_\alpha m^*_{,\alpha}d\Gamma+$$
$$+T^{(w)*}(\mathbf{x}^{bf})-T^{(w)*}(\mathbf{x}^{bs}).$$

Since on short segments of the boundary edge, we can write $\int_{\Gamma^{bs}\cup\Gamma^{bf}}n_\alpha m^*_{,\alpha}d\Gamma\approx\varepsilon\left(\frac{\partial m^*}{\partial\mathbf{n}}(\mathbf{x}^{bs})+\frac{\partial m^*}{\partial\mathbf{n}}(\mathbf{x}^{bf})\right)$, the b.c. $V^*\big|_{\partial\Omega}=0$ becomes

$$T^{(w)*}(\mathbf{x}^{bf})-T^{(w)*}(\mathbf{x}^{bs})+\varepsilon\left(\frac{\partial m^*}{\partial\mathbf{n}}(\mathbf{x}^{bs})+\frac{\partial m^*}{\partial\mathbf{n}}(\mathbf{x}^{bf})\right)=0. \tag{10}$$

This expression of boundary condition still involves the 2nd order derivatives in twisting moment terms.

In numerical solutions of boundary value problems, the governing equations can be considered either in the strong sense at interior nodes $\mathbf{x}^a\in\Omega$ as

$$\nabla^2 w^*(\mathbf{x}^a)-m^*(\mathbf{x}^a)=0,\quad \nabla^2 m^*(\mathbf{x}^a)=t_3^*(\mathbf{x}^a), \tag{11}$$

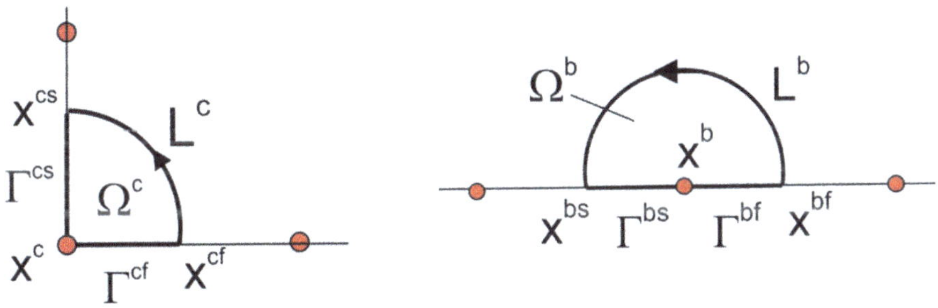

Figure 1: Illustration of boundary contours $\Gamma^c = \Gamma^{cs} \cup \Gamma^{cf}$ and L^c associated with a corner point $\mathbf{x}^c$ or a boundary node $\mathbf{x}^b$ on a smooth portion of the boundary.

or in the local weak sense on small subdomain $\Omega^a \ni \mathbf{x}^a$ around interior nodes $\mathbf{x}^a \in \Omega$ as

$$\int_{L^a} n_\alpha w^*_{,\alpha} d\Gamma - \int_{\Omega^a} m^* d\Omega = 0, \quad \int_{L^a} n_\alpha m^*_{,\alpha} d\Gamma - \int_{\Omega^a} t^*_3 d\Omega = 0. \tag{12}$$

It is seen that the 2nd order derivatives of primary field variables are required in the case strong formulation and the 1st order derivatives are sufficient in the weak formulation. On the other hand, these derivatives are needed only at nodal point in case of the strong formulation, while they are needed at certain integration points in case of weak formulation. Thus, the strong formulation requires higher order continuity of approximation than the weak formulation, while the computational efficiency is worse in the case of weak formulation.

3 MOVING FINITE ELEMENT APPROXIMATION

In contrast to the classical FEM, the analysed domain is not discretized into the mesh of finite elements, but only a mesh of nodes is utilized [4]. A finite element is associated with each node and this element is created automatically according to the position of the reference node on the bounded domain $\Omega \cup \partial\Omega$ [4]. In this paper, we shall consider the bi-quadratic Lagrange elements (with 9 nodes). Construction of the moving finite elements (FE) associated with particular nodes $\mathbf{x}^a$ follows the rules [4]: (i) if $\mathbf{x}^a \in \Omega$, $\mathbf{x}^a$ is the centre of the FE; (ii) if $\mathbf{x}^a \in \partial\Omega$ and $\partial\Omega$ is smooth at the node $\mathbf{x}^a$, this node is the mid-side node of the FE; (iii) if $\mathbf{x}^a$ lies at a corner on the boundary edge $\partial\Omega$, the FE is constructed uniquely. The approximation of the spatial variation of primary field variables, such as $u(\mathbf{x}) \in \left\{ w^*(\mathbf{x}), m^*(\mathbf{x}) \right\}$, is given by polynomial interpolation within the moving finite element

$$E^a \quad \text{as} \quad u(\mathbf{x})\big|_{E^a} \approx \sum_{\gamma=1}^{9} u^{a_\gamma} N^\gamma(\xi), \quad u^{a_\gamma} = u(\mathbf{x}^{a_\gamma}), \tag{13}$$

where ξ is the shortcut for intrinsic coordinates (ξ_1, ξ_2), $N^\gamma(\xi)$ stand for standard interpolation shape functions in Lagrange finite element [9], and a_γ is the global number of the node on the element E^a with the local number $\gamma \in \{1, 2, ..., 9\}$. Utilization of bi-quadratic Lagrange FE for interpolation enables us to apply also the formulations involving the 2nd order derivatives of field variables. Since the polynomial interpolation is in terms of intrinsic coordinates, the expression of derivatives is slightly complicated. Therefore we give them, but without details of derivation. The gradient of the field variable is given within the moving FE as

$$u_{,i}(\mathbf{x})\Big|_{E^a} = \frac{\partial u(\mathbf{x})}{\partial x_i}\Big|_{E^a} \approx \sum_{\gamma=1}^{9} u^{a_\gamma} b_i^{\gamma a}(\xi), \quad b_i^{\gamma a}(\xi) = \mathrm{Y}_{ik}^a \frac{\partial N^\gamma(\xi)}{\partial \xi_k}, \quad \mathrm{Y}_{ik}^a = \frac{\partial \xi_k}{\partial x_i}\Big|_{E^a}, \tag{14}$$

in which $[Y^a] = [J^a]^{-1}$ and

$$J_{ki}^a = \frac{\partial x_i}{\partial \xi_k}\Big|_{E^a} = \sum_{\gamma=1}^{9} x_i^{a_\gamma} \frac{\partial N^\gamma}{\partial \xi_k}, \quad Y_{ij}^a = \frac{\varepsilon_{ik3}\varepsilon_{jl3}}{\left|J^a\right|} J_{lk}^a, \quad \left|J^a\right| = \varepsilon_{mn3} J_{1m}^a J_{2n}^a, \tag{15}$$

since we have used the isoparametric elements for field variables and Cartesian coordinates

$$x_i\Big|_{E^a} \approx \sum_{\gamma=1}^{9} x_i^{a_\gamma} N^\gamma(\xi). \tag{16}$$

Similarly, for the 2nd order derivatives, we obtain

$$u_{,ij}(\mathbf{x})\Big|_{E^a} = \frac{\partial^2 u(\mathbf{x})}{\partial x_i \partial x_j}\Big|_{E^a} \approx \sum_{\gamma=1}^{n} u^{a_\gamma} \frac{\partial \xi_m}{\partial x_j} \frac{\partial}{\partial \xi_m} b_i^{\gamma a}(\xi)$$

$$= \sum_{\gamma=1}^{n} u^{a_\gamma} Y_{jm}^a \left[Y_{ik}^a N_{,km}^\gamma(\xi) - Y_{is}^a J_{sl,m}^a b_l^{\gamma a}(\xi) \right] = \sum_{\gamma=1}^{n} u^{a_\gamma} b_{ij}^{\gamma a}(\xi), \tag{17}$$

with $J_{sl,m}^a = \sum_{\gamma=1}^{n} x_l^{a_\gamma} \frac{\partial^2 N^\gamma}{\partial \xi_s \partial \xi_m} = J_{ml,s}^a$,

$$b_{ij}^{\gamma a}(\xi) := Y_{jm}^a \left[Y_{ik}^a N_{,km}^\gamma(\xi) - Y_{is}^a J_{sl,m}^a b_l^{\gamma a}(\xi) \right] = b_{ji}^{\gamma a}(\xi).$$

The numerical implementation of the strong formulation can be obtained very easily by using the expressions for the derivatives at nodal points in both the governing equations and boundary conditions. In the case of weak formulation, some additional implementation is needed for the geometry of local subdomains and boundary contours, and it can be done as shown in [4]. There is no restriction on the shape of the local subdomain around a nodal point, but the subdomain should lie within the moving FE associated with the considered nodal point. For numerical integrations, the most appropriate shape of subdomain is the circular subdomain in the intrinsic coordinate space.

4 NUMERICAL EXAMPLES

In order to test the accuracy, numerical stability and compare the computational efficiency, we have considered simple examples for bending of a thin elastic homogeneous plate with clamped and/or simply supported edges and subjected to stationary uniform transversal loading $t_3^* = 1$. The square plate $L \times L$ is considered with the thickness $h = L/50$ and constant Poisson ratio $\nu = 0.3$. For the governing equations (g.e.), we have used two alternatives – the strong as well as weak formulations, while the Dirichlet boundary conditions are considered in strong sense and the Neumann boundary conditions either in strong or weak sense. The convergence study with increasing the number of uniformly distributed nodal points is shown in Figs 2 and 3.

Practically the same results have been obtained for all the formulation variants applied to plate with CE: (a) weak form of g.e. and Neumann b.c. at all boundary nodes, (b) weak form of g.e. and Neumann b.c. only at corners, (c) strong form of g.e. and weak form of Neumann b.c. at all nodes. Convergence is achieved for rather large amount of nodes ($51 \times 51 = 2601$).

In case of the plate with SSE, the convergence is significantly faster ($11 \times 11 = 121$ nodes) and it is practically the same in both the considered formulation variants: (a) weak form of g.e., (b) strong form of g.e. The boundary conditions in plates with SSE are Dirichlet type and considered only in the strong sense.

From comparison of Figs 2 and 3, it is seen that the vanishing slope (normal derivative of deflections) on the CE affects significantly also the values of deflections at nodes far from the boundary edge, if the approximation elements are extensively large.

Hence, the convergence is much slower than in case of the plate with SSE. Moreover, in case of SSE plate, the analytical solution is available in the form of infinite series [8], which converges fast. On the other hand, the closed form solution is not available for CE plate. Therefore, we can present comparison of numerically obtained distribution of deflections with analytical one only for the SSE plate (Fig. 4).

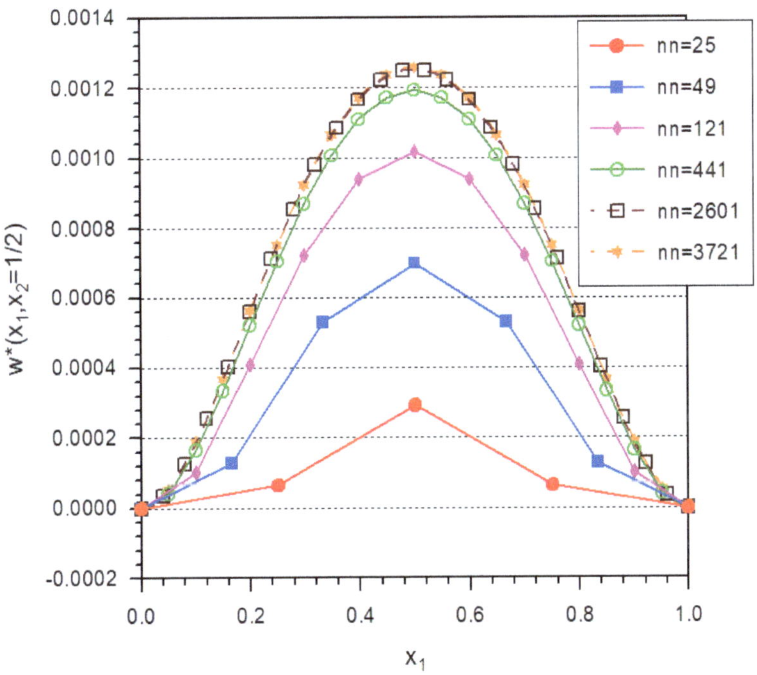

Figure 2: Convergence of numerical results for deflections in plate with CE.

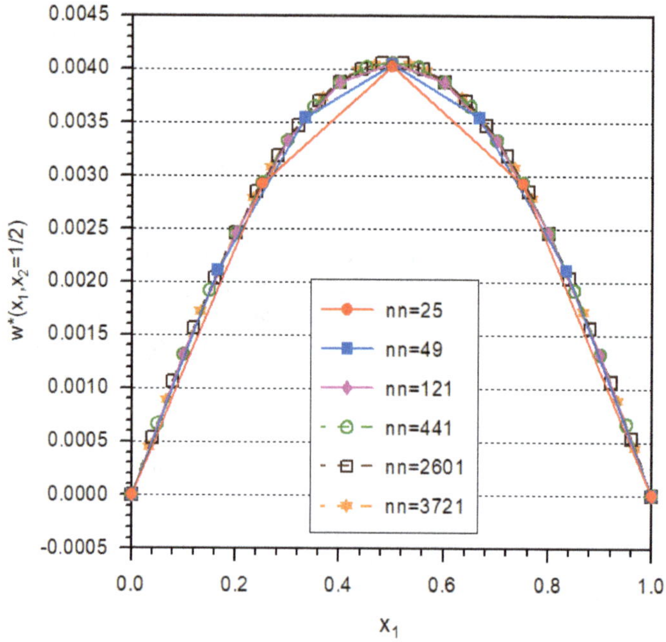

Figure 3: Convergence of numerical results for deflections in plate with SSE.

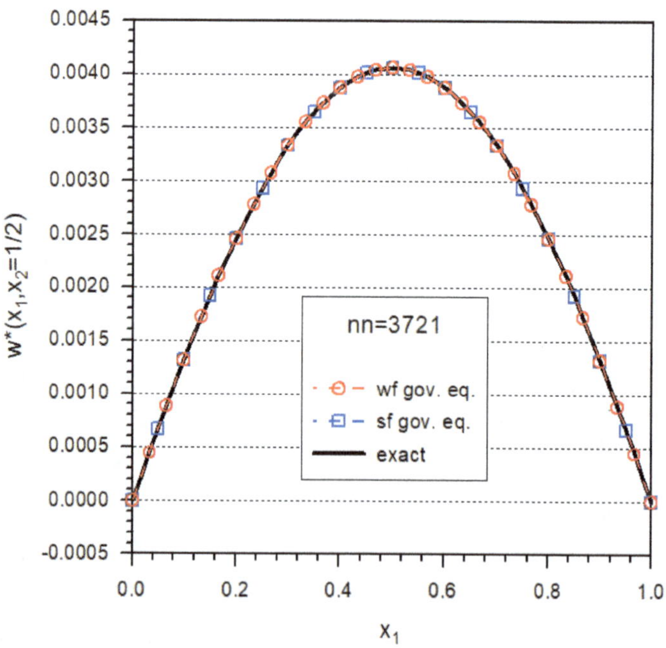

Figure 4: Comparison of numerically computed distribution of deflections with analytical one in SSE plate.

The deviation of numerically computed maximal deflection $w^*_{\max} = 0.4062227 \times 10^{-2} q^*$ from the exact value $w^*_{\max} = 0.4062353 \times 10^{-2} q^*$ is 0.0031% in case of the SSE plate, while in the CE plate the numerical value $w^*_{\max} = 0.1258785 \times 10^{-2} q^*$ deviates from the exact value $w^*_{\max} = 0.126 \times 10^{-2} q^*$ by 0.096%, provided that 5041 nodes were used in numerical computations.

Furthermore, for the SSE plate, we can evaluate also the average error as

$$\%\ \text{error of deflections} = \frac{1}{nn} \sum_{a=1}^{nn} \left| \frac{w^*(\mathbf{x}^a)}{w^*_{ex}(\mathbf{x}^a)} - 1 \right| \times 100\%,$$

and the convergence of the computational method can be measured via the dependence of the % error of deflections on the number of nodes (nn) (Fig. 5). The exact solution for the SSE plate [8] is given as

$$w^*_{ex} = \frac{16}{\pi^6} \frac{t_3^*}{L^4} \sum_{m=1,3,\ldots}^{\infty} \sum_{n=1,3,\ldots}^{\infty} \frac{\sin\dfrac{m\pi x_1}{L_1} \sin\dfrac{n\pi x_2}{L_2}}{mn\left[\left(\dfrac{m}{L_1}\right)^2 + \left(\dfrac{n}{L_2}\right)^2\right]^2}, \quad L_1 = L_2 = L.$$

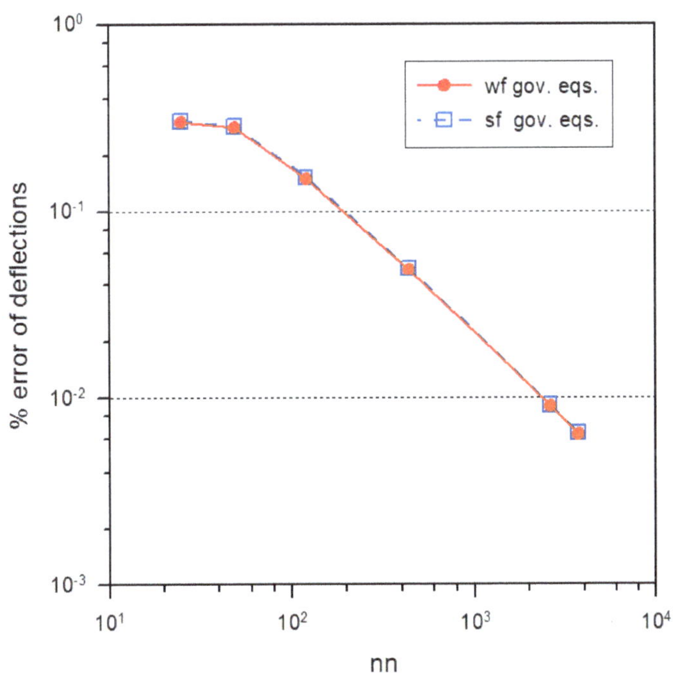

Figure 5: Dependence of the % error of deflections on the number of nodes in SSE plate.

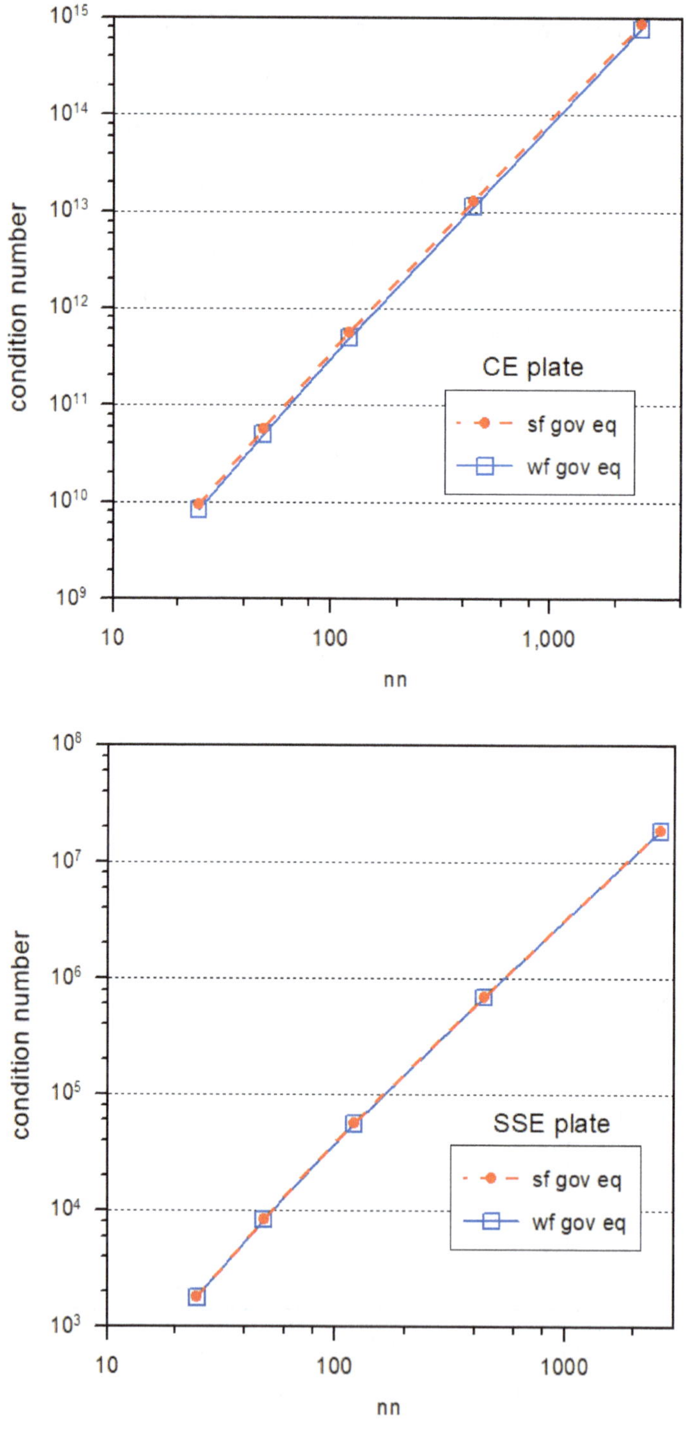

Figure 6: Condition number of discretized equations vs. number of nodes.

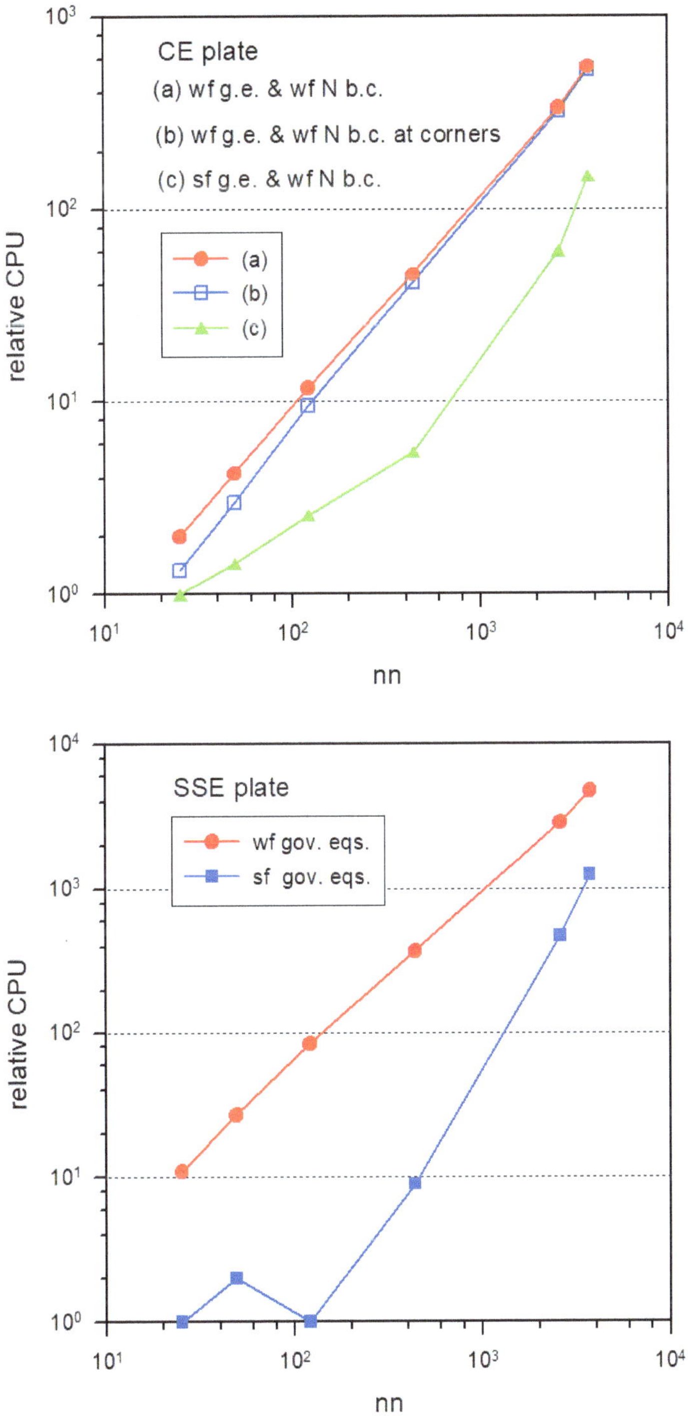

Figure 7: Relative CPU times vs. number of nodes for various techniques in CE and SSE plates.

Another characteristic of computational methods is the numerical stability, which is determined by the sensitivity of accuracy of numerical solution of discretized equations, i.e. linear system of algebraic equations, with respect to a change of the right hand side evoked for instance by inaccuracy of specification of external influence. For this purpose, the condition number is usually used, defined as the maximum ratio of the relative error in solution to the relative error in the r.h.s. of the system of algebraic equations. The dependences of the condition number on the amount of nodes for CE as well as SSE plates are shown in Fig. 6. Significantly better conditioning of the system of discretized equations is achieved in the SSE plate than in the CE plate.

Finally, for comparison of the computational efficiency of approaches employed in this study, we have used the relative CPU defined as the ratio of the actual CPU measured in seconds and the normalization value. The normalization value is selected as the smallest value of the CPU which is 0.140625 (sec) in case of CE plate and 0.15625 (sec) in SSE plate with using 25 nodes. From Fig. 7, it can be seen that the numerical integrations involved in weak formulations result in higher values of the CPU than in strong formulations. These differences between the CPU times are decreasing with increasing the number of nodes, since the time needed for solution of the system of discretized equations is becoming dominant and the time needed for numerical integrations is becoming less significant in approaches incorporating weak formulations.

5 CONCLUSIONS

The Moving Finite Element approximation has been developed and implemented for the solution of bending boundary value problems considered within the classical Kirchhoff–Love theory. Both the strong and weak formulations for governing equations as well as proper formulations of boundary conditions have been derived and verified in numerical test examples with studying: (i) convergence; (ii) accuracy; (iii) numerical stability of solution of the system of discretized equations; and (iv) computational efficiency of various computational approaches. In numerical studies the dependence on increasing number of nodes is involved.

ACKNOWLEDGEMENT

The financial support by the Slovak Research and Development Agency through grant SK-CN-RD-18-0005 is gratefully acknowledged.

REFERENCES

[1] Sladek, V., Sladek, J. & Zhang, C., Local integro-differential equations with domain elements for the numerical solution of partial differential equations with variable coefficients. *Jour. Eng. Math.*, **51**, pp. 261–282, 2005.

[2] Sladek, V., Sladek, J. & Zhang, C., Computation of stresses in non-homogeneous elastic solids by local integral equation method. *Comput. Mechanics*, **41**, pp. 827–845, 2008.

[3] Sladek, V., Sladek, J. & Zhang, C., The use of finite elements for approximation of field variables on local subdomains in a mesh-free way. *Composites with Micro-and Nano-Structure*, ed. V. Kompis, Springer, pp. 87–106, 2008.

[4] Sladek, V., Repka, M. & Sladek, J., Moving finite element method. *WIT Transactions on Engineering Sciences*, vol. 122, WIT Press: Southampton and Boston, 2019. DOI: 10.2495/BE410111.

[5] Gao, X.W., Gao, L.F., Zhang, Y., Cui, M. & Lv, J., Free element collocation method: A new method combining advantages of finite element and mesh free methods. *Computers and Structures*, **215**, pp. 10–26, 2019.

[6] Repka, M., Sladek, V. & Sladek, J., Numerical study of size effects in micro/nano plates by moving finite elements. *Composite Structures*, **212**, pp. 291–303, 2019.
[7] Sator, L., Sladek, V. & Sladek, J., Coupling effects in elastic analysis of FGM composite plates by mesh-free methods. *Composite Structures*, **115**, pp. 100–110, 2014.
[8] Timoshenko, S. & Woinowsky-Krieger, S., *Theory of Plates and Shells*, McGraw Hill: New York, 1959.
[9] Hughes, T.J.R., *The Finite Element Method, Linear Static and Dynamic Finite Element Analysis*, Prentice-Hall: Englewood Cliffs, 1987.

NUMERICAL ANALYSIS OF BURIED VIBRATION PROTECTION DEVICES USING THE METHOD OF FUNDAMENTAL SOLUTIONS

CARLOS ALBINO[1], LUÍS GODINHO[1], DANIEL DIAS-DA-COSTA[1,2] & PAULO AMADO-MENDES[1]
[1]ISISE, Department of Civil Engineering, University of Coimbra, Portugal
[2]School of Civil Engineering, The University of Sydney, Australia

ABSTRACT

Buried structures may be used to control elastic wave propagation in soils and help to reduce vibrations in sensitive structures. The analysis of these structures using numerical tools is of high importance and is usually a demanding computational task. In the present work, the authors explore the possibility of using a meshless method for such simulations, namely the method of fundamental solutions (MFS). In many applications, the MFS has proved to be a worthy and more efficient alternative to classic methods, such as the FEM or even the BEM. Here, the authors present a study on the application of the MFS to simulate the propagation of elastic waves in a soil with multiple buried inclusions. Application examples are also presented, in which the vibration reduction provided by different sets of buried inclusions is analysed and discussed.
Keywords: elastodynamics, MFS, numerical modelling, periodic structures, vibration filtering.

1 INTRODUCTION

Vibration induced in soils by dynamic sources, particularly transportation means, can affect sensitive constructions, and can be nowadays a strong limitation to the implementation of new infrastructures, such as railways. Mitigation of vibrations affecting sensitive constructions has been discussed since the middle of last century when high-speed trains – with speeds above 200 km/h – emerged as regular intercity transport [1]. Nowadays, with the railways crossing urban areas and with an increase in the quality of life, people increasingly complain about the noise and vibrations caused by the passage of trains [2], with the noise and vibration exposure leading to an increased number of awakenings and even cardiac problems [3]. These vibrations induced by the activity of rail transport systems can be mitigated by reducing vibration at source (train, train-track interaction, railway sleepers, etc.) or at a distance from this source, intervening in the host medium, reducing the propagation of waves (trenches, buried structures, etc.). Different authors studied how to reduce the vibration emitted by the source [4] and investigated the mitigation of the vibrations through structures constructed in the host medium [5]. However, simulating these structures requires demanding computational resources if traditional numerical techniques, such as the Finite Element Method (FEM) or the Boundary Element Method (BEM), are to be used. For the FEM, the large finite element mesh required to simulate wave propagation problems can be a significant limitation for its use. The BEM [6], [7] seems to be a very interesting alternative to the FEM, since it allows a very compact description of the propagation medium, requiring only discretization of the interfaces between different materials, and so it has indeed received great interest from the scientific community [8], [9]. However, the BEM also exhibits some limitations, since it produces fully filled equation systems, requires the previous knowledge of the fundamental solutions for the physical problem, and exhibits some additional mathematical complexity due to the need of calculating singular or quasi-singular integrals.

In the last two decades meshless methods emerged trying to overcome some of the difficulties of the conventional methods. Among other advantages, in many cases those

WIT Transactions on Engineering Sciences, Vol 126, © 2019 WIT Press
www.witpress.com, ISSN 1743-3533 (on-line)
doi:10.2495/BE420201

methods allow reducing the computational cost [10], and can be quite simple to implement. An interesting and accepted approach that found applications in many domains is the method of fundamental solutions (MFS). In the case of wave propagation problems, the MFS displacement field is simulated by the superposition of fields generated by a number of virtual sources located outside the domain. The errors of the method can be controlled and minimized through the right choice of the virtual sources and collocation points used to define the solution and enforce the boundary conditions, although this is still an active research topic [11], [12]. In a recent work [13], the singular boundary method (SBM) was also proposed, based on the method of fundamental solutions (MFS). In the SBM, unlike the MFS, virtual sources coinciding with collocation points in the physical boundary of the problem are used to simulate the displacement field and an inverse interpolation technique regularises the singularity of the fundamental solution. In an attempt to combine the advantages of different methods, some researchers have also developed strategies to combine and couple different methods for some specific problems (see for example [14], [15]).

In the present work, the authors make use of the MFS to simulate the propagation of waves in the presence of a set of periodically spaced elastic inclusions buried in a half-space. The main goal is to better understand the possibility of using such periodic systems for the mitigation of propagating vibrations induced by different sources. In what follows, first, a brief description of the governing equations is given, followed by an overview of the MFS formulation and of the involved Green's functions. Then, a study on the accuracy of the method is given. Finally, some numerical examples are presented to illustrate the main results computed in the scope of this work.

2 MATHEMATICAL FORMULATION

2.1 Wave propagation

In the time domain, wave propagation in a homogeneous elastic medium follows the equilibrium equation [9]:

$$\sigma_{ij,j} = \rho \frac{\partial^2 \hat{u}_i}{\partial t^2}, \tag{1}$$

where σ_{ij}, $\hat{u}_i$ and ρ are the stress tensor, the mechanical displacements and the volumetric mass density of the elastic medium, respectively, and the indices i and j can denote either x or y. The comma followed by the index denotes the partial differentiation with respect to the spatial coordinate x_i. The elastic stress tensor is defined as

$$\sigma_{ij} = \lambda \delta_{ij} u_{k,k} + \mu(u_{i,j} + u_{j,i}), \tag{2}$$

where λ and μ are the Lamé material constants, and δ_{ij} is the Kronecker delta function. Following eqns (1) and (2), it can be written that

$$(\lambda + 2\mu)\nabla(\nabla \cdot \hat{\mathbf{u}}) - \mu\nabla \times (\nabla \times \hat{\mathbf{u}}) - \rho \frac{\partial^2 \hat{\mathbf{u}}}{\partial t^2} = 0, \tag{3}$$

where $\hat{\mathbf{u}}$ is the displacement vector. The application of a time Fourier transformation leads to the following equation, when null initial conditions are assumed:

$$(\lambda + 2\mu)\nabla\nabla \cdot \mathbf{u} - \mu\nabla \times \nabla \times \mathbf{u} + \omega^2 \rho \mathbf{u} = 0, \tag{4}$$

in which ω is the angular frequency and $\mathbf{u}$ is the displacement vector in the frequency domain. Here, the particular case of a plane strain problem is addressed. The vector equation can be solved using the MFS which approximates the solution within a given elastic domain as a combination of fundamental solutions related to a set of virtual sources positioned outside the domain to avoid singularities.

2.2 General formulation of the MFS

Consider a solid inclusion Ω_i, with density ρ_i – where compressional waves (P) and shear waves (S) propagate with α_i and β_i velocities, respectively, embedded in an elastic host medium Ω_h, with density ρ_h which allows P and S waves to propagate with α_h and β_h velocities. The host medium is excited by a dynamic load F at position $P_0(x_0, y_0)$ oscillating with a frequency ω, according to the Fig. 1.

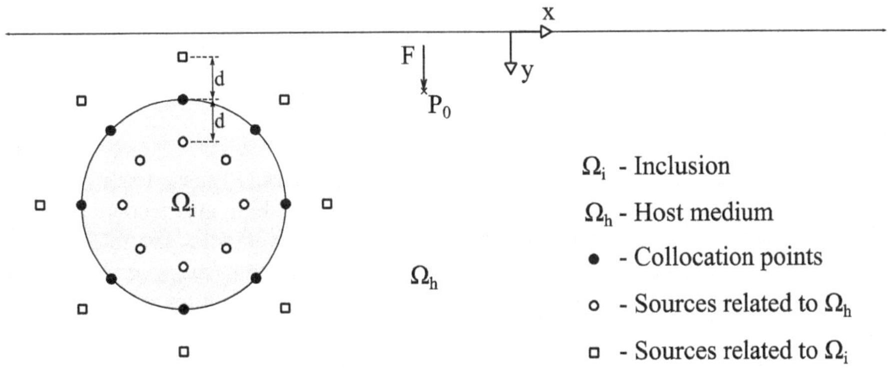

Figure 1: Schematic representation of MFS model.

The wave field can be approximated using the MFS by linearly superimposing the effects of a number of virtual sources located outside the propagation domain [12], [13]. In the present case, two sets of NC virtual sources placed on either side of that interface at a distance d are considered. At each position, virtual loads must be considered acting in both directions x and y. Sources inside the inclusion (related to the host medium) have unknown amplitudes P_{nj} whereas those placed in the host medium (related to the inclusion) have unknown amplitudes Q_{nj}; n is the source order number and j is the direction in which the load acts on this source. Accordingly, in the host medium, the displacements are given by:

$$u_i^h(x,y) = \sum_{n=1}^{NC}\sum_{j=1}^{2} P_{nj} G_{ij}^h(x_n^h, y_n^h, x, y, \rho_h, \alpha_h, \beta_h), \tag{5}$$

while inside the inclusion:

$$u_i^m(x,y) = \sum_{n=1}^{NC}\sum_{j=1}^{2} Q_{nj} G_{ij}^m(x_n^m, y_n^m, x, y, \rho_m, \alpha_m, \beta_m), \tag{6}$$

where $G_{ij}^m(\cdot)$ is the displacement at point (x, y), in medium m, along direction i, generated by a load acting along j at position (x_n^m, y_n^m), computed using the adequate Green's functions [16], [17].

To determine the unknown amplitudes, it is necessary to impose the boundary conditions of continuity of tangential and normal displacements and stresses along the interface, at NC collocation points, establishing a linear equation system with $4NC$ equations and $4NC$ unknowns. After solving these equations, the displacements at any point (x, y) in the propagation domain can be determined as a summation of the contribution of their respective virtual sources.

In the analysed scenarios, the system of equations is generated from two sets of fundamental solutions: one corresponding to the Green's functions for an elastic half-space simulating the host medium and another being the Green's functions for an elastic full-space simulating the field within the inclusion.

2.3 Green's functions

Both sets of fundamental solutions, which relate the field variables (stresses or displacements) at some location in the domain caused by a dynamic source placed elsewhere in the medium, are extensively described in works by Tadeu et al. [16] and Tadeu and Kausel [17]. To avoid the discretization of the horizontal surface of the half-space, Green's functions that take its presence into account are used, following the works of these latter authors. For a half-space medium, the total wave field can be expressed by taking into account the incident field generated by the source (source terms), and the terms generated at the surface (surface terms). The source terms can be written making use of the equations proposed by Tadeu and Kausel [17] for 2D loads, while surface terms can be represented by a set of one dilatational, ϕ, and another shear, ψ, potentials with unknown amplitude values. The continuous integrals of the effects of plane waves which express both the source and the surface terms, can be discretized into summations of discrete terms, assuming the existence of an infinite number of virtual sources placed along the x direction at equal intervals, L_x. The distance separating them is large enough to prevent the virtual loads from contaminating the response. Considering the contribution of $2N + 1$ terms, the potentials that define the surface terms for loads acting along the x direction are

$$\phi_0^x = E_a \sum_{n=-N}^{n=+N} \left(\frac{k_n}{\upsilon_n} E_{b0} A_n^x\right) E_d; \; \psi_0^x = -E_a \sum_{n=-N}^{n=+N} (E_{c0} C_n^x) E_d, \tag{7}$$

and for loads acting along the y direction

$$\phi_0^y = E_a \sum_{n=-N}^{n=+N} (E_{b0} A_n^y) E_d; \; \psi_0^y = E_a \sum_{n=-N}^{n=+N} \left(\frac{k_n}{\gamma_n} E_{c0} B_n^y\right) E_d, \tag{8}$$

where A_n^x, A_n^y, B_n^y and C_n^x, are the unknown coefficient to be determined from appropriate boundary conditions: $E_a = \frac{1}{2\rho_1 \omega^2 L_x}$; $E_{b0} = e^{-i\upsilon_n y}$; $E_{c0} = e^{-i\gamma_n y}$; $E_d = e^{-ik_n(x-x_0)}$; $i = \sqrt{-1}$; $k_n = 2\pi n/L_x$ (the horizontal wavenumber along x); $k_{p1} = \omega/\alpha_1$; $k_{s1} = \omega/\beta_1$; $\gamma_n = \sqrt{k_{s1}^2 - k_n^2}$ (where $Im(\gamma_n) \leq 0$); $\upsilon_n = \sqrt{k_{p1}^2 - k_n^2}$ (where $Im(\upsilon_n) \leq 0$); ω is the excitation frequency.

Within the inclusions, the base functions for the displacements field can be simply described by the full-space Green's functions [17], such that

$$G_{xx}^i(x_0, y_0, x, y, \rho_i, \alpha_i, \beta_i) = G_{xx}^{full}(x_0, y_0, x, y, \rho_i, \alpha_i, \beta_i), \tag{9}$$

and

$$G^i_{yx}(x_0, y_0, x, y, \rho_i, \alpha_i, \beta_i) = G^{full}_{yx}(x_0, y_0, x, y, \rho_i, \alpha_i, \beta_i), \tag{10}$$

where $G^{full}_{ij}(\cdot)$ represents the displacement generated along direction i, due to a load acting along j, by a dynamic source in an infinite elastic medium.

In the case of a half-space it is necessary to impose the following boundary conditions: null tangential and normal stresses at the half-space free surface. Thus, the systems of equations can be established for each value of k_n that allow the unknown amplitude factors to be calculated. As an example, taking the specific case of a load acting along x, after solving the relevant equation systems for a full sequence of values of k_n, in the range $\left[-\frac{2\pi}{L_x}N;\ +\frac{2\pi}{L_x}N\right]$, the final displacements contains an additional summation [17] due to the influence of the surface in order to satisfy the free-surface conditions:

$$G^h_{xx}(x_0, y_0, x, y, \rho_h, \alpha_h, \beta_h) =$$
$$G^{full}_{xx}(x_0, y_0, x, y, \rho_h, \alpha_h, \beta_h) + E_a \sum_{n=-N}^{n=+N} \left(A^x_n \frac{-ik_n^2}{v_n} E_{b0} - i\gamma_n E_{c0} C^x_n \right) E_d, \tag{11}$$

and

$$G^h_{yx}(x_0, y_0, x, y, \rho_h, \alpha_h, \beta_h) =$$
$$G^{full}_{yx}(x_0, y_0, x, y, \rho_h, \alpha_h, \beta_h) + E_a \sum_{n=-N}^{n=+N} (-ik_n^2 A^x_n E_{b0} + ik_n C^x_n E_{c0}) E_d. \tag{12}$$

3 NUMERICAL IMPLEMENTATION AND VERIFICATION

A comprehensive study was made in order to understand the accuracy and stability of the MFS in the analysis of different scenarios. A test case was analysed considering 6 solid circular inclusions with radius $R = 0.5$ m. For the host medium, a shear modulus $G_m = 43.52 \times 10^6$ Pa was considered. Both host medium and inclusion had a density $\rho_m = \rho_i = 1700\ \text{kg/m}^3$ and Poisson's ratio $v_m = v_i = 0.33$. The centre of the upper inclusions was placed at a depth of 2 m and spaced from each other 2 m, either vertically or horizontally, as shown in Fig. 2.

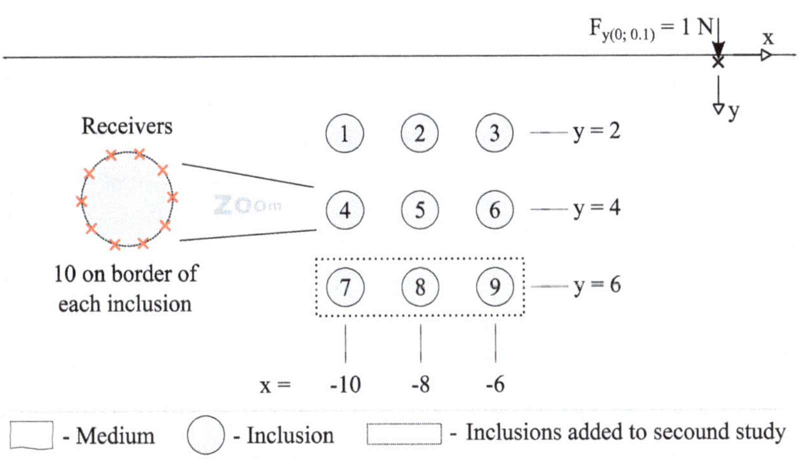

Figure 2: Schematic representation of multiple inclusion model.

The response, in terms of horizontal (x) and vertical (y) displacements, is computed in NE sets of 10 receivers evenly distributed along each inclusion border. Since receivers are placed over the interface between the inclusions and the host medium, the computation is performed by two possible ways: first using the result of eqn (5), as if receivers were in the host medium; and then the result of eqn (6), as if receivers were inside the inclusion.

Results were first computed for 6 inclusions with the same material properties of the host medium and excited by a frequency of 20 Hz, with virtual sources positioned at $0.5R$ from the boundary. For this model (shown in Fig. 2) different collocation points were considered. For the case of 40 points, the imaginary (a) and real (b) parts of response are schematically represented in Fig. 3, where "•" markers are receivers on the original position; "○" markers represent the analytical response computed directly with the Green's functions (since the properties of the inner and outer materials are the same); "×" markers represent the result computed as if the receivers were inside the inclusion using eqn (6); and "+" markers highlight results computed as if the receivers were in the host medium and calculated using the half-space Green's functions depicted in eqn (5). As can be seen in the figure, the numerical results match very well those calculated analytically, using the half-space Green's functions. The relative errors with respect to the analytical response are shown in Fig. 4 which presents the imaginary (a) and real (b) parts of relative errors for all different numbers of collocation points (between 20 and 80). Observing the presented plots, it can be concluded that the relative error tends to decrease with increasing numbers of collocation points, indicating a progressive convergence to the correct solution. Using 40 collocation points the relative error is already lower than 10^{-5}, indicating excellent solution accuracy. Observing Fig. 4(c), it can be seen that the system matrix becomes progressively more ill-conditioned, although the quality of the calculations does not seem to be degraded due to this issue. It is indeed known that when using the MFS, and this parameter should be monitored to avoid numerical errors due to badly conditioned matrices.

An additional check was performed for different contrasts between the properties of the inclusion and those of the host medium. Results are shown in Fig. 5 through the representation of the deformed shape of the inclusion with the same stiffness ("○" markers), half stiffness ("×" markers) and twice the stiffness ("+" markers) of the host medium. This figure also compares these results with those achieved using the BEM under the same

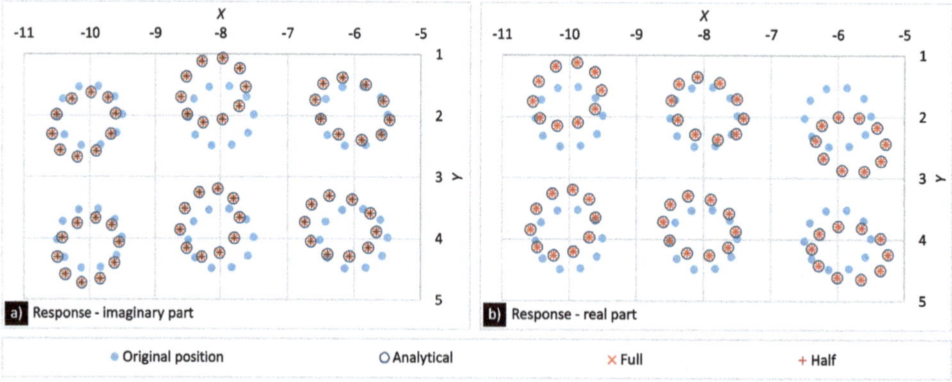

Figure 3: Deformed shape of the inclusions when 40 collocation points are used. (a) Imaginary; and (b) Real parts of the response, amplified by a factor of 1.5×10^8.

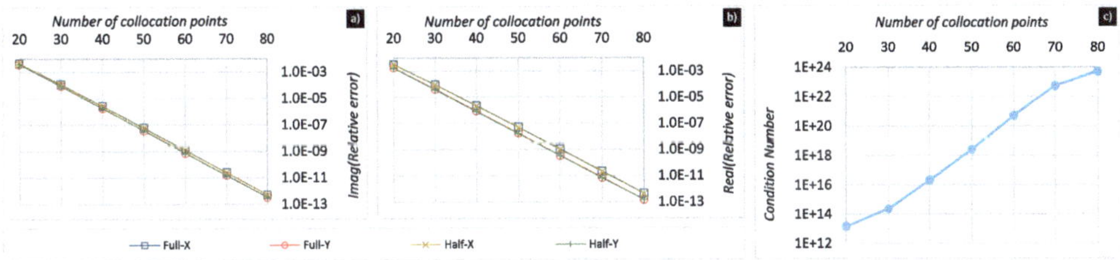

Figure 4: (a) Imaginary; (b) Real parts of relative errors; and (c) System matrix condition number.

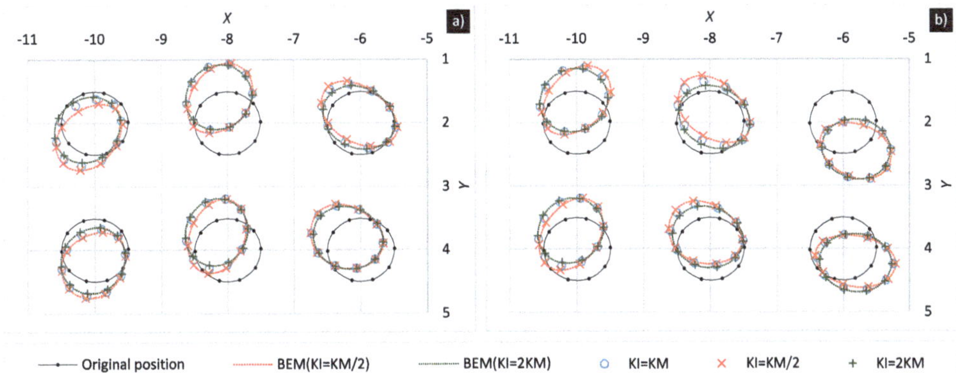

Figure 5: Modified positions of receivers. (a) Imaginary; and (b) Real parts of the responses, amplified by a factor of 1.5×10^8.

conditions. As can be seen, an excellent agreement of the MFS and BEM is obtained. Although not shown here, equally good results were obtained for larger sets of inclusions.

4 NUMERICAL ANALYSIS

To evaluate the effect of the presence of inclusions in the vibrations registered in receivers, it is evaluated a reduction in vibration levels. This reduction is computed in terms of an insertion loss, IL, that is defined as a ratio of the displacement vibration levels obtained in the presence of inclusions to the displacement vibration levels obtained without the inclusions. This ratio is given in dB by the following equation:

$$IL_i = -20log\frac{|u_i|}{|G_{ij}|}, \tag{13}$$

where u_i is the displacement field, along direction i, computed at receivers in presence of inclusions and G_{ij} represents the displacement field, along direction i and generated by a load acting along direction j, computed without the inclusions. According to eqn (13), positive values correspond to a reduction of the displacement vibration levels in the presence of inclusions and negative values of the insertion loss stand for losing protective solutions efficiency.

In this numerical study 4 different groups of solid circular inclusions will be compared, as shown in Fig. 6, buried in an elastic half-space at two depths (from the centre of the upper inclusion), (1 and 3) m, and subjected to two frequencies, (15 and 120) Hz resulting from the dynamic vertical load $F_y = 1$ N acting near the surface at point $(0, 0.1)$ m. All inclusions have a radius of 0.3 m and are made by a homogenous material with elastic properties of $\rho_i = 2700\,\text{kg/m}^3$, $G_i = 1840 \times 10^6$ Pa and $\nu_i = 0.20$, while the hosting soil is characterized by $\rho_m = 1700\,\text{kg/m}^3$, $G_m = 43.61 \times 10^6$ Pa and $\nu_m = 0.33$. The inclusions are placed along the system axes equally spaced between them by a distance which is twice their diameter, and the centre of each group along horizontal direction is kept at $x_G = -15$ m.

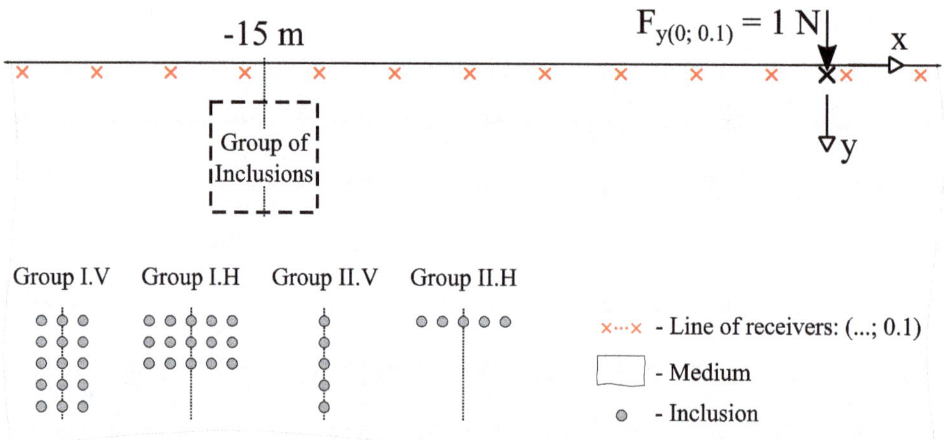

Figure 6: Model for the numerical application related to insertion loss.

All inclusions have a radius of 0.3 m and are made by a homogenous material with elastic properties of $\rho_i = 2700\,\text{kg/m}^3$, $G_i = 1840 \times 10^6$ Pa and $\nu_i = 0.20$, while the hosting soil is characterized by $\rho_m = 1700\,\text{kg/m}^3$, $G_m = 43.61 \times 10^6$ Pa and $\nu_m = 0.33$.

The response in terms of horizontal, x, and vertical, y, displacements is computed along a line of receivers placed 0.1 m below the surface and 0.1 m horizontally spaced in a range of $\{-30, ..., 30\}$ m. The computations are performed using the MFS model, with 30 collocation points distributed along the boundary of each inclusion, and with two sets of 30 virtual sources placed at either side of that interface, at a distance of 0.24 m. Fig. 7 illustrates the insertion loss, in dB, along the line of receivers for x and y components of the displacements and shows the influence in the response of the depth (1 m or 3 m) of inclusions Group I.V (see Fig. 6) and Group I.H, and for frequencies of 15 Hz and 120 Hz. Considering Group I.V (Fig. 7(a)), it can be concluded that for low frequencies (15 Hz) the influence of the depth of the inclusions is not very perceptible, and low attenuation values are always registered. However, at a higher frequency (120 Hz) the effect of this geometric parameter reveals to be determinant in the effect provided by the presence of the inclusions. At receivers placed between the inclusions zone and the sensible structures (downstream zone), considerable attenuation values (positive IL values) are identified, for the lowest depth values. Considering now Group I.H (rotating the inclusions Group I.V by 90°), the computed results at the same line of receivers are presented in Fig. 7(b). Again, it is clear that for the lower frequency, with larger wavelengths, very small attenuations are registered, while a much stronger attenuation effect is seen in the downstream zone for the higher frequency, and for inclusions closer to the surface. Comparing the results computed for both cases, it can be seen that a stronger attenuation is observed when a larger number of inclusions is distributed along the horizontal direction, thus allowing a stronger interference with waves traveling closer to the surface.

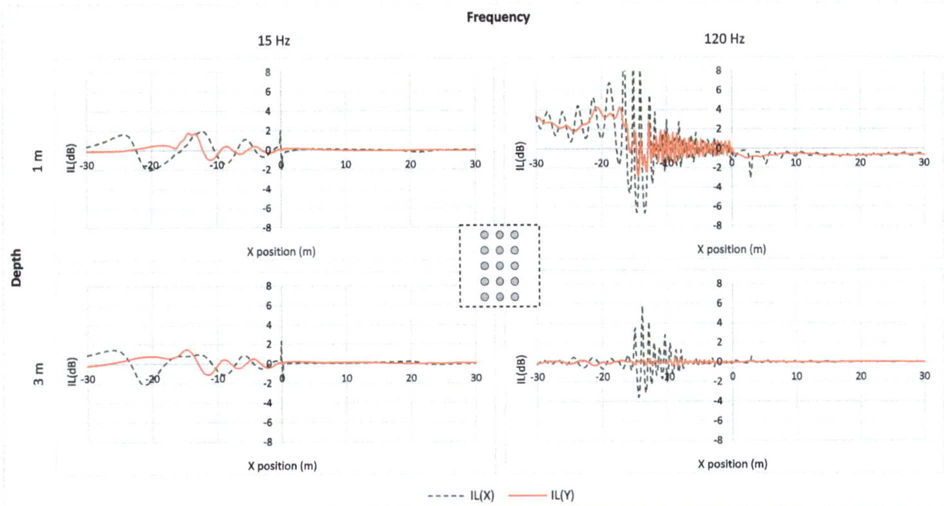

(a)

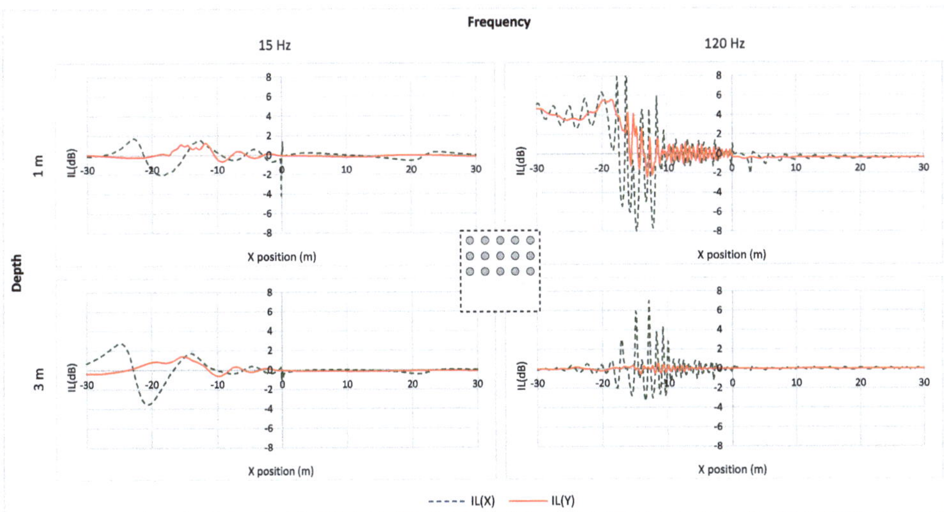

(b)

Figure 7: Attenuation provided by (a) Group I.V; and (b) Group I.H.

In order to identify the effect of varying number of inclusions, relative to the previous model, consider now just the presence of one line (Group II.H) with 5 inclusions, whose results can be seen in Fig. 8. Analysing this set of results, it becomes clear that, for the higher frequency, the effect provided by just one line of inclusions is very similar to that obtained when three horizontal lines are included. When a lower frequency is considered, the larger wavelengths involved lead again to a smaller attenuation when only one line of inclusions is used.

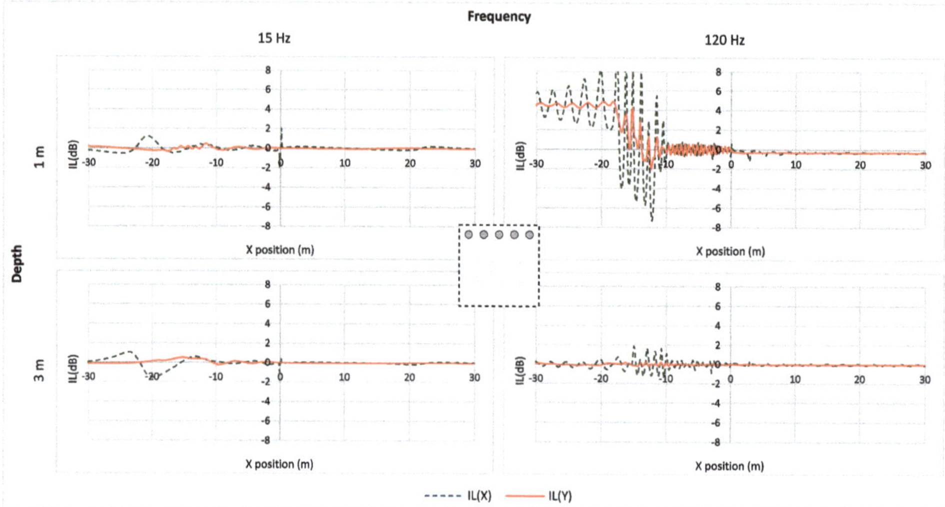

Figure 8: Attenuation provided by Group II.H.

To better asses the attenuation provided by the different configurations, Fig. 9 illustrates the insertion loss (in dB) for each of the inclusions' groups in one third octave bands and computed in the receivers placed on the downstream zone. Clearly, when only one vertical line of inclusions is considered very low insertion loss values are registered throughout the frequency range analysed. When more inclusions are considered along the horizontal direction, again very low attenuations are registered at lower frequencies, but starting on the 63 Hz band a much stronger insertion loss is visible. Indeed, considering the propagation velocities allowed by the host medium (around 247 m/s for P waves, 160 m/s for S waves and 150 m/s for Rayleigh waves), this frequency band coincides with the so-called "phononic crystal effect", generated by the Bragg interference phenomenon occurring between inclusions, which occurs approximately around c/2d (c being the velocity and d the spacing between inclusions). As more inclusions are considered along the horizontal direction, this effect becomes stronger, and leads to stronger attenuations, as is clear in the provided results. This results confirms that this type of mitigation strategy may be designed to fit a given frequency of the propagating vibration field, thus originating a good attenuation and providing effective protection for sensible structures positioned the beyond the inclusions.

5 CONCLUSIONS

In this work, a general MFS strategy was presented to allow the analysis of wave propagation in the presence of buried inclusions in a half-space. The proposed strategy is general and allows the simulation of any number of inclusions filled with distinct materials. The presented numerical results evidence the stability and convergence of solutions. The accuracy of the MFS is quite good, particularly when compared with BEM results. This paper only reported findings concerning the 2D case, and further in-depth studies need to be conducted to verify and improve the proposed strategy, in particular for 2.5D and 3D cases. Conceptually, the

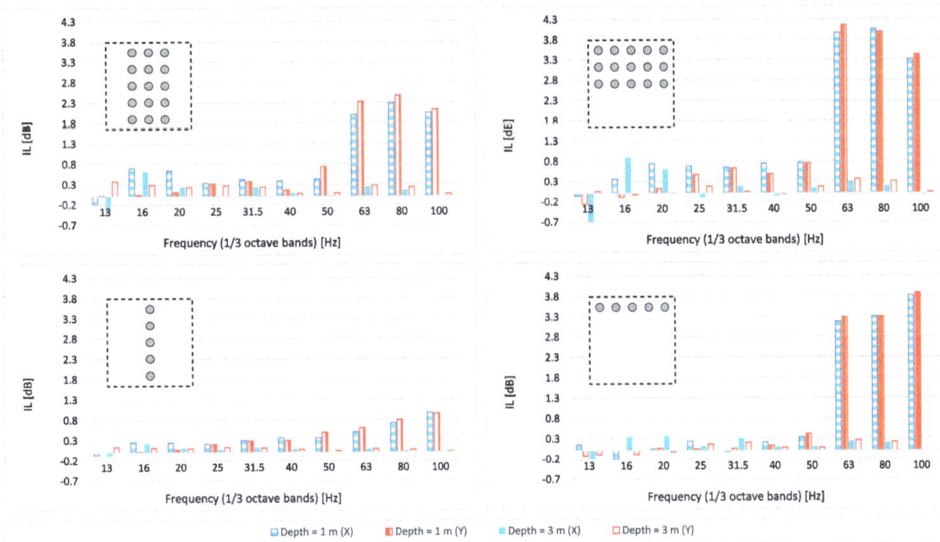

Figure 9: Insertion loss in 1/3 octave bands.

inclusions have a significant effect on vibration levels reduction, and it was observed that their efficiency peaks at a specific frequency range, which is strongly related to the Bragg effect occurring due to the multiple interferences between reflections at the equally spaced structures.

ACKNOWLEDGEMENTS

This work was financially supported by: Project PTDC/ECM-COM/1364/2014 – POCI-01-0145-FEDER-016783 (METASHIELD) – funded by FEDER funds through COMPETE2020 – Programa Operacional Competitividade e Internacionalização (POCI) and by national funds through FCT – Fundação para a Ciência e a Tecnologia, I.P. This work was partly financed by FEDER funds through the Competitivity Factors Operational Programme – COMPETE and by national funds through FCT – Foundation for Science and Technology within the scope of the project POCI-01-0145-FEDER-007633 (ISISE) and through the Regional Operational Programme CENTRO2020 within the scope of the project CENTRO-01-0145-FEDER-000006. D. Dias-da-Costa would like to acknowledge the support from the Australian Research Council through its Discovery Early Career Researcher Award (DE150101703) and ARC Projects (DP140100529, LP140100591).

REFERENCES

[1] Yang, Y.B. & Hung, H.H., *Wave Propagation for Train-Induced Vibrations. A Finite/Infinite Element Approach*, World Scientific Publishing: Singapore, 2009.

[2] Al Suhairy, S., *Prediction of Ground Vibration From Railways*, SP Swedish National Testing and Research Institute: Sweden, 2000.

[3] Croy, I., Smith, M.G. & Waye, K.P., Effects of train noise and vibration on human heart rate during sleep: An experimental study. *BMJ Open*, 2013. DOI: 10.1136/bmjopen-2013-002655.

[4] Lombaert, G., Degrande, G., Vanhauwere, B., Vandeborght, B. & François, S., The control of ground-borne vibrations from railway traffic by means of continuous floating slabs. *Journal of Sound and Vibration*, **297**(3–5), pp. 946–961, 2006.

[5] Andersen, L., Frigaard, P. & Augustesen, A., Mitigation of ground vibration by double sheet-pile walls. *Geotechnical Engineering/Construction Technology and Management: Proceedings of the 8th International Congress on Advances in Civil Engineering, ACE 2008*, Famagusta, North Cyprus, 2008.

[6] Ang, W.T., *A Beginner's Course in Boundary Element Methods*, Universal Publishers: Boca Raton, USA, 2007.

[7] Brebbia, C.A. & Dominguez, J., Boundary element methods for potential problems. *Applied Mathematical Modelling*, **1**(7), pp. 372–378, 1977.

[8] Tsiatas, G.C. & Yiotis, A.J., A BEM-based meshless solution to buckling and vibration problems of orthotropic plates. *Engineering Analysis with Boundary Elements*, **37**(3), pp. 579–584, 2013.

[9] Godinho, L., Amado-Mendes, P. & Tadeu, A., Meshless analysis of soil–structure interaction using an MFS–MLPG coupled approach. *Engineering Analysis with Boundary Elements*, **55**, pp. 80–92, 2015.

[10] Fries, T.-P. & Matthies, H.-G., *Classification and Overview of Meshfree Methods*, Institute of Scientific Computing, Technical University Braunschweig: Brunswick, Germany, 2004.

[11] Godinho, L.M.C., Costa, E.G.A., Pereira, A.S.C. & Santiago, J.A.F., Some observations on the behavior of the method of fundamental solutions in 3D acoustic problems. *International Journal of Computational Methods*, **9**(04), 1250049, 2012.

[12] Tadeu, A., António, J. & Godinho, L., Defining an accurate MFS solution for 2.5D acoustic and elastic wave propagation. *Engineering Analysis with Boundary Elements*, **33**(12), pp. 1383–1395, 2009.

[13] Gu, Y., Chen, W. & Zhang, C.-Z., Singular boundary method for solving plane strain elastostatic problems. *International Journal of Solids and Structures*, **48**(18), pp. 2549–2556, 2011.

[14] Soares, Jr. D. & Godinho, L., An overview of recent advances in the iterative analysis of coupled models for wave propagation. *Journal of Applied Mathematics*, 426283 2014.

[15] Godinho, L., Amado-Mendes, P., Pereira, A. & Soares, Jr. D., A coupled MFS–FEM model for 2-D dynamic soil–structure interaction in the frequency domain. *Computers and Structures*, **129**, pp. 74–85, 2013.

[16] Tadeu, A., António, J. & Godinho, L., Green's function for two-and-a-half dimensional elastodynamic problems in a half-space. *Computational Mechanics,* **27**, pp. 484–491, 2001.

[17] Tadeu, A. & Kausel, E., Green's functions for two-and-a-half dimensional elastodynamic problems. *Journal of Engineering Mechanics*, **126**(10), pp. 1093–1097, 2000.

COUPLING OF THE FAST BOUNDARY-DOMAIN INTEGRAL METHOD WITH THE STOCHASTIC COLLOCATION METHOD FOR FLUID FLOW SIMULATIONS

JURE RAVNIK[1], ANNA ŠUŠNJARA[2], JAN TIBAUT[1], DRAGAN POLJAK[2] & MARIO CVETKOVIĆ[2]
[1]Faculty of Mechanical Engineering, University of Maribor, Smetanova Ulica 17, SI-2000 Maribor, Slovenia
[2]University of Split, Faculty of Electrical Engineering, Mechanical Engineering and Naval Architecture, Croatia

ABSTRACT

In this paper we present the development of a Fast Boundary-Domain Integral Method for the solution of fluid flow and heat transfer problems. Adaptive cross approximation of integral matrices is performed on a hierarchically decomposed matrix structure. We couple the developed flow simulation algorithm with stochastic modelling of input parameters. In order to assess the influence of input parameters on the simulation results, we employ the stochastic collocation method as a wrapper around the deterministic code. We apply the developed numerical tool for the simulation of flow and heat transfer of nanofluids.
Keywords: Boundary-Domain Integral Method, stochastic collocation method, sensitivity analysis, nanofluid, natural convection.

1 INTRODUCTION

The Boundary-Domain Integral Method [1] has been successfully used to solve several kinds of engineering problems. Its main drawback is the large computational resources required due to the appearance of fully populated matrices during the discretization procedure. In this paper we present the development of a Fast Boundary-Domain Integral Method [2], [3] for the solution of fluid fl ow an d he at tr ansfer pr oblems, wi th th e aim to reduce computational requirements of the algorithm. The Navier–Stokes equations in velocity-vorticity formulation are considered. Diffusion and modified Helmholtz fundamental solutions are used to derive a Boundary-Domain Integral representation of the governing equations. We employ two discretization approaches. The vorticity transport equation is solved using a domain decomposition approach. This leads to a finite element type sparse structure of integral matrices and enables efficient use of computational resources. The kinematics equation, on the other hand, is solved on a single domain, which leads to full matrices of integrals. To mitigate the high computational resources required, we propose the use of a hierarchical decomposition of the domain, which leads to $\mathcal{H}$-matrix representation of integral matrices. Admissible matrix blocks are approximated using the adaptive cross approximation technique. Efficiency of the proposed acceleration strategy is discussed and an *optimal compression ratio* is found to exist. We define the optimal approximation ratio as the highest compression ratio, where the error introduced by the approximation is of the same order as other errors of the numerical scheme.

The reduced computational demands of the algorithm enables us to couple the developed flow simulation algorithm with stochastic modelling of input parameters. In order to assess the influence of input parameters on the simulation results, we employ the stochastic collocation method (SCM, [4], [5]) as a wrapper around the deterministic flow simulation code. In this way, we are able to propagate the uncertainty from input to output parameters. First, we identify the most important parameters using the "one-at-a-time" (OAT) principle and then, the full tensor SCM was used to assess the stochastic mean, variance and Sobol-like indices for sensitivity analysis [6].

WIT Transactions on Engineering Sciences, Vol 126, © 2019 WIT Press
www.witpress.com, ISSN 1743-3533 (on-line)
doi:10.2495/BE420211

Finally, we apply the developed numerical tool for the simulation of flow and heat transfer of nanofluids. A nanofluid is a term describing a dilute dispersion of particles in a fluid. The diameter of particles is in the order of ten nanometres. The particles are made of metal oxides, which enhances the thermal properties of the suspension. Natural convection in a cubical, differentially heated cavity is considered as a model problem for stochastic analysis.

2 METHODS

The velocity vorticity formulation of Navier–Stokes equations as a model for nanofluid flow and heat transfer was presented in Ravnik et al. [7]. In nondimensional setting, it reads:

$$\nabla^2\vec{v} + \vec{\nabla} \times \vec{\omega} = 0, \tag{1}$$

$$\frac{\partial\vec{\omega}}{\partial t} + (\vec{v}\cdot\vec{\nabla})\vec{\omega} = (\vec{\omega}\cdot\vec{\nabla})\vec{v} + Pr\frac{\mu_{nf}}{\mu_f}\frac{\rho_f}{\rho_{nf}}\nabla^2\vec{\omega} - PrRa\frac{\beta_{nf}}{\beta_f}\vec{\nabla}\times T\vec{g}, \tag{2}$$

$$\frac{\partial T}{\partial t} + (\vec{v}\cdot\vec{\nabla})T = \frac{k_{nf}}{k_f}\frac{(\rho c_p)_f}{(\rho c_p)_{nf}}\nabla^2 T, \tag{3}$$

where Pr and Ra are the Prandtl and Rayleigh numbers respectively. The nanofluid is modelled using effective properties, such as: density ρ_{nf}, viscosity μ_{nf}, heat capacitance $(c_p)_{nf}$, thermal expansion coefficient β_{nf} and thermal conductivity k_{nf}. Models for these are listed in [8]–[9]. The solution of vorticity and energy transport eqns (2) and (3) used the sub-domain Boundary Domain Integral Methods, as proposed by Ravnik et al. [10].

For the solution of the kinematics eqn (1) for unknown boundary vorticity values, we use the following integral statement:

$$c(\vec{\xi})\vec{v}(\vec{\xi}) + \int_\Gamma \vec{v}\,\vec{\nabla}u^\star\cdot\vec{n}d\Gamma = \int_\Gamma \vec{v}\times(\vec{n}\times\vec{\nabla})u^\star d\Gamma + \int_\Omega(\vec{\omega}\times\vec{\nabla}u^\star)d\Omega, \tag{4}$$

where $u^\star(\vec{r}) = \frac{1}{4\pi|\vec{r}-\vec{\xi}|}$ is the fundamental solution of the Laplace equation, Ω the domain, Γ the boundary and $\vec{\xi}$ is the source point. Placing $\vec{\xi}$ into all boundary nodes and performing discretization leads to a full system linear equations. We approximate the matrices using the adaptive cross approximation (ACA) technique [11]. First, the matrices are subdivided using a recursive algorithm, which gives a hierarchical $\mathcal{H}$ matrix structure (Fig. 1) [12]. We defined an optimal compression ratio, where maximum compression is achieved at a negligible effect of simulation accuracy. Further details of the approach are given in [2], [3], [13].

In order to asses the influence of nanofluid parameters (density, viscosity, heat capacitance, nanofluid concentration and thermal conductivity) on the simulation results, we propose the use of a stochastic model, which consists of d input parameters modelled as random variables (RV), organised in input vector: $\mathbf{X} = [x_1, \ldots x_d]$. The SCM is based on the polynomial approximation of the considered output Y in the d dimensional stochastic space:

$$\hat{Y}(\mathbf{X}) = \sum_{k=1}^{N} L_x(\mathbf{X})\cdot Y^{(k)}, \tag{5}$$

where $L_k(\mathbf{X})$ is Lagrange basis function and $Y^{(k)}$ is the output realization for the kth input point. The interpolation points (SC points) in each dimension are calculated according to Gauss–Legendre quadrature rule. The interpolation in the multivariate dimension space is done by using the tensor product of basis functions. The stochastic mean and variance are calculated by following the formula from the statistics.

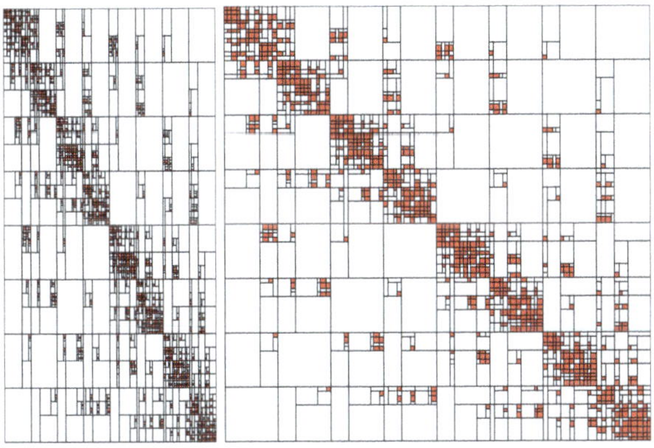

Figure 1: An example of domain-boundary (left) and boundary-boundary (right) $\mathcal{H}$ matrix structure. Red areas are inadmissible parts, where approximation is not performed, [2].

Sensitivity analysis based on variance decomposition in which the variance of a model is decomposed into terms depending on the input factors and their mutual interactions allows the computation of sensitivity indices of first and high order [5]. In practice, only first order indices are used:

$$S_i = \frac{V_{X_i}[E_{X_{\sim i}}(Y|X_i)]}{V(Y)}. \tag{6}$$

To capture the contribution of all terms which include the variable X_i, a total effect index is defined as follows:

$$S_{T_i} = 1 - \frac{V_{X_{\sim i}}[E_{X_i}(Y|X_{\sim i})]}{V(Y)}. \tag{7}$$

A more simple way is a "one-at-a-time" (OAT) approach in which one input parameter is modelled as RV while the others are kept at nominal values. The impact of variables is determined by comparing their respective variances.

3 RESULTS

As a model problem we consider the development of 3D natural convection in a differentially heated cavity. Water with aluminium nanoparticles was the chosen nanofluid. We simulated the problem at different nanoparticle concentrations and temperatures differences for a steady state solution.

In the first step, 12 input parameters are modelled as random with uniform distribution set as 10% variation from nominal value: volume fraction, temperature difference, fluid viscosity, heat capacity, density, thermal conductivity, thermal expansion coefficient, characteristic dimension, particle heat capacity, density, thermal conductivity, thermal expansion coefficient. To reduce the dimensionality of a problem 12 univariate test cases (TC) were done following the logic of OAT principle. Stochastic simulations with 3 and 5 SC points in each dimension were run leading to total of 73 deterministic simulations. The convergence of the method was satisfactory. After comparing the variances from 12 TCs, five input variables are recognized as the important ones, while the others will be neglected in further computations. Those 5 input variables are: volume fraction, temperature difference,

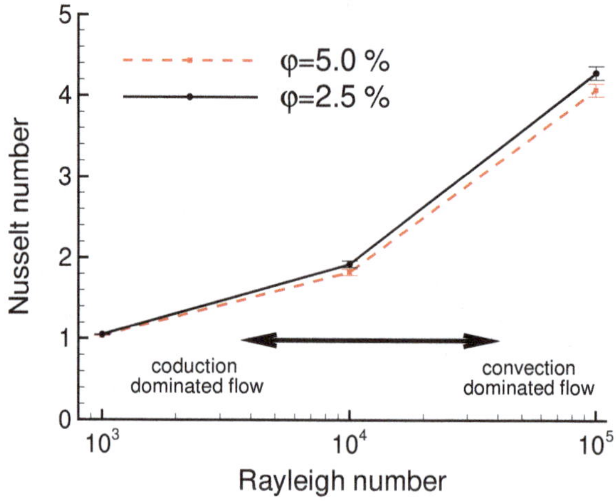

Figure 2: Heat flux variation due to uncertainty in input parameters. Results for two nanoparticle concentrations are shown.

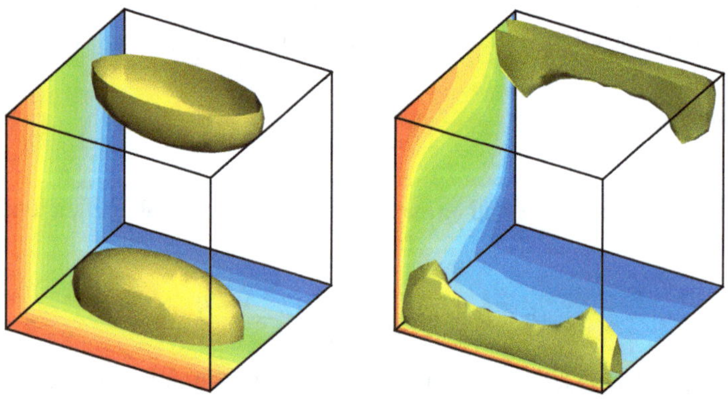

Figure 3: Isosurfaces indicating the volume where the greatest variation in temperature field occurs. Left panel: conduction dominated flow regime, $Ra = 10^3$, right panel: convection dominated flow regime, $Ra = 10^5$.

fluid viscosity, heat capacity and, thermal conductivity. In the second step a full tensor model with five random inputs and accuracy of 3SC was done.

In Fig. 2 we present the heat flux (expressed as Nusslet number) as a function of Rayleigh number. We observe that the uncertainty in input parameters is expressed more prominently in the convection dominated flow regime. Fig. 3 reveals areas in the flow domain where the greatest variation in the temperature field occurs. They are located at the top and bottom wall of the domain. In convection dominated flow regimes the areas of greatest variations are found along the edges of the domain.

4 SUMMARY

We presented an adapted cross approximation accelerated Boundary-Domain Integral Method for the simulation of nanofluids using the solution of the velocity-vorticity formulation of Navier–Stokes equations. The developed algorithm was coupled with a stochastic input parameter sensitivity analysis. We have successfully shown that acceleration of the boundary domain integral method enables performing larger simulation sets, which are needed to consider stochastic analyses.

REFERENCES

[1] Škerget, L., Hriberšek, M. & Kuhn, G., Computational fluid dynamics by boundary — domain integral method. *International Journal for Numerical Methods in Engineering*, **46**(8), pp. 1291–1311, 1999.

[2] Ravnik, J. & Tibuat, J., Fast boundary-domain integral method for unsteady convection-diffusion equation with variable diffusivity using the modified Helmholtz fundamental solution. *Numerical Algorithms*, 2019. https://doi.org/10.1007/s11075-019-00664-3.

[3] Tibaut, J. & Ravnik, J., Fast boundary-domain integral method for heat transfer simulations. *Engineering Analysis with Boundary Elements*, **99**, pp. 222–232, 2019.

[4] Poljak, D., Šesnić, S., Cvetković, M., Šušnjara, A., Dodig, H., Lalléchère, S. & El Khamlichi Drissi, K., Stochastic collocation applications in computational electromagnetics. *Mathematical Problems in Engineering*, 2018. https://doi.org/10.1155/2018/1917439.

[5] Saltelli, A., Annoni, P., Azzini, I., Campolongo, F., Ratto, M. & Tarantola, S., Variance based sensitivity analysis of model output. Design and estimator for the total sensitivity index. *Computer Physics Communications*, **181**(2), pp. 259–270, 2010.

[6] Ravnik, J., Susnjara, A., Tibaut, J., Poljak, D. & Cvetkovic, M., Stochastic modelling of nanofluid heat transfer. *4th Workshop UNEMA*, Split, Croatia, 2018.

[7] Ravnik, J., Skerget, L. & Hribersek, M., Analysis of three-dimensional natural convection of nanofluids by BEM. *Engineering Analysis with Boundary Elements*, **34**(12), pp. 1018–1030, 2010.

[8] Khanafer, K., Vafai, K. & Lightstone, M., Buoyancy-driven heat transfer enhancement in a two-dimensional enclosure utilizing nanofluids. *Int J Heat Mass Transfer*, **46**, pp. 3639–3653, 2003.

[9] Oztop, H.F. & Abu-Nada, E., Natural convection of water-based nanofluids in an inclined enclosure with a heat source. *Int J Heat Fluid Flow*, **29**, pp. 1326–1336, 2008.

[10] Ravnik, J., Škerget, L. & Žunič, Z., Velocity-vorticity formulation for 3D natural convection in an inclined enclosure by BEM. *International Journal of Heat and Mass Transfer*, **51**(17–18), pp. 4517–4527, 2008.

[11] Bebendorf, M., Approximation of boundary element matrices. *Numerische Mathematik*, **86**(4), pp. 565–589, 2000.

[12] Hackbusch, W., A sparse matrix arithmetic based on H-matrices. Part I: Introduction to H-matrices. *Computing*, **108**, pp. 89–108, 1999.

[13] Tibaut, J., Škerget, L. & Ravnik, J., Acceleration of a BEM based solution of the velocity-vorticity formulation of the Navier–Stokes equations by the cross approximation method. *Engineering Analysis with Boundary Elements*, **82**, pp. 17–26, 2017.

IMMERSED BOUNDARY METHOD APPLICATION AS A WAY TO BUILD A SIMPLIFIED FLUID-STRUCTURE MODEL

JONATAS EMMANUEL BORGES[1], MARCOS ANTONIO DE SOUZA LOURENÇO[2], ELIE LUIS MARTÍNEZ PADILLA[3] & CHRISTOPHER MICALLEF[4]
[1]Federal University of Mato Grosso, Brazil
[2]Federal University of Technology – Paraná, Brazil
[3]Federal University of Uberlândia, Brazil
[4]University of Malta, Malta

ABSTRACT

A simplified fluid-structure interaction model, consisting of a cylinder tethered by a spring system interacting dynamically with an incompressible two-dimensional lid-driven cavity flow, is solved using the immersed boundary method. Results show that when the spring forces are weaker than the fluid drag force, the springs stretch freely and the cylinder motion is the direct result of the fluid dynamics action. For higher values of spring forces, the cylinder motion reaches a maximum displacement, and the spring forces induce the cylinder to an oscillatory movement damped by the fluid drag forces. Subsequently the amplitude of the displacement decreases. The cylinder motion is restricted within the mainstream fluid flow, where the maximum displacement reduces as the Reynolds number increases.
Keywords: computational fluid dynamics, immersed boundary method, fluid-structure interaction.

1 INTRODUCTION

A lot of research is going on in the field of computational fluid dynamics (CFD) in order to balance the need to model increasingly complex boundary conditions and achieve highly accurate results at a minimum amount of computational time and resources [1]. The great majority of engineering fluid flow problems are characterized by complex geometries, which are often associated with the presence of solid, moving or flexible walls.

Most situations encountered in engineering practice involve complex geometries which are not fit for Cartesian grids and the choice of grid is not at all trivial. In order to deal with complex geometries, whole families of numerical methods have been developed. Body-fitted coordinate grids, such as curvilinear grids, non-orthogonal grids and non-structured grids, are usually employed. Unfortunately grid generation for complex geometries is generally an issue since it consumes a lot of user time especially when commercial codes are not employed.

One alternative to body-fitting meshes is the increasingly popular immersed boundary (IB) method. The immersed boundary method was pioneered by Peskin [2] in a paper on simulation of blood flowing through cardiac valves, where biological deformable tissue interact with the viscous flow. Since then, the method has experienced successive refinements and modifications, attracting considerable interest as an important tool for solving general fluid–fluid or fluid–solid interaction problems [3]–[6].

The distinguishing quality of an IB method is to simulate an interface inside a flow by adding a force field to discretised Navier–Stokes equations. The embedded interface is represented by an arbitrary Lagrangian mesh whereas the flow domain is usually discretised by a simple Eulerian orthogonal grid. An interpolation function transfers the information from one domain to another and back.

Borges et al. [7] applied the direct forcing method [4] in order to investigate Newtonian laminar flow through a three-dimensional sudden contraction using Cartesian meshes. To

WIT Transactions on Engineering Sciences, Vol 126, © 2019 WIT Press
www.witpress.com, ISSN 1743-3533 (on-line)
doi:10.2495/BE420221

compute the transfer functions between the Eulerian and Lagrangian grids, the moving-least-square (MLS) approximation [5] was employed. Comparison between numerical and experimental velocity profiles has shown good agreement.

The advantages of IB method are interesting for fluid-structure interaction where the body motion or body deformation does not imply the need for grid restructuring at every single time step. Fluid-structure interaction (FSI) problems occur in many areas of engineering and represent one of the greatest challenges of our time.

Fluid-structure interaction problems can be solved either monolithically, by solving the flow and structure equations simultaneously, or in a partitioned way, having separate solvers for the fluid flow equations and the structural equations. The main advantage of the monolithic approach is that no coupling iterations within the time step are required. On the other hand, existing flow solvers and structural solvers can be employed in a partitioned simulation, reducing software development efforts and improving manageability. Furthermore, this approach can solve FSI problems that do not require matching meshes at the interfaces between the solid and the fluid [8]. Degroote et al. [9], [10] compared the monolithic and partitioned approaches and reported that the performance of the two approaches was problem-dependent.

Nowadays, FSI simulations are widely used in industry. In the aerospace industry, FSI is used in order to study flutter on control surfaces, lift devices and also turbo machinery blades. On the other hand FSI is used in the renewable energy industry to investigate wind turbine behavior and also wave-energy converters. Parachute dynamics is another interesting field as large deformations appear. The design and analysis of artificial heart valves, the prediction of the rupture of aneurysms or of the outcome of surgery also rely extensively on FSI simulations in patient-specific geometries [9].

The difficulty of developing mathematical models which adequately represent the complexity of fluid-structure interaction problems, has led to simplified physical models in order to initiate the study and for a better understanding of the phenomena involved. One of the strategies for the development of simplified models was the proposal where the structure is represented by a body, spring and damper assembly. As an example, important knowledge has been acquired on the vortex-induced vortex vibration problem (VIV) considering the flow on oscillating cylinders [11]–[13].

Campregher [14] employed the physical virtual model originally proposed by Lima e Silva et al. [3] to study the flow around a stationary cylinder. Results were validated against well reported benchmarks in published literature. The study was then extended to a fluid-structure interaction problem using the partitioned approach. The dynamic system chosen was composed by a cylinder tethered by springs, immersed in the flow. This study produced interesting results.

This work presents a simplified fluid-structure problem where the cylinder, represented by the multi-direct forcing of Wang et al. [6], is tethered by springs immersed in an incompressible two-dimensional Newtonian lid-driven cavity flow, where fluid-structure interaction takes places.

2 MATHEMATICAL DESCRIPTION

2.1 Governing equations for fluid flow

This work is restricted to incompressible Newtonian fluid flows which are governed by Navier–Stokes and continuity equations. In the physical space and relative to an inertial reference frame these equations are

$$\frac{\partial u_i}{\partial t} + \frac{\partial(u_i u_j)}{\partial x_j} = -\frac{1}{\rho}\frac{\partial p}{\partial x_i} + \frac{\partial}{\partial x_j}\left[\upsilon\left(\frac{\partial u_i}{\partial x_j} + \frac{\partial u_j}{\partial x_i}\right)\right] + \frac{f_i}{\rho}, \quad (1)$$

$$\frac{\partial u_j}{\partial x_j} = 0, \quad (2)$$

where $u_i(\vec{x},t)$ and $p(\vec{x},t)$ are the velocity and pressure fields, respectively; $f_i(\vec{x},t)$ is the force field, ρ is the density and ν is the kinematic viscosity. The term $f_i(\vec{x},t)$ is the external force exerted on the flow field which is the mutual interaction force between fluid and immersed boundary expressed as follows

$$f_i(\vec{x},t) = \int_\Gamma F_k(\overrightarrow{x_k},t)\,\delta(\vec{x} - \overrightarrow{x_k})dx_k, \quad (3)$$

where $\delta(\overrightarrow{x_k} - \vec{x})$ is a distribution function, $\overrightarrow{x_k}$ is the position of the Lagrangian markers set at the immersed boundary, $\vec{x}$ is the position of the computational Eulerian mesh and $F_k(\overrightarrow{x_k},t)$ is the force exerted on the Lagrangian marker $\overrightarrow{x_k}$. Fig. 1 illustrates the distribution of the Lagrangian force to adjacent Eurelian volumes.

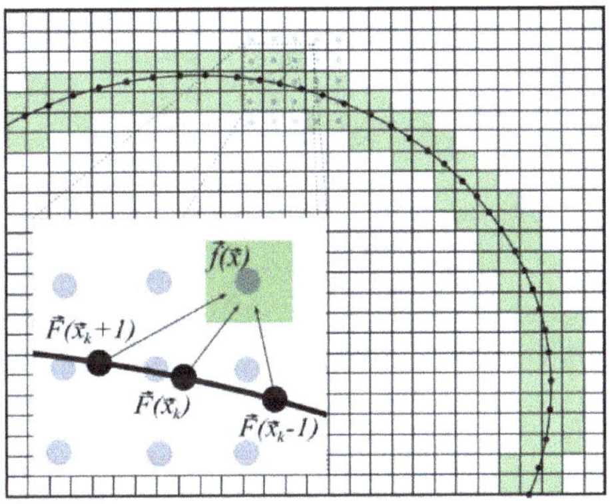

Figure 1: Distribution of the Lagrangian force to adjacent Eulerian volumes (shaded).

3 NUMERICAL METHODS

The governing eqns (1) and (2) are discretized by the finite volume method, where the advective and diffusive terms are discretised using a central differencing scheme, in a staggered arrangement, as proposed by Patankar [15]. In a simple rectilinear mesh discretization, the components u and v are positioned in the volume's normal faces in the x and y directions respectively. Scalar values, as the pressure, are located at the volume center. The discretization in time was performed by a Runge–Kutta second order scheme. Pressure–velocity coupling uses the two-step fractional method of Kim and Moin [16]. The solution of the Poisson equation is performed by Successive Over Relaxation (SOR) iterative method.

The velocity field is initially estimated as u_i^*, in eqn (4)

$$u_i^* = u_i^t + \left(-\frac{1}{\rho}\frac{\delta p^t}{\delta x_i} - \frac{\delta(u_i u_j)}{\delta x_j} + \upsilon\frac{\delta^2 u_{k\,i}}{\delta x_{k\,j}\delta x_{k\,j}}\right)^l \frac{\Delta t}{2-l} \qquad l = 0{,}1. \quad (4)$$

The velocity $u_i^{t+\Delta t}$ is estimated using

$$u_i^{t+\Delta t} = u_i^* - \Delta t\,\frac{\delta p'}{\delta x_i}, \quad (5)$$

where the pressure fluctuation p' is determined by the solution of a Poisson equation in the form of

$$\frac{\delta}{\delta x_i}\left(\frac{\delta p'}{\delta x_i}\right) = \frac{\rho}{\Delta t}\frac{\delta u_i^*}{\delta x_i}, \tag{6}$$

leading to the pressure $p^{t+\Delta t}$ estimation through

$$p^{t+\Delta t} = p' + p^t. \tag{7}$$

3.1 Mathematical model for the immersed interface

The Lagrangian force field is calculated by the direct forcing method proposed by Uhlmann [4]. This method enables the modeling of the wall boundary condition in the immersed interface without introducing any ad hoc constants. The Lagrangian force, $F_i(\overrightarrow{x_k}, t)$, is calculated through the momentum equation (eqn (8)) over a fluid particle in the fluid–solid interface. A force $F_i(\overrightarrow{x_k}, t)$ is imposed on the Lagrangian marker to equate its velocity to the desired velocity $u_{i\,IBM}$ at the immersed boundary

$$\frac{F_i(\overrightarrow{x_k},t)}{\rho} = \frac{\partial u_{k\,i}}{\partial t} + \frac{\partial(u_{k\,i}u_{k\,j})}{\partial x_{k\,j}} + \frac{1}{\rho}\frac{\partial p_k}{\partial x_{k\,i}} - \upsilon\frac{\partial^2 u_{k\,i}}{\partial x_{k\,j}\partial x_{k\,j}}. \tag{8}$$

The Lagrangian force for the marker x_k is determined using

$$\frac{F_i(\overrightarrow{x_k},t)}{\rho} = \frac{u_{k\,i}^{t+\Delta t} - u_{k\,i}^{\sim} + u_{k\,i}^{\sim} - u_{k\,i}^{t}}{\Delta t} + \left[\frac{\partial(u_{k\,i}u_{k\,j})}{\partial x_{k\,j}} + \frac{1}{\rho}\frac{\partial p_k}{\partial x_{k\,i}} - \upsilon\frac{\partial^2 u_{k\,i}}{\partial x_{k\,j}\partial x_{k\,j}}\right]^t, \tag{9}$$

where $u^{\sim}$ is a temporary parameter (Wang et al. [6]) and Δt is the time step. Eqn (9) can be decomposed into two complementary equations

$$\frac{u_{k\,i}^{\sim} - u_{k\,i}^{t}}{\Delta t} + \left[\frac{\partial(u_{k\,i}u_{k\,j})}{\partial x_{k\,j}} + \frac{1}{\rho}\frac{\partial p_k}{\partial x_{k\,i}} - \upsilon\frac{\partial^2 u_{k\,i}}{\partial x_{k\,j}\partial x_{k\,j}}\right]^t = 0, \tag{10}$$

$$\frac{F_i(\overrightarrow{x_k},t)}{\rho} = \frac{u_{k\,i}^{t+\Delta t} - u_{k\,i}^{\sim}}{\Delta t} = \frac{u_{i\,IBM} - u_{k\,i}^{\sim}}{\Delta t}, \tag{11}$$

where $u_{i\,IBM}$ is the immersed boundary velocity obtained from the fluid-structure interaction. Under the effect of the force, the velocity on the Lagrangian marker x_k at time $t + \Delta t$ $\left(u_{k\,i}^{t+\Delta t}\right)$ can be modified to the desired velocity $u_{i\,IBM}$ (Wang et al. [6]).

A distribution function D_h is applied to spread the two-way coupling between the Eulerian grid and the Lagrangian markers at the immersed boundary x_k. This distribution function is determined from its surrounding Eulerian grid nodes x_i. The velocity on the Lagrangian marker is calculated as

$$u_{k\,i}^{\sim} = \sum u_i^* D_h(\overrightarrow{x_{k\,i}} - \overrightarrow{x_i})h^2, \tag{12}$$

$$D_h(\vec{x}_k - \vec{x}) = \frac{1}{h^2}\left[w_g(r_x)w_g(r_y)\right], \tag{13}$$

$$w_g(r) = \begin{cases} \dfrac{3-2|r|+\sqrt{1+4|r|-4|r|^2}}{8} & for\ 0 < |r| \le 1, \\ \dfrac{5-2|r|-\sqrt{-7+12|r|-4|r|^2}}{8} & for\ 1 < |r| \le 2, \\ 0\ for\ |r| > 2, \end{cases} \tag{14}$$

where D_h is the distribution function, h is the characteristic length associated to the Eulerian mesh size, $r_x = \frac{x_k - x}{h}$ and $r_y = \frac{y_k - y}{h}$.

The effect of the force on the Lagrangian markers which spreads into the Eulerian grid is expressed as

$$f_i(\vec{x},t) = \sum F_i(\overrightarrow{x_k},t) D_h(\overrightarrow{x_i} - \overrightarrow{x_{ki}}) h'^2, \tag{15}$$

where h' is the characteristic length associated to the Lagrangian mesh size.

The estimated velocity field from the first step of the two-step fractional method (eqn (4)), is updated to take into account the body presence. It can be written as

$$u_i^* = u_i^* + \Delta t \frac{f_i}{\rho}. \tag{16}$$

During the process of interpolation to obtain the simulated velocity on the Lagrangian markers and extrapolation to spread the effect of the force on its surrounding Eulerian grids, the velocities on the Lagrangian markers may not be satisfied entirely. Therefore, the multi-direct forcing technique, proposed by Wang et al. [6], is applied. This iterative technique is employed until the difference between the velocity simulated on the Lagrangian markers and the desired velocity approaches zero. The norm L_2 is used as the criterion for the loop process at the time $t + \Delta t$. The total Lagrangian force $F_{i\,lag}^{t+\Delta t}$, is the sum of all Lagrangian markers forces in all loops required to achieve the norm L_2 criterion at the time $t + \Delta t$.

3.2 Structural model of the spring tethered system

With reference to Fig. 2, assuming that the springs are perfectly elastic having negligible mass and drag properties and neglecting the cylinder's inertia, applying Newton's second law and Hooke's law, the following equations, for the resultant forces in each direction, result

$$\sum F_{X\,tot}^{t+\Delta t} = F_{mx}^{t} - F_{x\,lag}^{t+\Delta t}, \tag{17}$$

$$\sum F_{Y\,tot}^{t+\Delta t} = F_{my}^{t} - F_{y\,lag}^{t+\Delta t}, \tag{18}$$

where

$$F_{mx}^{t} = -F_1^t \cos(\beta_1{}^t) + F_2^t \cos(\beta_2{}^t) - F_3^t \sin(\beta_3{}^t), \tag{19}$$

$$F_{my}^{t} = -F_1^t \sin(\beta_1{}^t) - F_2^t \sin(\beta_2{}^t) + F_3^t \cos(\beta_3{}^t), \tag{20}$$

where $F_{i\,lag}$ is the total Lagrangian force acting on the cylinder at the time $t + \Delta t$ and F_{mi} represents the resultant springs force at the time t. F_1, F_2 and F_3 are calculated by the Hooke's law at the time t. The cylinder center position at the time t is given by $(x, y) = (x^t, y^t)$ and the geometrical domain center is given by $(x, y) = (x_{cen}, y_{cen})$, thus, the angles values and the springs forces are calculated as follows

$$\beta_1^t = \tan^{-1}\left(\frac{y^t}{x^t}\right), \tag{21}$$

$$\beta_2^t = \tan^{-1}\left(\frac{y^t}{2x_{cen} - x^t}\right), \tag{22}$$

$$\beta_3^t = \tan^{-1}\left(\frac{x^t - x_{cen}}{2y_{cen} - y^t}\right), \tag{23}$$

$$F_1^t = k\, l_1^t, \tag{24}$$

$$F_2^t = k\, l_2^t, \tag{25}$$

$$F_3^t = k\, l_3^t, \tag{26}$$

where the displacement by the springs, l_i^t, is calculated by the cylinder center position at the time t and the springs equilibrium position which coincides with the geometrical domain center (illustrated in Section 4), calculated as follows

$$l_1^t = \sqrt{(x^t)^2 + (y^t)^2} - \sqrt{(x_{cen})^2 + (y_{cen})^2}, \tag{27}$$

$$l_2^t = \sqrt{(2x_{cen} - x^t)^2 + (y^t)^2} - \sqrt{(x_{cen})^2 + (y_{cen})^2}, \tag{28}$$

$$l_3^t = \sqrt{(x^t - x_{cen})^2 + (2y_{cen} - y^t)^2} - \sqrt{(y_{cen})^2}. \tag{29}$$

After calculating the total force acting on the cylinder, it is possible to determine the components of acceleration

$$\ddot{x}^{t+\Delta t} = \frac{\sum F_{X\,tot}^{t+\Delta t}}{m}, \tag{30}$$

$$\ddot{y}^{t+\Delta t} = \frac{\sum F_{Y\,tot}^{t+\Delta t}}{m}. \tag{31}$$

These accelerations are integrated to estimate the velocities of the body

$$u_{IBM} = \dot{x}^{t+\Delta t} = \dot{x}^t + \ddot{x}^{t+\Delta t}\Delta t, \tag{32}$$

$$v_{IBM} = \dot{y}^{t+\Delta t} = \dot{y}^t + \ddot{y}^{t+\Delta t}\Delta t. \tag{33}$$

Integrated again to determine the new cylinder center

$$x^{t+\Delta t} = x^t + \dot{x}^{t+\Delta t}\Delta t, \tag{34}$$

$$y^{t+\Delta t} = y^t + \dot{y}^{t+\Delta t}\Delta t. \tag{35}$$

The computational algorithm is described in the flow chart on the following pages.

4 PROBLEM DESCRIPTION

The problem consists of a square two-dimensional cavity of width l_x and height l_y, with $l_x = l_y = L = 1$ m, where the upper wall moves from left to right with a given velocity $\mathrm{U} = 1.0\ \mathrm{m/s}$ which drives the flow and $U = V = 0\ \mathrm{m/s}$ on the remaining three sides. The cylinder with diameter $D = 0.1\ m$ is sustained by a set of three springs, as depicted in Fig. 2. Initially the fluid is at rest and the cylinder is in the equilibrium position at the center of the domain. The Reynolds number is defined based on the characteristic length, $L = 1$ m and the characteristic velocity $\mathrm{U} = 1.0\ \mathrm{m/s}$. The density of the fluid, $\rho = 1\ \mathrm{kg/m^3}$ and the kinematic viscosity is the variable responsible to change the Reynolds number ($\nu = U\,L/Re$).

5 RESULTS

The numerical platform was validated by the simulations of the classical lid-driven cavity flow without the presence of the immersed body. The spring system was then validated. The final step involved the fluid-structure interaction simulation.

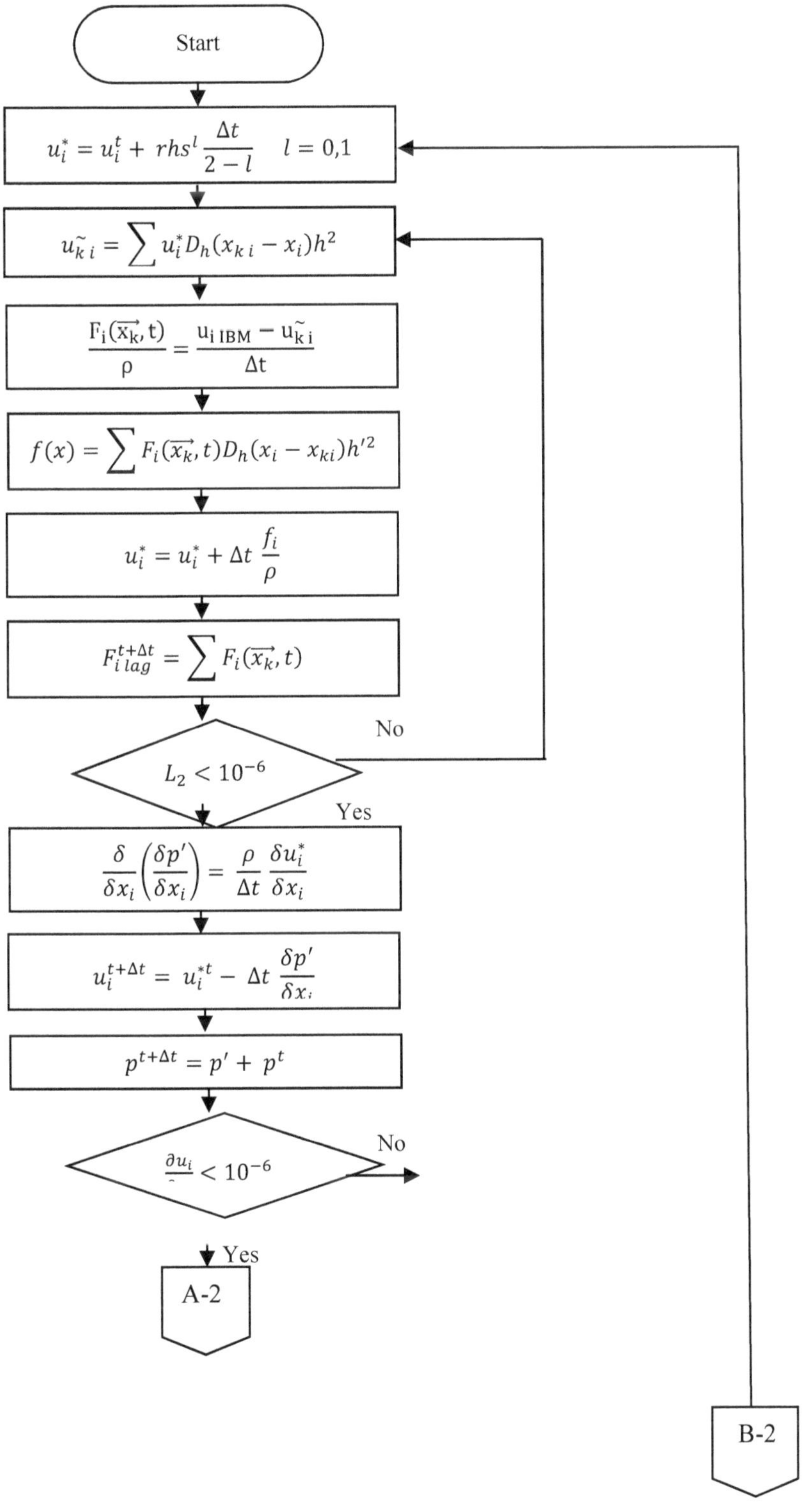
Start
$u_i^* = u_i^t + rhs^l \frac{\Delta t}{2-l} \quad l = 0,1$
$\tilde{u_{k\,i}} = \sum u_i^* D_h(x_{k\,i} - x_i)h^2$
$\frac{F_i(\overrightarrow{x_k}, t)}{\rho} = \frac{u_{i\,IBM} - \tilde{u_{k\,i}}}{\Delta t}$
$f(x) = \sum F_i(\overrightarrow{x_k}, t) D_h(x_i - x_{ki}) h'^2$
$u_i^* = u_i^* + \Delta t \frac{f_i}{\rho}$
$F_{i\,lag}^{t+\Delta t} = \sum F_i(\overrightarrow{x_k}, t)$
$L_2 < 10^{-6}$
No
Yes
$\frac{\delta}{\delta x_i}\left(\frac{\delta p'}{\delta x_i}\right) = \frac{\rho}{\Delta t}\frac{\delta u_i^*}{\delta x_i}$
$u_i^{t+\Delta t} = u_i^{*t} - \Delta t \frac{\delta p'}{\delta x_i}$
$p^{t+\Delta t} = p' + p^t$
$\frac{\partial u_i}{-} < 10^{-6}$
No
Yes
A-2
B-2

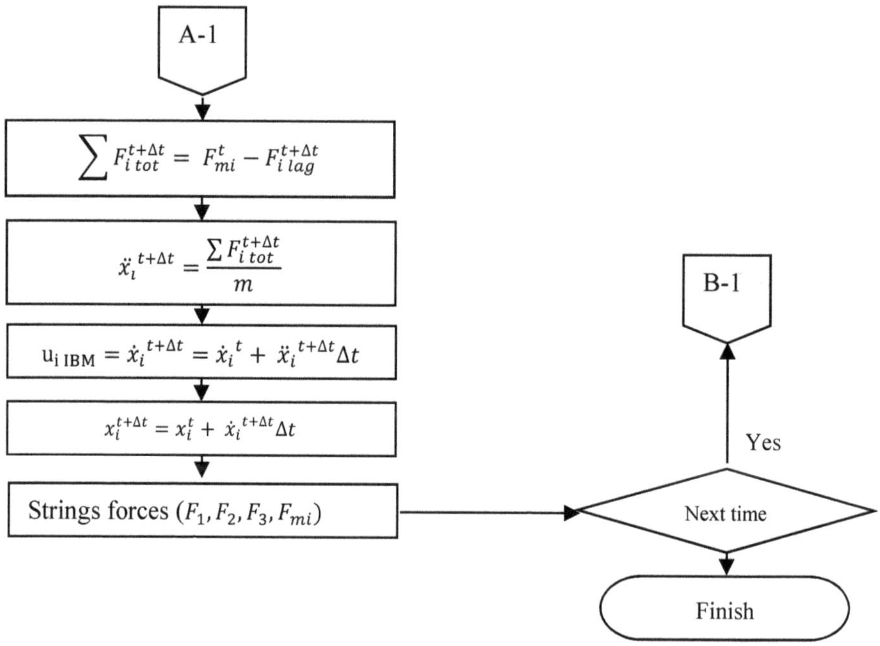

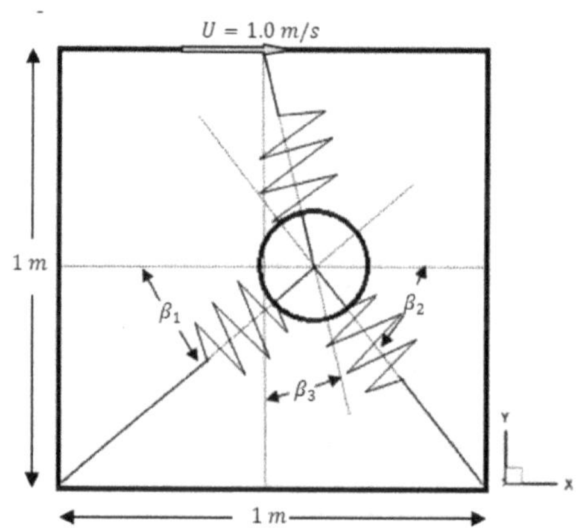

Figure 2: Problem proposal scheme with the cylinder in an unbalanced position.

For the lid-driven cavity flow without the presence of the immersed body, the simulations were performed for Reynolds numbers of 100, 400 and 1,000. Uniform rectilinear meshes of 129 x 129, 141 x 141 and 160 x 160 were used. Ghia et al. [17] reported the first benchmark results for the classical problem of the two-dimensional, incompressible lid-driven cavity flow. The results include the flow patterns, structures identification, minimum and maximums of several fields and velocity profiles passing through the geometric center of the cavity. This data was used to compare and validate the present numerical model.

The characteristics and topology of the Lid-driven cavity flow depend on the Reynolds number. For $Re = 100$ the flow structure is characterized by a primary eddy centred slightly to the upper right side of the cavity. In addition to this eddy, there are two smaller eddies at the bottom left and at the bottom right corners of the cavity. As the Reynolds number increases the secondary eddies grow in size and the primary eddy adapts to this change, shifting towards the center of the cavity. Fig. 3 shows the streamlines pattern for primary and secondary eddies associated with the velocity vector magnitude ($\sqrt{u^2 + v^2}$). As the Reynolds number increases the flow becomes more energized and the primary eddy is circumvented by this region.

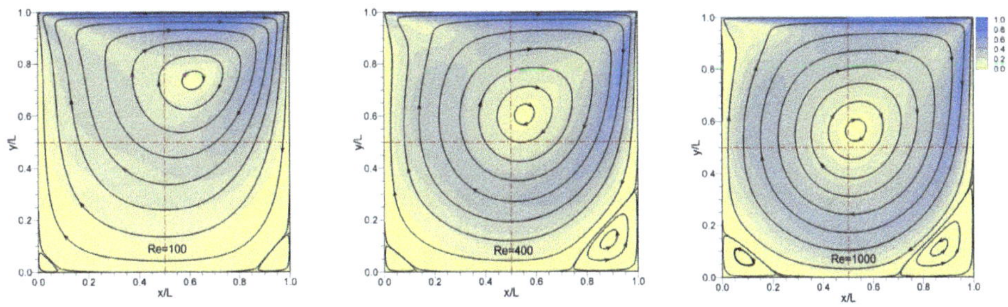

(a) (b) (c)

Figure 3: Streamlines pattern for primary and secondary eddies associated with the velocity vector magnitude for: (a) Re = 100; (b) Re = 400; and (c) Re = 1000.

In order to validate the spring system equations, the cylinder was initially placed in an arbitrary position, such as $x/L = 0.4$ and $y/L = 0.6$. The cylinder's center position was then tracked against dimensionless time ($t* = t/(L/U)$) as shown in Fig. 4. Since no damping is present, the cylinder continues to move indefinitely.

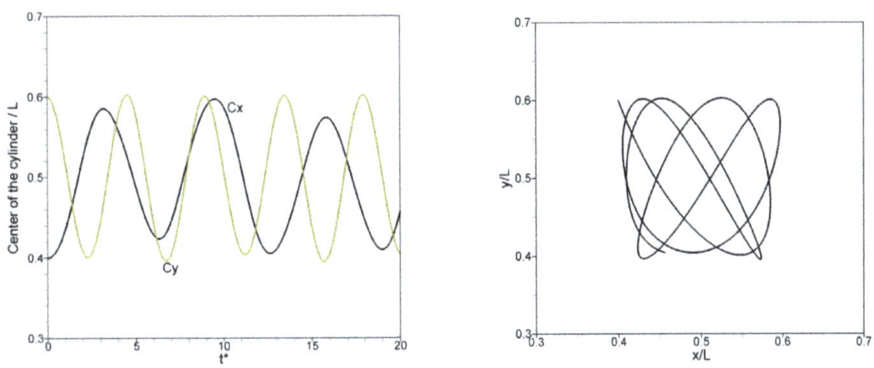

(a) (b)

Figure 4: History of the cylinder center position at k = 0.1 N/m. (a) With respect to time; and (b) With respect to space.

For the fluid-structure interaction, the square cavity was discretized using a mesh size of 160x160 volumes. The simulations were run at Reynolds numbers of Re = 100, 400 and

1,000 and spring constants $k = 0.01$, 0.1 and 1.0 N/m. The cylinder diameter was fixed at D = 0.1m. With the fluid initially at rest and the cylinder in the equilibrium position ($x/L = y/L = 0.5$), the simulations were run for 100s.

Results show that the transient cylinder center position depend on the values of k and Re. As shown in Fig. 5, there are two distinct results related to the k value. When the spring constant is small, $k = 0.01$ N/m, the resultant springs force is much weaker than the fluid dynamics forces (Fig. 5(a)) and the springs stretch freely resulting in the cylinder motion mainly affected by the fluid dynamics action. On the other hand, when the spring constant is larger, a clear response from the springs forces interacting with the fluid dynamics is evident.

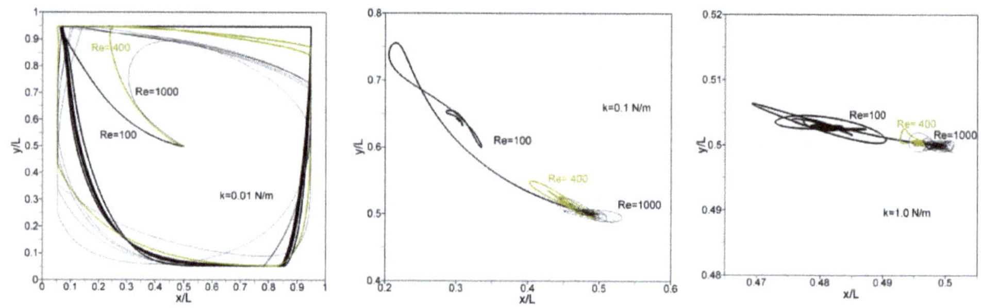

(a) (b) (c)

Figure 5: The cylinder center position along the time for spring constant. (a) k = 0.01 N/m; (b) k = 0.1 N/m; and (c) k = 1 N/m.

6 CONCLUSIONS

In this paper, a simplified fluid-structure interaction model consisting of a cylinder tethered by a spring system interacting dynamically with the incompressible two-dimensional lid-driven cavity flow using the immersed boundary (IB) method was investigated. A structured Cartesian grid was employed for the Eulerian domain while the boundaries were represented through Lagrangian markers. The advantage of this method is that bodies of almost any arbitrary shape can be added without grid restructuring, a procedure which is often time-consuming. The numerical implementation of a Cartesian grid is simpler to implement than an alternative body-fitted coordinates grid.

Using the IB method, the motion of the cylinder does not impose grid restructuring at each time step when FSI is present. Additionally the partitioned approach used in this study shows an efficient way to couple the flow and structural solver.

From the cases studied, cases having large spring constants, strings forces induce the cylinder to an oscillatory motion damped by the fluid viscous forces, consequently the amplitude of the displacements decrease.

It can therefore be concluded that the immersed boundary method is a good tool to model fluid-structure interaction problems. Additionally this method can also accommodate complex geometry without the need of complex grid creation algorithms.

ACKNOWLEDGEMENTS

The authors thank the College of Mechanical Engineering (FEMEC) of Federal University of Uberlandia (UFU), Federal University of Mato Grosso (UFMT), Federal University of Technology – Paraná (UFTPR) and University of Malta for the support.

REFERENCES

[1] Ferziger, J.H. & Peric, M., *Computational Methods for Fluid Dynamics*, 3rd ed., Springer-Verlag: Berlin and New York, 2002.

[2] Peskin, C., Flow patterns around heart valves: A numerical method. *Journal of Computational Physics*, **10**, pp. 252–271, 1972.

[3] Lima e Silva, A.L.F., Silveira-Neto, A. & Damasceno, J., Numerical simulation of two-dimensional flows over a circular cylinder using the immersed boundary method. *Journal of Computational Physics*, **198**, pp. 351–370, 2003.

[4] Uhlmann, M., An immersed boundary method with direct forcing for the simulation of particulate flows. *Journal of Computational Physics*, **209**, pp. 448–476, 2005.

[5] Vanella, M. & Balaras E., A moving-least-squares reconstruction for embedded boundary formulations. *Journal of Computational Physics*, **228**(18), pp. 6617–6628, 2009.

[6] Wang, Z., Fan, J. & Luo, K., Combined multi-direct forcing and immersed boundary method for simulating flows with moving particles. *International Journal of Multiphase Flow*, **34**, pp. 283–302, 2007.

[7] Borges, J.E., Lourenço, M.A.S., Padilla, E.M. & Micallef, C., Immersed boundary method application as a way to deal with the three-dimensional sudden contraction. *Computation*, **6**(3), p. 50, 2018.

[8] Ha, S.T., Ngo, L.C., Saeed M., Jeon, B.J. & Choi, H., A comparative study between partitioned and monolithic methods for the problems with 3D fluid-structure interaction of blood vessels. *Journal of Mechanical Science and Technology*, **31**, pp. 281–287, 2017.

[9] Degroote, J., Haelterman, R., Annerel, P., Bruggeman, P. & Vierendeels, J., Performance of partitioned procedures in fluid-structure interaction. *Computers and Structures*, **88**, pp. 446–457, 2010.

[10] Degroote, J., Bathe, K. & Vierendeels, J., Performance of a new partitioned procedure versus a monolithic procedure in fluid-structure interaction. *Computers and Structures*, **87**(11–12), pp. 793–801, 2009.

[11] Chern, M.J., Kuan, Y.H., Nugroho, G., Lu, G.T. & Horng, T.L., Direct-forcing immersed boundary modeling of vortex-induced vibration of circular cylinder. *Journal of Wind Engineering and Industrial Aerodynamics*, **134**, pp. 109–121, 2014.

[12] Dettmer, W. & Perić, D., A computational framework for fluid–rigid body interaction: Finite element formulation and applications. *Computer Methods in Applied Mechanics and Engineering*, **195**(13–16), pp. 1633–1666, 2006.

[13] Williamson, C.H.K. & Govardhan, R., Vortex-induced vibrations. *Annual Review of Fluid Mechanics*, **36**, pp. 413–455, 2004.

[14] Campregher, R., Mathematical modeling for three-dimensional fluid-structure interaction problems. Doctoral thesis, University Federal of Uberlandia, 2005.

[15] Patankar, S., *Numerical Heat Transfer and Fluid Flow, Series in Computational and Physical Processes in Mechanics and Thermal Sciences*, 1st ed., Hemisphere Publishing Company, 1980.

[16] Kim, J. & Moin, P., Application of a fractional step method to incompressible Navier–Stokes equations. *Journal of Computational Physics,* **59**, pp. 308–323, 1985.

[17] Ghia, U., Ghia, K.N. & Shin, C.T., High-re solutions for incompressible flow using the Navier–Stokes equations and a multigrid method. *Journal of Computational Physics*, **48**, pp. 387–411, 1982.

FAST METHODS FOR VORTEX INFLUENCE COMPUTATION IN MESHLESS LAGRANGIAN VORTEX METHODS FOR 2D INCOMPRESSIBLE FLOWS SIMULATION

DARIA LEONOVA[1], ILIA MARCHEVSKY[1,2] & EVGENIYA RYATINA[1,2]
[1]Bauman Moscow State Technical University, Russia
[2]Ivannikov Institute for System Programming of the RAS, Russia

ABSTRACT

Vortex methods are a powerful tool for solving engineering problems of incompressible flow simulation around airfoils. The vorticity is considered as a primary computed variable. According to the Navier–Stokes equations written down in Helmholtz-type form (for 2D case), new vorticity is generated only on the surface line of an airfoil. Its intensity is unknown and can be found from the solution of the boundary integral equation resulting from the no-slip boundary condition satisfaction. The right-hand side of the integral equation in the simplest case depends on the incident flow velocity and vorticity distribution in the flow domain. For the velocity field computation, it is necessary to take into account the influence of all the vortices, which simulate the vorticity distribution in the flow. These vortices are moving in the flow with velocities calculated as sums of point-to-point vortex influences, so the computation complexity of such operation is proportional to N^2 where N is the number of vortices. In practice, N can have the order of tens or hundreds of thousands, up to a million, so the application of the "direct" method for velocities calculation becomes impossible. In this paper, two fast methods having logarithmic (proportional to $N\log N$) computational complexity are implemented and investigated. The first method is an analogue of the Barnes–Hut fast method for the N-body problem. The second one is based on fast solution of the Poisson's equation with respect to the stream function on rather coarse mesh by using the fast Fourier transform technique with some special procedure for the results correction. Numerical complexity estimations for both methods are derived. Their sequential and parallel implementations are developed. Numerical experiments show that the FFT-based method is more efficient. Total acceleration compared with the "direct" method is over 1000 times for 500,000 vortex elements.

Keywords: vortex methods, viscous incompressible media, vortex influence calculation, fast methods, Barnes–Hut-type method, Poisson equation, fast Fourier transform.

1 INTRODUCTION

In many engineering applications the problem of two-dimensional outer gas and fluid flows simulation with small subsonic speeds appears. If the flow compressibility can be neglected, the so-called vortex methods [1], [2] can be very efficient in comparison to well-known mesh methods for solving such problems, especially when we deal with large body displacements. We consider pure Lagrangian meshless modification of vortex methods, namely Viscous Vortex Domains method [3]; its main idea is considering the vorticity as a primary computed variable. The vorticity in the flow moves with the velocity which is superposition of the convective velocity and diffusive one caused by viscosity influence. It means that at every time step there are no vorticity generation in the flow domain (only the motion of existing one); new vorticity is generated on the airfoil surface line. Vorticity distribution in 2D vortex methods is simulated by a set of elementary vorticity carriers – vortex elements. Each of them is described by the position in the flow domain $\vec{r}$, small constant radius ε and circulation Γ ,which is also assumed to be constant. The airfoil surface line K can be modelled with the

WIT Transactions on Engineering Sciences, Vol 126, © 2019 WIT Press
www.witpress.com, ISSN 1743-3533 (on-line)
doi:10.2495/BE420231

free vortex sheet of unknown intensity [4], so the problem is reduced to the solution of the boundary integral equation with respect to the vortex sheet intensity $\gamma(\vec{r})$ following from the no-slip boundary condition satisfaction for the tangential velocities components [5]

$$\oint_K \frac{\vec{n}(\vec{r})\cdot(\vec{r}-\vec{\xi})}{2\pi|\vec{r}-\vec{\xi}|^2}\gamma(\vec{\xi})dl_\xi - \frac{\gamma(\vec{r})}{2} = f_\tau(\vec{r}) \qquad \vec{r}\in K, \tag{1}$$

where $\vec{n}(\vec{r})$ is unit normal vector to the airfoil surface line K. This equation has to be solved at every time step. The right-hand side of eqn (1) is a known function which depends only on the incident flow velocity $\vec{V}_\infty$ and vorticity distribution $\Omega(\vec{r})$ in the flow domain in the simplest case of immovable and non-deformable airfoil:

$$f_\tau(\vec{r}) = -\vec{\tau}(\vec{r})\cdot\left(\vec{V}_\infty + \int_S \frac{\vec{k}\times(\vec{r}-\vec{\xi})}{2\pi|\vec{r}-\vec{\xi}|^2}\Omega(\vec{\xi})dS_\xi\right), \tag{2}$$

where $\vec{\tau}(\vec{r})$ is unit tangent vector to the airfoil surface line, $\vec{k}$ is unit vector orthogonal to the flow plane, $\vec{k} = \vec{n}(\vec{r})\times\vec{\tau}(\vec{r})$. According to the mentioned vorticity distribution representation through a set of n vortex elements with circulations Γ_i, positions $\vec{r}_i$ and radius ε, the integral overflow domain in eqn (2) can be replaced by a sum over all vortex elements:

$$\int_S \frac{\vec{k}\times(\vec{r}-\vec{\xi})}{2\pi|\vec{r}-\vec{\xi}|^2}\Omega(\vec{\xi})dS_\xi \approx \sum_{i=1}^{n} \frac{\vec{k}\times(\vec{r}-\vec{r}_i)}{2\pi\left\{|\vec{r}-\vec{r}_i|^2,\, \varepsilon^2\right\}}\Gamma_i \qquad \vec{r}\in K, \tag{3}$$

After solving the integral eqn (1), all the vortex elements in the flow should be moved to new positions. Their movement is described by the system of differential equations

$$\frac{d\vec{r}_i}{dt} = \vec{V}(\vec{r}_i) + \vec{W}(\vec{r}_i),$$

where $\vec{W}(\vec{r})$ is the diffusive velocity, proportional to the viscosity coefficient [3]. This system is normally solved by Euler's explicit method. The convective velocity $\vec{V}(\vec{r}_i)$ is calculated in the similar way as the right-hand side eqn (3), but now it is necessary to calculate such a sum for all positions $\vec{r}_i$ of the vortex elements:

$$\vec{V}(\vec{r}_i) = \sum_{j=1}^{n} \frac{\vec{k}\times(\vec{r}_i-\vec{r}_j)}{2\pi\left\{|\vec{r}_i-\vec{r}_j|^2,\, \varepsilon^2\right\}}\Gamma_j, \qquad i=1,\ldots,n. \tag{4}$$

Vortex elements influences are inversely proportional to the distances to the corresponding vortex elements. So, at every time step it is necessary to calculate the vortex influence according to eqn (4) for every vortex element in the flow domain, that leads to the computational complexity proportional to n^2, where n is the number of vortex elements. This problem is similar to the n-body gravitational problem, where the calculating of the attractive forces between all the bodies is necessary.

It should be noted that vortex influence calculation is the most time-consuming operation in the whole algorithm of the vortex method; in some cases it can reach about 70% of the total execute time. In order to provide flow simulation with rather high accuracy, it is necessary to consider large number of vortex elements – up to hundreds of thousands. Thus, the "direct" method of vortex influence calculation becomes impossible. The usage of multiprocessor systems and Nvidia graphic accelerators allows for solving this problem only partly. In Kuzmina et al. [6] it is shown that parallel technologies provide significant acceleration of the whole algorithm, especially using combined technologies MPI+CUDA. However, for some problems time of computations remains unacceptably high, for example, in case when the number of vortices exceeds about 100 000.

The aim of this research is implementation of the fast algorithms for vortex influence calculation permitting one to reduce the computational complexity of this operation. Two fast methods of logarithmic (proportional to $n \log n$) computational complexity for vortex influence calculation are considered: the Barnes–Hut-type method and the method based on fast solution of the Poisson's equation with respect to the stream function on rather coarse mesh using fast Fourier transform technique with further solution correction.

2 THE BARNES–HUT-TYPE METHOD

The Barnes–Hut method [7] initially had been developed for the gravitational n-body problem and later adapted to vortex methods by Dynnikova [8]. The main idea of this method is that the influence of the groups of closely adjacent vortex elements on another such groups located far apart, can be calculated approximately using linearized formulae [8].

For this purpose, in the flow domain the hierarchical tree-structure of rectangular space domains (cells), containing all vortex elements, is constructed. The zero-level cell contains all vortex elements, and then it is divided across its long side into two first-level cells, each is reduced horizontally and vertically according to its vortices in order to exclude empty area. Next, similarly the second-level cells are constructed, etc. (Fig. 1). Such procedure continues until the target level is achieved or the cell contains single vortex.

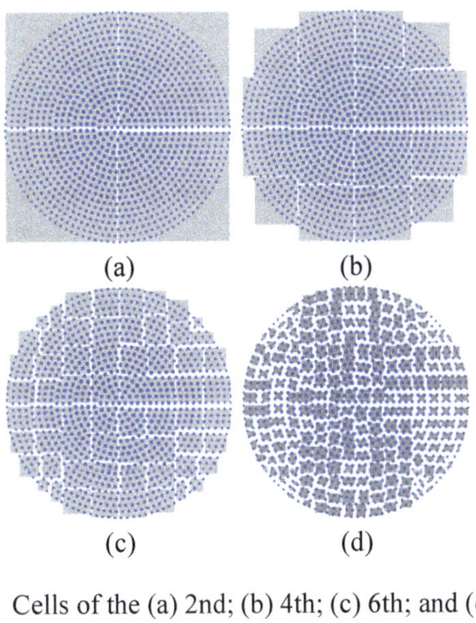

Figure 1: Cells of the (a) 2nd; (b) 4th; (c) 6th; and (d) 8th levels.

The set of terminal vertices of the resulting tree contains all the vortex elements and for each cell the tree traversal is carried out according to the scheme in Fig. 2. Moreover, the transition to lower level cells (red arrows) is carried out only if the "proximity criterion" for the current cell is satisfied (the distance between cells centers is compared with the sum of their sides).

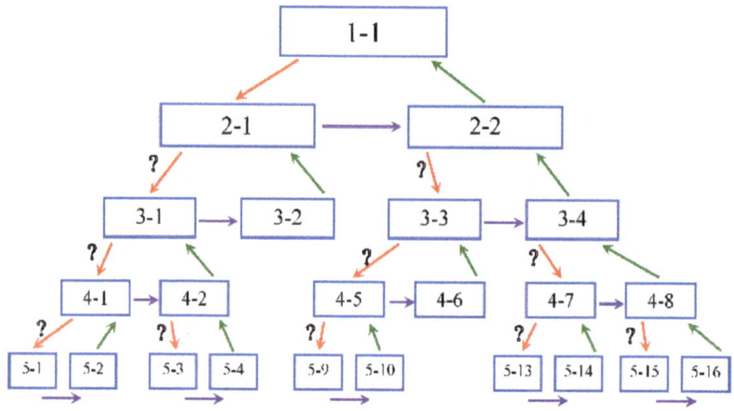

Figure 2: Tree traversal scheme.

So, the Barnes–Hut-type algorithm of vortex influence computation at every time step of flow simulation consists of the following stages:

1. Zero-level cell formation for all vortex elements existing at the current time step.
2. Tree structure construction; determination of the terminal vertices.
3. Calculation of the properties for every tree-cell such as centers of positive and negative vorticity and total circulations (also positive and negative).
4. For every terminal tree cell 3 operations are performed:
   a. tree traversal in order to determine far-spaced cells from the current one and calculation of the coefficients in linearized representation of the vortex influence;
   b. calculation of the vortex influence from closely adjacent vortex elements (placed in the so-called "neighboring" zone) according to the direct eqn (4);
   c. accumulation of the influences calculated approximately (a) and exactly (b).

The "proximity criterion" used for "neighboring" cells determination depends on the adjustable parameter θ (usually $0<\theta\leq 1$) which determines the accuracy of the method. For small values of this parameter ($\theta \ll 1$) the Barnes–Hut method provides high accuracy, but it also has high computational complexity; increase of the parameter θ leads to decreasing both the computational complexity and accuracy of the method. Numerical experiments show, that the value of the parameter θ should not exceed 0.2…0.4 in order to provide the relative error at the level of 0.2%.

It should be noted, that number of tree levels k is also an adjustable parameter; it practically does not influence the accuracy of the method, but the computational complexity of the algorithm depends on its value significantly – deviation on 2–3 levels from the optimal value leads to increase of calculation time in 2 or more times. In Table 1, the relative time of the Barnes–Hut-type algorithm execution is given for the problems with large number of vortex elements (n = 100,000, n = 250,000, n = 500,000, n = 1,000,000), where the ratio of times of computations is given for non-optimal and optimal numbers of tree levels (denoted as k^*); in brackets the acceleration of the algorithm compared to the "direct" Biot–Savart method (eqn (4)) is shown. Note, that hereinafter the algorithm execution time means the vortex influence calculation at one time step.

Table 1: Time of computations ratio for the Barnes–Hut method at optimal and non-optimal number of tree levels and different number of vortices.

Relative calculation time t_k/t_{k^*} (k^* is an optimal number of tree levels)							
n = 100,000 k^* = 14		n = 250,000 k^* = 15		n = 500,000 k^* = 16		n = 1,000,000 k^* = 17	
K	t_k/t_{k^*}	k	t_k/t_{k^*}	k	t_k/t_{k^*}	K	t_k/t_{k^*}
12	1.76	13	2.02	14	1.78	15	1.91
13	1.20	14	1.19	15	1.23	16	1.10
14	1.00 (22.3)	15	1.00 (40.5)	16	1.00 (64.8)	17	1.00 (92.8)
15	1.36	16	1.17	17	1.25	18	1.24
16	2.18	17	1.92	18	2.08	19	2.11

In order to estimate parallel properties of the algorithm (its scalability), taking in mind the possibility of its parallel version development, in Table 2 the contributions in percent of the above mentioned main operations to the total execution time of the Barnes–Hut algorithm are shown for the model problem with n = 1,000,000 vortex elements and different numbers of tree levels k. Here the operation *1* is the operation of tree construction (stage 2), operation *2* is coefficients calculation (stage 4a), operation *3* is direct influence calculation in neighboring zone (stage 4b), t_1 is calculation time for the Barnes–Hut method, t_2 is calculation time for the direct method, s is the achieved acceleration.

Table 2: Contributions of the main operations in the Barnes–Hut algorithm at different number of tree levels.

Levels	Operation *1*	Operation *2*	Operation *3*	Other ops.
15	0.95%	2.82	96.22	0.01
16	1.69%	11.58	86.70	0.03
17	1.90%	30.59	67.44	0.07
18	1.57%	60.61	37.77	0.05
19	1.00%	78.53	20.43	0.04

It is seen that with increasing number of levels the point-to-point stage of calculations (Operation *3*) is decreased, while that the stage of coefficients calculation (Operation *2*) increases.

The parallel version of the Barnes–Hut method for vortex influence computation was implemented using both MPI (Message Passing Interface, for cluster systems) and OpenMP (Open Multi-Processing, for systems with shared memory) technologies. Taking into account that stages 4a–4c of the Barnes–Hut algorithm can be executed independently for the terminal tree cells, the simplest way of parallelization is to split the terminal cells onto MPI-processes, including additional parallelization using OpenMP technology for each process. Due to small contribution of the tree construction procedure (less than 2%, according to Table 2), it is not parallelized and executed simultaneously by all the MPI processes. The achieved values of acceleration are shown in Table 3 for the problem with large number of vortex elements (n = 1,000,000) for one, two and three cluster nodes. It is seen, that the most efficient is hybrid (MPI + OpenMP) parallelization.

Table 3: Acceleration results using MPI and OpenMP technologies for the Barnes–Hut algorithm.

Number of nodes	1	2	3
1 OpenMP thread per node (per MPI process)			
Time, sec	106.13	54.02	36.67
Acceleration, times	1.00	1.96	2.89
4 OpenMP threads per node (per MPI process)			
Time, sec	30.84	16.82	12.11
Acceleration, times	3.44	6.31	8.76
4 MPI processes per node			
Time, sec	33.58	18.20	13.17
Acceleration, times	3.16	5.83	8.06

Taking into account, that the optimal variant of the Barnes–Hut method (for 17 tree levels) provides 92,8 times acceleration in comparison to the direct method in sequential mode (Table 2), now we obtain, that its parallel version, being run on 3 nodes, provides total acceleration of about 810 times. Thus, the total time of computation can be reduced from approximately 165 minutes (2.75 hours) to about 12 seconds.

However, despite the similarity the problem of vortex influence calculation and n-body gravitational problem, there is a significant difference. In gravitational problem due to 3D problem statement, the attractive force, caused by one body decreases proportionally to the squared distance, whilst the two-dimensional problem statement in vortex methods leads to influence decreasing with the first degree of the distance only, so the efficiency of the Barnes–Hut algorithm in vortex method is lower in comparison to the original gravitational problem. Thus, it is interesting to investigate other possible ways to approximate fast vortex influence calculation.

3 THE FFT-BASED METHOD

This method is based on the similar idea as the Barnes–Hut method: the influence of the vortex elements, placed if far-spaced regions can be calculated approximately. Such influence can be calculated fast due to the possibility of the Poisson's equation fast solution with respect to the stream-function using fast Fourier transform (FFT) technique [9].

The velocity can be represented through the stream-function as $\vec{V}(\vec{r}) = \nabla \times \left(\psi(\vec{r})\vec{k}\right)$, where the stream-function ψ satisfies the Poisson's equation with known right-hand side

$$\Delta\psi = -\Omega(\vec{r}). \qquad (5)$$

The solution of such equation can be represented as the convolution of the Green's function $G(\vec{r})$ and the right-hand side, so for the velocity $\vec{V}(\vec{r})$ we obtain the following eqn:

$$\vec{V}(\vec{r}) = \nabla \times \left(\psi(\vec{r})\vec{k}\right) = \int_S \left(\nabla \times G(\vec{r}-\vec{\xi})\vec{k}\right)\Omega(\vec{\xi})dS_\xi =$$
$$= \int_S \frac{\vec{k}\times(\vec{r}-\vec{\xi})}{2\pi\,|\,\vec{r}-\vec{\xi}\,|^2}\Omega(\vec{\xi})dS_\xi = \int_S \vec{Q}(\vec{r}-\vec{\xi})\,\Omega(\vec{\xi})dS_\xi, \tag{6}$$

If rectangular uniform mesh is introduced, which contains $M_x \times M_y$ nodes in horizontal and vertical directions, respectively, the integral in eqn (6) can be approximately replaced with a sum over all the mesh nodes:

$$\int_S \vec{Q}(\vec{r}-\vec{\xi})\Omega(\vec{\xi})dS_\xi \approx \sum_{i=0}^{M_x-1}\sum_{j=0}^{M_y-1}\vec{Q}(\vec{r}-\vec{r}_{ij})\Gamma_{ij}, \tag{7}$$

where Γ_{ij} are the nodal values of vorticity, i.e., the vorticity should be projected somehow onto the nodes in such a way, that the following condition is satisfied:

$$\int_S \Omega(\vec{\xi})dS_\xi = \sum_{i=0}^{M_x-1}\sum_{j=0}^{M_y-1}\Gamma_{ij}. \tag{8}$$

We consider rather coarse mesh, that means $M_x M_y \ll n$, and now describe briefly the main stages of the FFT-based fast method of vortex influence calculation:

1. To calculate the nodal circulations Γ_{ij} using some smoothing operator, for example, introduced in Monaghan [11].
2. To calculate vector kernel components
$$\vec{Q}(\vec{r}) = \frac{\vec{k}\times\vec{r}}{2\pi\,|\,\vec{r}\,|^2} = \{Q_x, Q_y\}$$
at mesh nodes. For the origin point $\vec{r} = \{0, 0\}$ we assume $\vec{Q}(\vec{r}) = \{0, 0\}$. Nodal values of Q_x and Q_y form matrices Q^x and Q^y, respectively.
3. To quadruple the mesh domain is by its extension to the right and upper directions (reflection technique) in order to satisfy the boundary conditions (zero velocity on infinity). In additional nodes the matrices Q^x and Q^y are extended symmetrically and skew-symmetrically [10], circulation values Γ_{ij} are assumed to be zero. All following stages should be performed for obtained extended matrices.
4. To perform two-dimensional fast Fourier transform [9] for the abovementioned matrices Q^x, Q^y, Γ; the resulting complex-valued images are $\hat{Q}^x$, $\hat{Q}^y$, $\hat{\Gamma}$.
5. According to the well-known feature of the FFT, the nodal values of eqn (7) image form the matrix, which components are calculated as elementwise multiplications of two matrices, so as a result we obtain two matrices, consist of the images of the velocity components nodal values: $\hat{V}^x = \hat{Q}^x \circ \hat{\Gamma}$ and $\hat{V}^y = \hat{Q}^y \circ \hat{\Gamma}$.
6. To perform inverse fast Fourier transform procedure and obtain matrices of nodal values of horizontal and vertical velocity components V^x and V^y.
7. To interpolate the velocity nodal values onto vortices placed in the corresponding mesh cells.

Due to coarse mesh usage, a significant error occurs. In particular cell it is caused mainly by significant error in influence calculation from the vortex elements, placed in this cell itself and in the neighboring zone, so the correction procedure should be performed. It is necessary to exclude the “incorrect” influence calculated at the Poisson’s equation solution only from the vortices from neighboring zone and add the "correct" influence, calculated point-to-point using eqn (4). So, the following algorithm [10] should be performed for each mesh cell.

1. Calculation of the correction matrix:
   a) the vortex of unit circulation is placed in arbitrary mesh cell;
   b) using the same smoothing operator, which have been used for nodal values Γ_{ij} calculation, its circulation is redistributed onto 16 nearest nodes (black points in Fig. 3), the resulting nodal circulations vector is $\{\Gamma\}$;

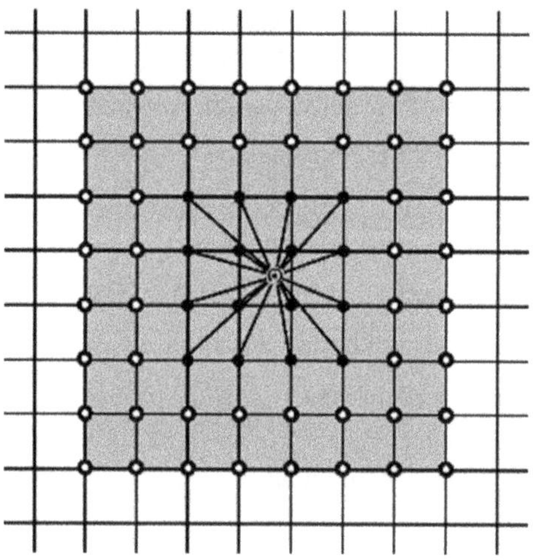

Figure 3: Neighboring zone of the cell.

   c) the steps 4–6 of the basic algorithm are performed for this "sub-problem"; as a result we obtain the velocities at 64 nearest nodes (black and white points in Fig. 3). Due to 2D problem statement, the resulting vector $\{V\}$ has 128 components;
   d) based on assumption of linear dependence between velocity and circulation vectors we obtain the matrix equality $\{V\}=[C]\{\Gamma\}$, where $[C]$ is correction matrix having size 128×16;
   e) the 15 similar sub-problems are considered in order to construct the system of equations for correction matrix components; note, that for every sub-problem the unit vortex is placed in different position to avoid a singular matrix.
2. After constructing a correction matrix $[C]$, the influences of the vortices placed in the neighboring zone of each cell can be excluded. For this purpose, for each cell the circulations of the vortices contained in it are redistributed onto 16 nearest nodes; the resulting vector is $\{V\}=[C]\{\Gamma\}$ which means the current cell influence on its neighboring zone. Corresponding nodal velocities are kept save for each cell and they are summarized with analogous influences from other cells in neighboring zone (for the cells, which do not refer to the neighboring zone such influence will be equal to zero). Thus, for all the cells we can exclude incorrectly calculated influence of the vortices places in the neighboring zones.
3. For each cell now it is necessary to add the correctly (exactly) calculated vortex influence from its neighboring zone according to the Biot–Savart law (eqn (4)):

$$\vec{V}_{neigh}(\vec{r}_i) = \sum_{j \in B_i \cup B_i'} \frac{\vec{k} \times (\vec{r}_i - \vec{r}_j)}{2\pi \left\{ |\vec{r}_i - \vec{r}_j|^2, \varepsilon^2 \right\}} \Gamma_j \qquad i \in B_i, \tag{9}$$

where B_i means the set of vortices in the i-th cell; B_i' means the set of vortices in its neighboring zone to the i-th cell (not including the i-th cell itself).

3.1 Numerical experiment

The model problem with 500 vortex elements was considered (Fig. 4). The mesh 16 x 16 was considered for the numerical solution of the Poisson's equation vortices circulations were chosen arbitrary ($-0.01 \le \Gamma_i \le 0.01$).

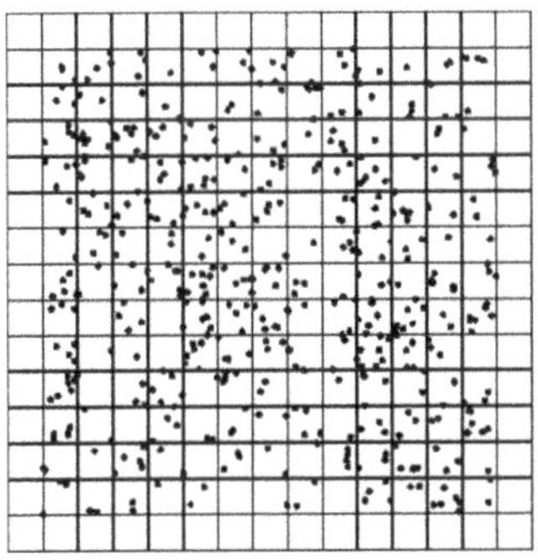

Figure 4: Model problem statement: vortex elements distributed uniformly in the unit square.

At Fig. 5 the streamlines of the velocity field are shown for the central mesh cell before the correction procedure performing (Fig.5 (a)) and after this procedure (Fig. 5(b)). Here the streamlines, obtained using the direct eqn (4), are shown in red color; the streamlines obtained using the FFT-based method are shown in blue color. It is apparent that velocities obtained using the fast solution of the Poisson's equation (without correction) are quite different from the real ones calculated directly point-to-point, while the correction permits to achieve quite accurate results.

Calculation for the vortices in the central cell:

$$\delta V_i = \frac{|\vec{V}_{i,exact} - \vec{V}_{i,FFT}|}{\max\limits_j |\vec{V}_{j,exact}|} \qquad i, j = 1, \ldots, n_{cell}. \tag{10}$$

A quantitative empirical estimation was obtained also for the accuracy of velocities where δV_i is relative error the velocities calculation; $\vec{V}_{i,exact}$ – exact value of the velocity of the i-th vortex element, calculated "directly" according to eqn (4); $\vec{V}_{i,FFT}$ – approximate value, calculated using the FFT-based method. The maximum in the denominator is calculated over the vortices in the central cell. Numerical results for different sizes of the neighboring zone, i.e., numbers of cell layers around each cell, are shown in Table 4. Zero number of cell layers corresponds to the algorithm without correction stage.

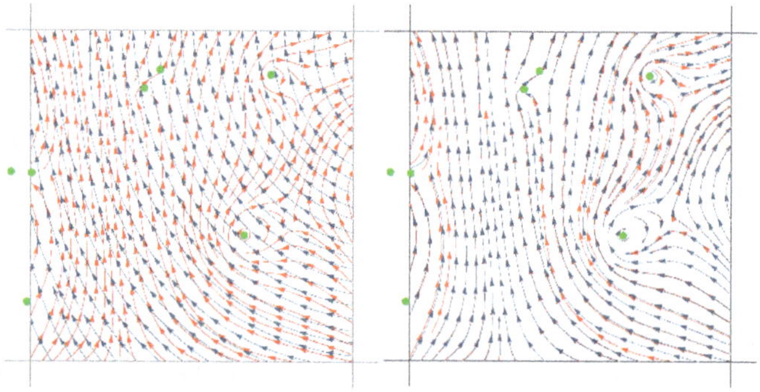

(a) (b)

Figure 5: Velocities streamlines in the central mesh cell. (a) Before correction procedure; and (b) After it.

Table 4: Relative error of the FFT-based method.

Number of cell layers	0	1	2	3	4
max δV_i	0.2840	0.0326	0.0024	0.0017	0.0003

It is seen, that the error of the algorithm without correction exceeds 20%, which makes the algorithm inapplicable in practice. Three cell layers in the neighboring zone (i.e., the neighboring zone in this case consists of 49 cells including the central cell itself) provides the error level less than 0.2%. Further neighboring zone size increase provides smaller error: 0.03% for 4 cell layers, however the neighboring zone here consists of 81 mesh cells that leads to significant growth of the algorithm computational complexity.

At the same time number of numerical experiments, performed in order to investigate the properties of the vortex methods algorithms, shows the error level of 0.2% is acceptable for the majority of problems, being solved by vortex methods. This level of the error is comparable with the error, arising from the boundary integral equation discretization, which determines the vorticity generation on the airfoil surface line. Therefore, it seems to be optimal from point of view of the ratio between the method accuracy and computational time to choose the size of the neighboring zone to be equal 3 cell layers, as it is shown in Fig. 3.

3.2 Complexity estimation of the algorithm

Note, that the accuracy of the FFT-based method practically does not depend on the mesh size, used for solving of the Poisson's problem, however it influences significantly the computational complexity of the method in the whole.

In order to estimate the computational complexity, we assume that n vortex elements are distributed uniformly in the square domain, mesh cells are squares, and there are equal numbers of the mesh nodes (equal to M) in horizontal and vertical directions. We assume additionally, that there are no vortex elements in the outer layer of the cells, similarly to shown in Fig. 4: this permits to use known smoothing kernel M_4 [11] for vorticity redistribution from the vortex elements onto mesh nodes and satisfy eqn (8).

Taking into account only the operations of multiplication and division, it is possible to estimate the complexities of the following algorithm stages:

1. Circulations calculation at mesh nodes (redistribution) and correction procedure.
2. Fast Fourier Transform and inverse Fourier transform; we take into account, that the FFT-based algorithm is implemented at every time step, so for the fixed mesh it is possible to compute matrices Q^x, Q^y and perform 2D FFT for them (resulting matrices $\hat{Q}^x$, $\hat{Q}^y$) only once. Thus, at every time step the FFT should performed only for matrix Γ and two inverse transformations should be performed for matrices $\hat{V}^x$ and $\hat{V}^y$ in order to obtain nodal values of velocity components.
3. Interpolation procedure from the mesh nodes onto the vortex elements.

The theoretical complexities estimations are the following:

$$\left.\begin{aligned} Q_1 &= 192n + 2048(M-2)^2 \\ Q_2 &= 8M^2(7 + 6\log_2 M) \\ Q_3 &= 9n + 245n^2(M-2)^{-2} \end{aligned}\right\}. \tag{11}$$

It is also important to estimate the mentioned operations contributions to the method complexity; it significantly depends on the number of mesh nodes. With nodes number increasing the contributions Q_1 and Q_2 decrease, while Q_3 increases significantly. As an example, we consider the problem with $n = 1\,000\,000$ vortex elements and numbers of mesh nodes $M = 128$ and $M = 512$. The nodes numbers are chosen equal to power of two (2^k) because that makes it possible to perform the FFT procedure optimally. The comparison of the numerical results (blue columns) and theoretical estimations (red columns) of the operations Q_1, Q_2 and Q_3 contributions are shown in the diagrams in Fig. 6. A good agreement between the numerical results and theoretical estimations is seen.

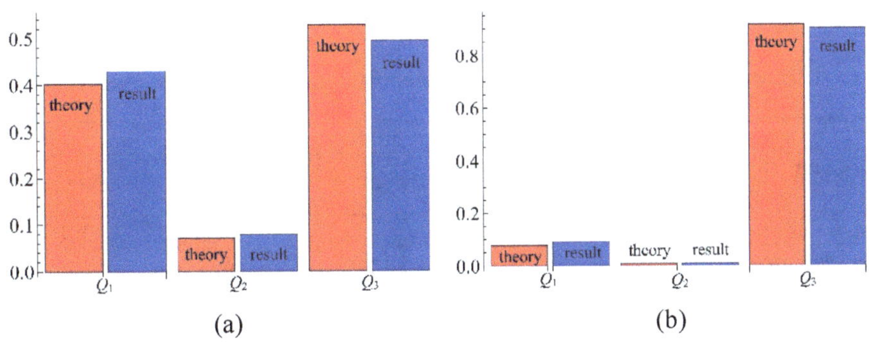

Figure 6: The main operations contributions in the FFT-based algorithm for the meshes consist of $M = 128$ nodes (a) and M = 512 nodes (b) in both directions.

The parallel version of the FFT-based algorithm was implemented using both MPI and OpenMP technologies, similarly to the above discussed parallelization scheme for the Barnes–Hut method. Note, that in the present implementation the FFT transformation (direct and inverse) procedure is performed sequentially, by using the Eigen library (http://eigen.tuxfamily.org). The achieved values of acceleration are shown in Table 5 for the problem with large number of vortex elements (n = 1,000,000) for one, two and three cluster nodes.

Table 5: Acceleration using MPI and OpenMP technologies for the FFT-based algorithm.

Nodes number	1	2	3
1 OpenMP thread per node (per MPI process)			
Time, sec	4.52	2.82	2.24
Acceleration, times	1.00	1.60	2.02
4 OpenMP threads per node (per MPI process)			
Time, sec	2.12	1,64	1.41
Acceleration, times	2.13	2.76	3.21
4 MPI processes per node			
Time, sec	1.79	1.48	1.40
Acceleration, times	2.53	3.05	3.23

Thus, the parallel implementation of the FFT-based method, being run on 3 nodes provides total acceleration of about 7000 times, that means that the total time of computation can be reduced from approximately 165 minutes (2.75 hours) to about 1.4 seconds. Note, that MPI parallelization is more efficient in comparison to hybrid (MPI + OpenMP) approach.

4 COMPARISON OF THE TWO FAST METHODS

In order to compare the efficiency of two fast methods for vortex influence calculation, four different problems with large number of vortex elements (n_1 = 100,000, n_2 = 250,000, n_3 = 500,000, n_4 = 1,000,000) were considered. The values of the adjustable parameters of both methods (coefficient θ in proximity criterion and number of tree levels k in the Barnes–Hut method; number of cell layers in the neighboring zone and number of nodes in the mesh in the FFT-based method) were chosen optimally; the relative error (10) in all cases was less than 0.2%. In Table 6 time of computations in seconds is given for all the mentioned problems (all the computations are performed in sequential mode).

Table 6: Time of computations (in sec.) for all the methods for different number of vortices.

	n_1 = 100,000	n_2 = 250,000	n_3 = 500,000	n_4 = 1,000,000
Biot-Savart (direct) method	98.5	615.6	2462.2	9848.9
Barnes–Hut method	4.43	15.21	38.02	106.13
FFT-based method	0.47	1.10	2.47	4.52

The achieved acceleration is shown in Table 7 in comparison to the direct method (computations according to the Biot–Savart law (eqn 4)). It is clearly seen that the FFT-based method is much more efficient in comparison to the Barnes–Hut-type method; for example, time of computations is nearly the same for $n_1 = 100\,000$ using the Barnes–Hut method and for $n_4 = 1\,000\,000$ using the FFT-based method.

Table 7: Fast methods acceleration in times.

	n_1 = 100,000	n_2 = 250,000	n_3 = 500,000	n_4 = 1,000,000
Barnes–Hut method	22	40	65	93
FFT-based method	210	560	997	2179

5 CONCLUSIONS

In the present paper two different methods of vortex influence computation were implemented and investigated. Both methods have the logarithmic (n log n) computational complexity and can be efficiently applied in case the number of vortex elements n is about few hundreds of thousands, up to one million. Both methods have some adjustable parameters, such as the coefficient in the proximity criterion in the Barnes–Hut method and the number of cell layers in the neighboring node in the FFT-based method, which effect the ratio between the accuracy and time of computations. The other parameters – number of tree levels in the Barnes–Hut method and mesh size in the FFT-based method practically do not affect the accuracy, but their values determine the computational complexity of the corresponding algorithms. It is found, that the FFT-based method is at least 10 times more efficient in comparison to the Barnes–Hut method. Parallel implementations of both methods are developed using OpenMP and MPI technologies; their efficiencies are investigated for a small number of processors/cores.

ACKNOWLEDGEMENTS

The research was supported by the Russian Foundation for Basic Research (project No. 18-31-20051, "Development of new algorithms based on vortex methods for flow around airfoils simulation and coupled hydroelastic problems solving and their efficient software implementation for multiprocessor computers of different architectures").

REFERENCES

[1] Lifanov, I.K., *Singular Integral Equations and Discrete Vortices*, Utrecht, p. 475, 1996.

[2] Cottet, G.H. & Koumoutsakos, P.D., *Vortex Methods: Theory and Practice*, Cambridge University Press, p. 328, 2000.

[3] Andronov, P.R., Guvernyuk, S.V. & Dynnikova, G.Y., *Vortex Calculation Methods of Non-Stationary Hydrodynamical Loadings*, Moscow University Press, p. 184, 2006.

[4] Tokaty, G.A., *A History and Philosophy of Fluid Mechanics*, Courier Corporation, p. 241, 1994.

[5] Kempka, S.N., Glass, M.W., Peery, J.S., Strickland J.H. & Ingber, M.S., Accuracy considerations for implementing velocity boundary conditions in vorticity formulations. Sandia report, Sand96-0583 UC-700, p. 52, 1996.

[6] Kuzmina, K.S., Marchevsky, I.K. & Ryatina, E.P., On CPU and GPU parallelization of VM2D code for 2D flows simulation using vortex method. *Proceedings of the 6th European Conference on Computational Mechanics (ECCM 6) and 7th European Conference on Computational Fluid Dynamics (ECFD 7)*, 2018, www.eccm-ecfd2018.org/admin/files/filePaper/p2045.pdf. Accessed on: 10 Mar. 2019.

[7] Barnes, J. & Hut, P., A hierarchical *O(N log N)* force-calculation algorithm. *Nature*, **324**(4), pp. 446–449, 1986.

[8] Dynnikova, G.Y., Fast technique for solving the N-body problem in flow simulation by vortex methods. *Computational Mathematics and Mathematical Physics*, **49**(8), pp. 1389–1396, 2009.

[9] Nussbaumer, H.J., *Fast Fourier Transform and Convolution Algorithms*, Springer Verlag, p. 286, 1982.

[10] Morgenthal, G. & Walther, J.H., An immersed interface method for the Vortex-In-Cell algorithm. *Computers and Structures*, **85**, pp. 712–726, 2007.

[11] Monaghan, J.J., Extrapolating B-splines for interpolation. *Journal of Computational Physics*, **60**(2), pp. 253–262, 1985.

NUMERICAL SCHEMES FOR VORTEX SHEET INTENSITY COMPUTATION IN VORTEX METHODS TAKING INTO ACCOUNT THE CURVILINEARITY OF THE AIRFOIL SURFACE LINE

KSENIIA KUZMINA[1,2] & ILIA MARCHEVSKY[1,2]
[1]Bauman Moscow State Technical University, Russia
[2]Ivannikov Institute for System Programming of the RAS, Russia

ABSTRACT
In vortex methods, vorticity is the primary computed variable. The problem of the accuracy improvement of vorticity generation simulation at the airfoil surface line in 2D vortex methods is considered. The generated vorticity is simulated by a thin vortex sheet at the airfoil surface line, and it is necessary to determine the intensity of this sheet at each time step. It can be found from the no-slip boundary condition, which leads to a vector boundary integral equation. There are two approaches to satisfy this equation: the first one leads to a singular integral equation of the 1st kind, while the second one leads to a Fredholm-type integral equation of the 2nd kind with bounded kernel for smooth airfoils. Usually, for numerical solution of the boundary integral equation, the airfoil surface line is replaced by a polygon, which consists of straight segments (panels). A discrete analogue of the integral equation can be obtained using the Galerkin method. Different families of basis and projection functions lead to numerical schemes with different complexity and accuracy. For example, a numerical scheme with piecewise-constant basis functions provides the first order of accuracy for vortex sheet intensity, and a numerical scheme with piecewise-linear functions gives the second order of accuracy. However, the velocity field near the airfoil surface line is also of interest. In the case of rectilinear airfoil surface line discretization, the accuracy of velocity field reconstruction has no more than the first order of accuracy for both, piecewise-constant and piecewise-linear numerical schemes. In order to obtain a higher order of accuracy for velocity field reconstruction, it is necessary to take into account the curvilinearity of the airfoil surface line. In this research, we have developed such an approach, which provides the second order of accuracy both, for vortex sheet intensity computation and velocity field reconstruction.
Keywords: vortex method, boundary integral equation, vortex sheet, curvilinear panel.

1 INTRODUCTION

There are many different methods for the numerical simulation of the interaction between structures and flow, the most popular are mesh methods. The most common of them are finite difference method, finite element method, finite volume method. Within the framework of such approaches, there are many modifications that allow one to consider models of diverse complexity, taking into account different physical effects. However, often in engineering applications there is a need to analyze a large number of possible design parameters. In this case, it is impossible to perform detailed investigation for all parameter values, so there is a need to involve approximate, but cost efficient methods for the primary rough analysis of the parameters. Meshless Lagrangian vortex methods [1]–[6] fall into the category of such approaches. Their range of applicability is limited by the model of incompressible flow, however, in this field, they can be much more efficient in comparison with mesh methods [7].

Vortex methods are based on the consideration of vorticity as the primary calculated value. There is no new vorticity generation in the flow region; new vorticity is generated only at the airfoil surface line: the generated vorticity is simulated by a vortex sheet. This vorticity then becomes part of the vortex wake as a separate vortex elements, which move in the flow

WIT Transactions on Engineering Sciences, Vol 126, © 2019 WIT Press
www.witpress.com, ISSN 1743-3533 (on-line)
doi:10.2495/BE420241

with some velocity. The intensity of this vortex sheet can be determined from the boundary condition at the airfoil surface line [8], which is expressed by the boundary integral equation.

Numerical schemes which are commonly used in vortex methods for approximate solution of boundary integral equation, imply the splitting the airfoil surface line into segments (usually called "panels"). Then, a system of linear algebraic equations approximating the integral equation is constructed. The accuracy of the simulation of the vorticity generation process directly affects the accuracy of the whole problem solution.

In the case of using the of well-known and well-studied method of discrete vortices (MDV) [2], which is applicable only for inviscid flow simulation, vorticity is being shed from the airfoil surface to the vortex wake only at some preliminary known points of separation, which are often taken at sharp edges or angle points (Fig. 1(a)). However, modern and more complicated modifications of vortex methods, such as the method of viscous vortex domains (VVD) [9], which allows one to simulate viscous flows, imply the vorticity shedding from the whole airfoil surface line (Fig. 1(b)). So for such modification, in addition to the accuracy of the vortex sheet intensity calculation, the accuracy of velocity field reconstruction in the neighborhood of the airfoil surface line is also of interest since it determines the accuracy of vorticity evolution simulation in the flow.

(a) (b)

Figure 1: Vortices separation in (a) MDV methods; and (b) VVD methods.

In the present research, the method of Viscous Vortex Domains is considered as a basic modification of vortex methods. In the framework of VVD, vorticity in the flow domain is represented by a set of vortex elements having constant (in time) circulation and moving with velocity $\vec{V}+\vec{W}$, where $\vec{V}$ is the flow velocity and $\vec{W}$ is the so-called "diffusive velocity" [9].

Normally, in vortex methods an airfoil surface line is approximated by rectilinear panels [10]. For this case, numerical schemes for a boundary integral equation approximate solution of different accuracy and the corresponding numerical complexity can be constructed. For example, in [11] several schemes based on the Galerkin method ideas are developed. In these schemes, the vortex sheet intensity is represented by a piecewise-constant or piecewise-linear function. Note that in [12] a piecewise-linear scheme was also constructed, however, for other reasons. In [11] it is shown that piecewise-constant and piecewise-linear schemes provide the 1st and the 2nd order of accuracy for the vortex sheet intensity, respectively. However, the piecewise-linear scheme does not permit one to provide the 2nd order of accuracy for the velocity field reconstruction in neighborhood of the airfoil surface line. This is due to the fact that, in the case of the airfoil surface line approximation with rectilinear panels instead of the original smooth curve, we deal with dummy angular points between the panels. Their presence effects significantly the velocity field generated by the vortex sheet which is located on such rectilinear panels. It leads to the loss of smoothness of the velocity field in proximity to the angle points, so only 1st order of accuracy can be achieved. Therefore, it seems that it is necessary to take into account the curvilinearity of the airfoil surface line to improve the order of accuracy of the velocity field reconstruction.

The aim of this research is the accuracy analysis for vortex sheet intensity computation and velocity field reconstruction when the airfoil surface line is approximated by rectilinear and curvilinear panels.

2 THE PROBLEM STATEMENT

In vortex methods, the vortex sheet intensity at the airfoil surface line K can be determined from the no-slip boundary condition. This condition, together with the generalized Helmholtz decomposition, leads to the vector boundary integral equation with respect to the vortex sheet intensity [8], [13]. This equation can be reduced to a scalar integral equation by projecting it onto a normal or a tangent vector at the airfoil surface line [8]. In the first case, we obtain a singular integral equation, in the second case – a Fredholm-type integral equation of the second kind. In [14], it is shown that the second approach allows one to obtain a solution with higher accuracy. According to this approach, the boundary integral equation with respect to the vortex sheet intensity $\gamma(r)$ is as follows [8], [11], [14]

$$\oint_K Q_\tau(\vec{r},\vec{\xi})\gamma(\vec{\xi})dl_\xi - \frac{1}{2}\gamma(\vec{r}) = f_\tau(\vec{r}), \qquad \vec{r} \in K. \tag{1}$$

Here,

$$Q_\tau(\vec{r},\vec{\xi}) = \frac{\vec{n}(\vec{r})\cdot(\vec{r}-\vec{\xi})}{2\pi|\vec{r}-\vec{\xi}|^2}, \tag{2}$$

$\vec{n}$ is unit outward normal vector to the airfoil surface line, and f_τ is the right-hand side which depends on the vorticity distribution in the flow region, the airfoil surface line velocity and the incident flow velocity. We assume that the right-hand side is known function, which corresponds to the case when the velocity of the airfoil surface line is known function or when the coupled hydroelastic problem is solved according to partitioned approach. In the more complex case where the coupled hydroelastic problem is solved using a monolithic approach, the airfoil velocity is also an unknown variable, and motion equation for the airfoil surface line should be solved together with the boundary integral equation.

Eqn (1) has infinite set of solutions; in order to select the unique one, the additional condition should be added

$$\oint_K \gamma(\vec{r})dl_r = \Gamma, \tag{3}$$

where Γ is given value of total circulation around the airfoil.

Note that the kernel $Q_\tau(\vec{r},\vec{\xi})$ of eqn (1) is bounded for smooth airfoils; if the surface line is C^2-smooth curve, it is easy to prove that

$$\lim_{|\vec{r}-\vec{\xi}|\to 0} |Q_\tau(\vec{r},\vec{\xi})| = \frac{\varkappa}{4\pi},$$

where $\varkappa$ is the curvature of the curve K at the corresponding point.

If there is angle poin at the airfoil surface line, the kernel is unbounded in this point, and exact solution for the vortex sheet intensity has a weak singularity [15]. If the initial airfoil surface line has an angular point, this singularity is physically justified and represents real flow specificity. However, when we discretize the airfoil surface line by rectilinear panels, we obtain a set of "dummy" angle points (Fig. 2). When the numerical solution is assumed to be piecewise-constant along the panels, or even piecewise-linear, but the neighboring panels have nearly the same lengths, airfoil approximation with the polygon is acceptable and the convergence of the numerical solution to the exact one takes place for vortex sheet intensity with the 1st and 2nd order of accuracy, respectively. For significantly different lengths of neighboring panels, the result of numerical solution can be far from exact solution, not only quantitatively, but even qualitatively.

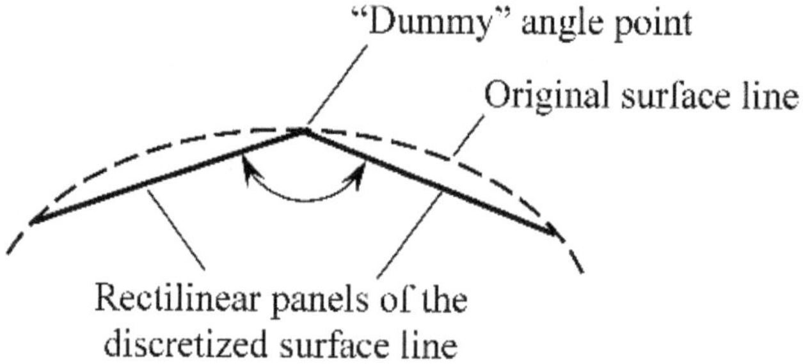

Figure 2: Dummy angle point formation between rectilinear panels.

Thus, in order to increase the accuracy vortex sheet intensity computation, it is necessary to avoid the appearance of artificial angle points, i.e. to take into account the curvilinearity of the airfoil surface line.

The flow velocity can be reconstructed using the Biot–Savart law through known vorticity distribution $\vec{\Omega}(\vec{r})$ in the flow domain S, the intensity of the vortex sheet $\vec{\gamma}(\vec{r})$ on the airfoil surface line K, and the velocity of airfoil surface line. In the considered particular case of the immovable airfoil, it takes the following form

$$\vec{V}(\vec{r}) = \vec{V}_\infty + \int_S \frac{\vec{\Omega}(\vec{r}) \times (\vec{r} - \vec{\xi})}{2\pi |\vec{r} - \vec{\xi}|^2} dS_\xi + \oint_K \frac{\vec{\gamma}(\vec{r}) \times (\vec{r} - \vec{\xi})}{2\pi |\vec{r} - \vec{\xi}|^2} dl_\xi,$$

where $\vec{V}_\infty$ is the incident flow velocity. Note that hereinafter scalar vorticity field $\Omega(\vec{r})$ and scalar vortex sheet intensity $\gamma(\vec{r})$ are considered

$$\vec{\Omega}(\vec{r}) = \Omega(\vec{r})\vec{k}, \quad \vec{\gamma}(\vec{r}) = \gamma(\vec{r})\vec{k},$$

where k is unit vector orthogonal to the flow plane.

More generally, when the airfoil surface line is movable, the flow velocity according to the generalized Helmholtz decomposition also depends on the airfoil surface line velocity [8].

3 THE NUMERICAL METHOD

For the numerical solution of the boundary integral equation, we use the basic ideas developed in [11], where the Galerkin approach is applied to numerical schemes construction. In [11], schemes with a rectilinear discretization of the airfoil surface line are described, an approximate solution is represented as piecewise-constant or piecewise-linear function. Here, we briefly describe such schemes and then generalize main ideas to the case of curvilinear panels [16].

3.1 Numerical scheme with rectilinear panels

Let the airfoil surface line be discretized by rectilinear panels K_i, $i=1,\dots,N$. Then, the system of basis functions $\varphi_i^q(\vec{r})$, $i=1,\dots,N$, $q=0,1$, is introduced such that

$$\varphi_i^0(\vec{r})=\begin{cases}1, & \vec{r}\in K_i,\\ 0, & \vec{r}\notin K_i;\end{cases}\qquad \varphi_i^1(\vec{r})=\begin{cases}\dfrac{(\vec{r}-\vec{c}_i)\cdot\vec{\tau}_i}{L_i}, & \vec{r}\in K_i,\\ 0, & \vec{r}\notin K_i,\end{cases} \tag{4}$$

where $\vec{c}_i$ is the center of the i-th panel; L_i is the length of the i-th panel.

Then, the piecewise-constant and piecewise-linear numerical solutions are represented as linear combinations of the basis functions with unknown coefficients $\{\gamma_i^0\}_{i=1}^N$ and $\{\gamma_i^1\}_{i=1}^N$

$$\gamma^c(\vec{r})=\sum_{i=1}^N \gamma_i^0\varphi_i^0(\vec{r}),\quad \gamma^l(\vec{r})=\sum_{i=1}^N(\gamma_i^0\varphi_i^0(\vec{r})+\gamma_i^1\varphi_i^1(\vec{r})),\quad \vec{r}\in K. \tag{5}$$

All the formulae below are presented for piecewise-linear solution, however, they can be easily adapted to the piecewise-constant case.

Then, according to Galerkin approach, unknown coefficients γ_i^0 and γ_i^1 can be found from the orthogonality condition of the boundary integral equation residual to the projection functions which we choose the same as the basis functions $\varphi_i^q(s)$, $i=1,\dots,N$, $q=0,1$ [11], [14]. The residual of eqn (1) on the i-th panel has the following form

$$z_i(\vec{r})=\sum_{j=1}^N\left(\gamma_j^0\int_{K_j}Q_\tau(\vec{r},\vec{\xi})\varphi_j^0(\vec{\xi})dl_\xi+\gamma_j^1\int_{K_j}Q_\tau(\vec{r},\vec{\xi})\varphi_j^1(\vec{\xi})dl_\xi\right)-\frac{1}{2}\left(\gamma_i^0\varphi_i^0(\vec{r})+\gamma_i^1\varphi_i^1(\vec{r})\right)-f_\tau(\vec{r}).$$

Its orthogonality condition to the basis functions leads to a system of linear equations

$$\int_{K_i}z_i(\vec{r})\varphi_i^p(\vec{r})dl_r=0,\quad i=1,\dots,N,\quad p=0,1,$$

which, in turn, can be written down in the following form

$$\sum_{j=1}^N\left(\gamma_j^0\int_{K_i}\varphi_i^p(\vec{r})dl_r\int_{K_j}Q_\tau(\vec{r},\vec{\xi})\varphi_j^0(\vec{\xi})dl_\xi+\gamma_j^1\int_{K_i}\varphi_i^p(\vec{r})dl_r\int_{K_j}Q_\tau(\vec{r},\vec{\xi})\varphi_j^1(\vec{\xi})dl_\xi\right)-$$
$$-\frac{1}{2}\left(\gamma_i^0\int_{K_i}\varphi_i^0(\vec{r})\varphi_i^p(\vec{r})dl_r+\gamma_i^1\int_{K_i}\varphi_i^1(\vec{r})\varphi_i^p(\vec{r})dl_r\right)=$$
$$=\int_{K_i}f_\tau(\vec{r})\varphi_i^p(\vec{r})dl_r,\quad i=1,\dots,N,\quad p=0,1. \tag{6}$$

Eqn (3) is approximated straightforwardly

$$\sum_{i=1}^N\int_{K_i}\left(\gamma_i^0\varphi_i^0(\vec{r})+\gamma_i^1\varphi_i^1(\vec{r})\right)dl_r=\Gamma. \tag{7}$$

The system which consists of linear eqns (6) and (7) is overdetermined; we regularize it similarly to [2] by adding the so-called regularization variable R to those of eqn (6) which correspond to $p=0$. In block-matrix form regularized linear system can be written down as

$$\begin{pmatrix} A^{00}+D^{00} & A^{01}+D^{01} & I \\ A^{10}+D^{10} & A^{11}+D^{11} & O \\ L^0 & L^1 & 0 \end{pmatrix}\begin{pmatrix} \gamma^0 \\ \gamma^1 \\ R \end{pmatrix}=\begin{pmatrix} b^0 \\ b^1 \\ \Gamma \end{pmatrix}. \tag{8}$$

Here A^{pq} are $N \times N$ square matrices; D^{pq} are diagonal matrices, $p,q=0,1$; I and O are vectors consist of ones and zeros, respectively; L^0 and L^1 are vectors consist of integrals from the corresponding basis functions along the curvilinear panels; $\gamma^0=(\gamma_1^0,\dots,\gamma_N^0)^T$ and $\gamma^1=(\gamma_1^1,\dots,\gamma_N^1)^T$ are vectors of unknown coefficients; R is regularization variable; b^0 and b^1 are right-hand side vectors

$$\begin{gathered} A_{ij}^{pq}=\int_{K_i}\left(\int_{K_j} Q_\tau(\vec{r},\vec{\xi})\varphi_j^q(\vec{\xi})dl_\xi\right)\varphi_i^p(\vec{r})dl_r, \quad D_{ii}^{pq}=-\frac{1}{2}\int_{K_i}\varphi_i^p(\vec{r})\varphi_i^q(\vec{r})dl_r, \\ L_i^p=\int_{K_i}\varphi_i^p(\vec{r})dl_r, \quad b_i^p=\int_{K_i} f_\tau(\vec{r})\varphi_i^p(\vec{r})dl_r, \qquad i,j=1,\dots,N, \quad p,q=0,1. \end{gathered} \tag{9}$$

The scheme (8), which corresponds to a piecewise-linear approximate solution, can be reduced to a scheme for a piecewise-constant solution

$$\begin{pmatrix} A^{00}+D^{00} & I \\ L^0 & 0 \end{pmatrix}\begin{pmatrix} \gamma \\ R \end{pmatrix}=\begin{pmatrix} b^0 \\ \Gamma \end{pmatrix}, \tag{10}$$

where the coefficients, again, have the form (9).

The coefficients (9) can be calculated analytically; all the formulae for their calculation are given in [17].

3.1 Numerical scheme with curvilinear panels

Let us suppose that the airfoil surface line is parameterized with the natural parameter s – the arc length of the curve. Then, eqn (1) takes the form

$$\int_K Q_\tau(s,\sigma)\gamma(\sigma)d\sigma-\frac{1}{2}\gamma(s)=f_\tau(s), \quad s\in[0,L].$$

Here, L is the length of the airfoil surface line.

Let the airfoil surface line now be discretized by N curvilinear panels whose endings and beginnings correspond to the values of the parameter s. Then, the parameter s at the i-th panel varies over the range $[s_{i-1},s_i]$.

The basis functions $\varphi_i^q(s)$, $i=1,\dots,N$, $q=0,1$, similar to (4), are introduced, but now they depend on the arc length

$$\varphi_i^0(s)=\begin{cases} 1, & s\in[s_{i-1},s_i], \\ 0, & s\notin[s_{i-1},s_i]; \end{cases} \qquad \varphi_i^1(s)=\begin{cases} \dfrac{s-s_i^*}{\delta_i}, & s\in[s_{i-1},s_i], \\ 0, & s\notin[s_{i-1},s_i]. \end{cases} \tag{11}$$

Here, $\delta_i=s_i-s_{i-1}$ is the length of the i-th curvilinear panel, the parameter value $s_i^*=\dfrac{s_i+s_{i-1}}{2}$ corresponds to the center of the i-th panel.

The approximate solution on the airfoil surface line has the same form as earlier (5); the orthogonality condition of the residual of the equation to the basis functions leads to the systems which coincide with (10) or (8), respectively; the coefficients have the following form

$$A_{ij}^{pq} = \int_{s_{i-1}}^{s_i} \left(\int_{s_{j-1}}^{s_j} Q_\tau(s,\sigma)\varphi_j^q(\sigma)d\sigma \right)\varphi_i^p(s)ds, \quad D_{ii}^{pq} = -\frac{1}{2}\int_{s_{i-1}}^{s_i} \varphi_i^p(s)\varphi_i^q(s)ds,$$
$$L_i^p = \int_{s_{i-1}}^{s_i} \varphi_i^p(s)ds, \quad b_i^p = \int_{s_{i-1}}^{s_i} f_\tau(s)\varphi_i^p(s)ds, \qquad i,j = 1,\dots,N, \quad p,q = 0,1. \tag{12}$$

As earlier, a piecewise-constant scheme can be constructed, similarly to (10).

Of course, the calculation of coefficients (12), which now requires the integration over the curvilinear panels, is quite complicated problem. In contrast with case of rectilinear panels, it is not possible to obtain exact formulae for their calculation. In [18] a methodology has been developed that allows to calculate them approximately. Using the expanding of the integrands into Taylor series, all the necessary approximate formulae were obtained for integrals calculation over curvilinear panels panels.

4 NUMERICAL EXPERIMENTS

In order to compare the described schemes with rectilinear and curvilinear airfoil surface line discretization and piecewise-constant and piecewise-linear vortex sheet intensity representation, we consider the model problem of potential flow simulation around elliptical airfoil (Fig. 3(a)).

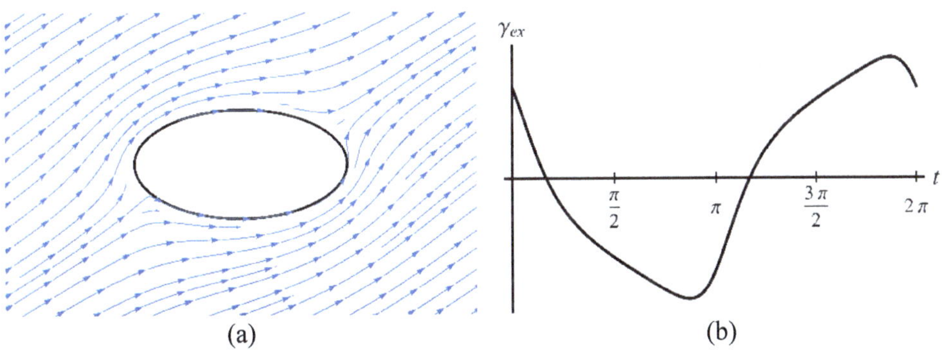

Figure 3: (a) Flow around elliptical airfoil; and (b) Exact solution for vortex sheet intensity.

Let us suppose that the surface line of elliptical airfoil with semiaxes a_1 and b_1 is given by the parametric relations

$$\begin{cases} x(t) = a_1 \cos t, \\ y(t) = b_1 \sin t, \end{cases} \qquad t \in [0;2\pi).$$

The incident flow has velocity $|\vec{V}_\infty|$ and angle of incidence β ; there is no vorticity in the flow region.

In this case, complex potential of the flow has the following form

$$f(z) = W_\infty^* \hat{\zeta}(z) + \frac{(a_1 + b_1)^2 W_\infty}{\hat{\zeta}(z)}, \quad z = x + iy,$$

where $W_\infty = \frac{1}{2} | V_\infty | e^{i\beta}$; $\hat{\zeta}(z) = z - \sqrt{z^2 - a^2}$; "$*$" means complex conjugate. From this complex potential, the velocity field can be calculated: $V^{ex} = f'(z)^*$.

Also, the exact solution for vortex sheet intensity for such problem can be derived

$$\gamma^{ex}(t) = \frac{| V_\infty | \sin(\beta - t)}{\sqrt{a_1^2 \sin^2 t + b_1^2 \cos^2 t}}, \qquad t \in [0; 2\pi).$$

The following values of parameters were chosen: incident flow velocity $| \vec{V}_\infty | = 1$, the angle of incidence $\beta = \pi / 6$, ellipse semiaxes $a = 1.0$ and $b = 0.5$. The exact solution for such parameters is shown in Fig. 3(b).

The airfoil surface line in numerical experiments was discretized into N panels with approximately the same lengths equal to δ.

The error of vortex sheet intensity computation is estimated using the L_1 norm

$$\| \Delta\gamma \|_{L_1} = \oint_K | \gamma^{ex}(\vec{r}) - \gamma(\vec{r}) | dl_r.$$

In [19], the method for this integral calculation is described: it is shown there how the approximate solution at the approximated airfoil can be projected onto the original airfoil surface line.

The error of the velocity field reconstruction in neighborhood of the airfoil can be estimated in the similar way

$$\| \Delta V \|_{L_1} = \oint_{K^h} | \vec{V}^{ex}(\vec{r}) - \vec{V}(\vec{r}) | dl_{K^h}.$$

Here, the integral is calculated over the curve K^h surrounding the boundary of the airfoil and spaced from the airfoil surface line by a value $\frac{1}{8}\delta$; $\vec{V}^{ex}(\vec{r})$ is the exact velocity field around the elliptical airfoil; $\vec{V}(\vec{r})$ is the velocity field reconstructed according to the Biot–Savart law through the numerical solution $\gamma(\vec{r})$.

The errors of vortex sheet intensity computation and velocity field reconstruction using the scheme with rectilinear panels and piecewise-constant and piecewise-linear solutions are presented in Table 1 for different number of panels. Also a'posteriori estimates for the order of accuracy are given in Table 1 as a logarithms of the errors ratio

$$m_\gamma = \log_{\frac{N_i}{N_{i-1}}} \frac{\| \Delta\gamma \|_{L_1}^{(N_{i-1})}}{\| \Delta\gamma \|_{L_1}^{(N_i)}}, \qquad m_V = \log_{\frac{N_i}{N_{i-1}}} \frac{\| \Delta V \|_{L_1}^{(N_{i-1})}}{\| \Delta V \|_{L_1}^{(N_i)}}.$$

In Table 1, the indices "c" and "l" correspond to the piecewise-constant and piecewise-linear numerical solutions, respectively.

Table 1 shows that the piecewise-constant and piecewise-linear schemes provide the first and second order of accuracy, respectively, for the vortex sheet intensity computation. It should be noted that the error of the piecewise-linear scheme is an order of magnitude lower compared to the piecewise-constant scheme.

Table 1: The errors for vortex sheet intensity computation and velocity field reconstruction using numerical schemes with rectilinear panels.

N	32	64	128	256	500	1000
$\Vert \Delta\gamma^c \Vert_{L_1}$	0.367716	0.182186	0.090671	0.045283	0.023180	0.011589
m_γ^c	–	1.01	1.00	1.00	1.00	1.00
$\Vert \Delta\gamma^l \Vert_{L_1}$	0.057839	0.017131	0.004532	0.001164	0.000309	0.000078
m_γ^l	–	1.76	1.92	1.96	1.98	1.99
$\Vert \Delta V^c \Vert_{L_1}$	0.188352	0.087041	0.042885	0.021439	0.011005	0.005460
m_V^c	–	1.11	1.02	1.00	1.00	1.01
$\Vert \Delta V^l \Vert_{L_1}$	0.137691	0.060351	0.028499	0.013966	0.007083	0.003496
m_V^l	–	1.18	1.08	1.03	1.01	1.02

However, both schemes provide only the first order of accuracy for the calculation of the velocity field near the airfoil surface line. Moreover, the use of the piecewise-linear scheme does not lead to significant decreasing of the error of velocity field calculation: the error decreases less than twice in comparison to piecewise-constant scheme.

The error values for the schemes with curvilinear panels are shown in Table 2.

Table 2: The errors for vortex sheet intensity computation and velocity field reconstruction using numerical schemes with curvilinear panels.

N	32	64	128	256	500	1000
$\Vert \Delta\gamma^c \Vert_{L_1}$	0.299910	0.150430	0.075123	0.037549	0.019225	0.009612
m_γ^c		1.00	1.00	1.00	1.00	1.00
$\Vert \Delta\gamma^l \Vert_{L_1}$	0.017236	0.004556	0.001132	0.000283	0.000074	0.000019
m_γ^l		1.92	2.01	2.00	2.00	2.00
$\Vert \Delta V^c \Vert_{L_1}$	0.149978	0.070900	0.034642	0.017184	0.008760	0.004343
m_V^c		1.08	1.03	1.01	1.01	1.01
$\Vert \Delta V^l \Vert_{L_1}$	0.011160	0.002802	0.000704	0.000176	0.000046	0.000012
m_V^l		1.99	1.99	2.00	2.00	2.01

From Table 2 it is seen that the piecewise-constant scheme, as earlier, provides the first order of accuracy both for the vortex sheet intensity and velocity field calculation. As for the piecewise-linear scheme, the usage of curvilinear panels makes it possible to improve significantly the accuracy of velocity field reconstruction: now the error has the 2nd order.

5 CONCLUSIONS

Different numerical schemes for boundary integral equation solution in vortex methods are considered. The unknown value in this integral equation is intensity of the vortex sheet at the airfoil surface line that simulates the generation of a new vorticity. The schemes differ in two factors: 1. different methods for approximate airfoil surface line description are used: by rectilinear and curvilinear panels; 2. there are two ways of numarical solution representation: in the form of piecewise-constant or piecewise-linear function.

These schemes were compared for a model problem of flow simulation around elliptical airfoil: in terms of accuracy of the vortex sheet intensity computation and in terms of the accuracy of the velocity field reconstruction near the airfoil surface line. It is shown that the piecewise-constant and piecewise-linear schemes provide the first and the second orders of accuracy, respectively, for the vortex sheet intensity both in the case of rectilinear and curvilinear discretization of the airfoil surface line. However, for the velocity field the piecewise-linear scheme with rectilinear panels does not allow one to obtain the second order of accuracy. To obtain the second order, it is necessary to take into account the curvilinearity of the airfoil surface line.

Of course, the calculation of the system coefficients in case of piecewise-linear vortex sheet intensity distribution and curvilinear panels is more complicated procedure, compared with schemes with rectilinear panels and piecewise-constant distribution. But the fact that this scheme provides results with much higher accuracy, allows one to use much fewer panels discretizing the airfoil surface line, and, therefore, decrease the matrix size. In the case when the airfoil is deformable, the decreasing of the matrix size significantly reduces the computational complexity of the algorithm, because in this case, fully filled matrix should be inversed at each time step. In addition, the nonuniform linear distribution of the vortex sheet intensity over the panel allows one to transform the vorticity from the vortex sheet to the vortex wake with higher accuracy (in comparison with discrete or piecewise-distribution of the vortex sheet intensity), since in this case several vortex elements with different intensities can be discharged from one panel in accordance with the distribution.

ACKNOWLEDGEMENT

The financial support for the present project was partly provided by the Russian Foundation for Basic Research (RFBR) (grant No. 18-31-00245).

REFERENCES

[1] Leonard, A., Vortex methods for flow simulation. *Journal of Computational Physics*, **37**, pp. 289–335, 1980.
[2] Lifanov, I., *Singular Integral Equations and Discrete Vortices*, VSP: Utrecht, 1996.
[3] Cottet, G.-H. & Koumoutsakos, P., *Vortex Methods: Theory and Practice*, Cambridge University Press: Cambridge, 2000.
[4] Lewis, R., *Vortex Element Methods for Fluid Dynamic Analysis of Engineering Systems*, Cambridge University Press: Cambridge, 2005.
[5] Yokota, R. & Obi, Sh., Vortex methods for the simulation of turbulent flows: Review. *Journal of Fluid Science and Technology*, **6**(1), pp. 14–29, 2011.
[6] Branlard, E., *Wind Turbine Aerodynamics and Vorticity-Based Methods: Fundamentals and Recent Applications*, Springer, 2017.
[7] Kraposhin, M., Kuzmina, K., Marchevsky, I. & Puzikova, V., Study of OpenFOAM efficiency for solving fluid: Structure interaction problems. *OpenFOAM Selected Papers of the 11th Workshop,* eds J. Nobrega & J. Jasak, Springer, pp. 465–479, 2019.

[8] Kempka, S.N., Glass, M.W., Peery, J.S., Strickland, J.H. & Ingber, M.S., Accuracy consideration for implementing velocity boundary conditions in vorticity formulations. *Sandia Report,* SAND96-0583 UC-700, p. 52, 1996.

[9] Dynnikova, G.Ya., The Lagrangian approach to solving the time-dependent Navier–Stokes equations. *Doklady Physics*, **49**(11), pp. 648–652, 2004.

[10] Katz, J. & Plotkin, A., *Low-Speed Aerodynamics. from Wing Theory to Panel Methods*, Singapore: McGraw-Hill, 1991.

[11] Kuzmina, K.S., Marchevskii, I.K., Moreva, V.S. & Ryatina, E.P., Numerical scheme of the second order of accuracy for vortex methods for incompressible flow simulation around airfoils. *Russian Aeronautics*, **60**(3), pp. 398–405, 2017.

[12] Morgenthal, G. & Walther, J.H., An immersed interface method for the vortex-in-cell algorithm. *Computers and Structures*, **85**, pp. 712–726, 2007.

[13] Wu, J.C. & Thompson, J.F., Numerical solutions of time-dependent incompressible Navier–Stokes equations using an integro-differential formulation. *Computers and Fluids*, **1**, pp. 197–215, 1973.

[14] Kuzmina, K.S., Marchevskii, I.K. & Moreva, V.S., Vortex sheet intensity computation in incompressible flow simulation around airfoil by using vortex methods. *Mathematical Models and Computer Simulations*, **10**(3), pp. 276–287, 2018.

[15] Kuzmina, K.S., Marchevsky, I.K. & Moreva, V.S., On vortex sheet intensity computation for airfoils with angle point in vortex methods. *International Journal of Mechanical Engineering and Technology*, **9**(2), pp. 799–809, 2018.

[16] Kuzmina, K. & Marchevsky, I., The boundary integral equation solution in vortex methods with the airfoil surface line discretization into curvilinear panels. *Proceedings of Topical Problems of Fluid Mechanics 2019,* Prague, pp. 131–138, 2019.

[17] Kuzmina, K.S., Marchevsky, I.K. & Ryatina, E.P., Exact analytical formulae for linearly distributed vortex and source sheets influence computation in 2D vortex methods. *Journal of Physics: Conference Series*, **918**, article 012013, 2017.

[18] Kuzmina, K.S., Marchevsky, I.K. & Soldatova, I.A., Improved algorithm of boundary integral equation approximation in 2D vortex method for flow simulation around curvilinear airfoil. *AIP Conference Proceedings*, **2027**, article 040048, 2018.

[19] Kuzmina, K.S. & Marchevsky, I.K., Exact solutions for flow simulation around elliptical and wing airfoils for accuracy estimation of vortex sheet generation simulation in 2D vortex methods. *Journal of Physics: Conference Series*, in press.

3D CAUCHY PROBLEM FOR AN ELASTIC LAYER: INTERFACIAL CRACKS DETECTION

ALEXANDER N. GALYBIN[1,2] & SERGEY M. AIZIKOVICH[2]
[1]The Schmidt Institute of Physics of the Earth (IPE RAS), Moscow, Russia
[2]Don State Technical University, Russian Federation

ABSTRACT

This study presents a Cauchy-type problem of 3D elasticity for an elastic layer that can be bonded to an infinite base (half-space) made of dissimilar elastic material. The initial conditions are given on one side of the layer and both stress and displacement vectors are assumed to be known simultaneously. No conditions are specified on the other side. In the case of this side being fully bonded to the base, the stress and displacement vectors are continuous across the interface. This fact introduces certain relationships that have to be imposed on the initial conditions in order to obey continuity. We use these in order to detect a possible appearance of delamination of the interface. By using the double Fourier transform and the general solution of 3D elasticity in terms of harmonic functions, the initial value problem is reduced to a system of Fredholm integral equations of the first kind. Solutions of such systems are usually unstable; therefore, a numerical approach is suggested to overcome this difficulty by using the SVD regularisation. A possibility of delamination detection is discussed.

Keywords: layered elastic structures, interfacial cracks, inverse problems, integral equations, regularisation.

1 INTRODUCTION

This study continues the previous research for a 2D elastic strip [1] and generalises it for the 3D case of an elastic layer that can also be bonded to an elastic base (half-space) made of a different material. The main focus of this study is the development of mathematical methods capable of crack (delamination) detection on the interface between the layer and the base by using the strain/displacement measurements on the stress-free surface of the layer. Such a problem belongs to a wide class of incorrectly posed inverse problems of elasticity. It is known that the problem with overdetermined boundary conditions on a part of the boundary is incorrectly posed and can be unsolvable or possess unstable solutions. There are many publications regarding 2D inverse problems of plane elasticity but not so many for the 3D case. Standard formulation of the Cauchy initial value problem of plane elasticity usually assumes that the displacements vector **u** is known together with the stress vector **p** on a part of the boundary (usually it is stress-free, i.e. **p**=0) and no boundary conditions are specified on the rest of the boundary. Shvab [2] referred to such mathematical formulations as the (**u**,**p**) problem. Some other formulations, (see survey in Bonnet and Constantinescu [3]), can be reduced to the (**u**,**p**) problem. It is evident that in the case of plane elasticity the (**u**,**p**) problem can be routinely solved by a finite difference method applied to the Cauchy type system of two first-order differential equations of equilibrium, three linear Hook's law equations and the three first order strain-displacement equations. In the case of 3D, the number of equations in the system increases but this would not present essential difference for the numerical analysis. However, such finite difference approaches require essential modifications due to numerical instability inherent for incorrectly posed formulations. Other possible numerical methods solving (**u**,**p**) problem in 2D are based on finite element or on boundary element techniques, which still necessitates the application of a regularisation technique to provide

WIT Transactions on Engineering Sciences, Vol 126, © 2019 WIT Press
www.witpress.com, ISSN 1743-3533 (on-line)
doi:10.2495/BE420251

stability of the solution. It is understood that the use of standard regularisation techniques (as Tikhonov's or SVD regularisations) is capable to stabilize the solution (e.g. [4]–[7]), but for large systems their application is limited. Therefore, the development of alternative techniques for 3D case is necessary. Shwab [8] has shown that the Cauchy problem of 3D elasticity can be reduced to the Moisil and Theodoresco system [9], that provides generalisation of the integral Cauchy theorem from 2D into 3D and proved the uniqueness theorem for the solution of the problem. He also noticed that the problem is conditionally ill-posed as it can possess unstable solutions and investigated the accuracy of the solution by employing a boundary element technique. Galybin and Irša [10] have developed an FEM type technique for solving the Cauchy type problem for 3D harmonic functions complemented by the LSQR algorithm [11] for large systems. This allowed one to perform reconstruction of the contact stresses under a rigid circular punch applied to the boundary of elastic isotropic half-space by using the data of strain monitoring on the free surface of the half-space. This paper is aimed at the development of methods for solving the (**u**,**p**) problem for layered structures, in particular for the bi-material structure consisting of a coating bonded to a substrate. The main focus is to derive the integral relationships that can be useful for detection of delamination on the interface between the coating and the substrate. The work employs the method of spacious harmonic functions and the Fourier transform, which is widely used for solving direct boundary value problems for layered structures or bodies with planar cracks (e.g. [12], [13]).

2 PROBLEM FORMULATION AND INTEGRAL EQUATIONS

2.1 General solution in terms of harmonic functions

We consider an elastic layer $0<x_3<H$ with the moduli G (shear modulus) and κ (Muskhelishvili's constant) that can be fully or partly bonded to the half-space $x_3<0$ that has the moduli G_0 and κ_0 as shown in Fig. 1.

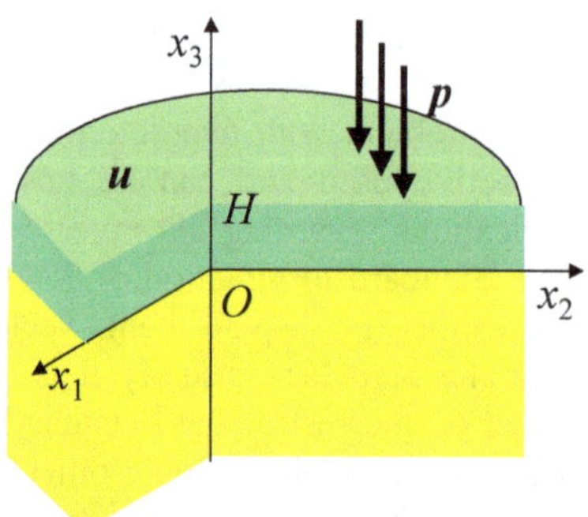

Figure 1: Geometry of the problem.

The boundary conditions are as follows:

BC1. The stress and displacement vectors are both known on the boundary $x_3=H$;

BC2. The stress and displacement vectors are both continuous across the interface $x_3=0$.

It is understood that this formulation is overdetermined if BC2 holds on the entire boundary $x_3=0$. However, if a delamination occurs on a part of the interface, BC2 can be used for the sake of its detection by solving the Cauchy initial value problem, IVP, for the layer with BC1. Therefore, we mainly focus on the Cauchy IVP and then discuss its application for the crack detection. In either case the aim is to construct a system of governing integral equations for the problem. Based on the formulas form Sih and Liebowitz [14] it has been shown [15], [16] that the following combination of the stress and displacement components can be expressed in terms of three spacious harmonic functions $\varphi=\varphi(x_1,x_2,x_3)$, $\psi=\psi(x_1,x_2,x_3)$ and $\chi=\chi(x_1,x_2,x_3)$ as follows:

$$\partial_1 u_2 - \partial_2 u_1 = 2(1-\nu)\chi, \tag{1}$$

$$\partial_1\sigma_{23} - \partial_2\sigma_{13} = 2G(1-\nu)\partial_3\chi, \tag{2}$$

$$\frac{\partial_1\sigma_{13}+\partial_2\sigma_{23}}{2G} = -\partial_{33}\psi - x_3\partial_{333}(\varphi+\psi) = (\partial_{11}+\partial_{22})\big(\psi + x_3\partial_3(\varphi+\psi)\big), \tag{3}$$

$$\begin{aligned}\partial_1 u_1 + \partial_2 u_2 &= -(1-2\nu)\partial_3\phi - 2(1-\nu)\partial_3\psi - x_3\partial_{33}(\phi+\psi) = \\ &= -\partial_3\big((1-2\nu)\phi + 2(1-\nu)\psi\big) + x_3\big(\partial_{11}+\partial_{22}\big)(\phi+\psi),\end{aligned} \tag{4}$$

$$u_3 = -2(1-\nu)\varphi - (1-2\nu)\psi + x_3\partial_3(\varphi+\psi), \tag{5}$$

$$\frac{\sigma_{33}}{2G} = -\partial_3\varphi + x_3\partial_{33}(\varphi+\psi). \tag{6}$$

Here, the symbols ∂_k stand for the derivative with respect to the k-th variable and since the functions φ, ψ and χ are harmonic we have used the following obvious relationship that is valid for any spacious harmonic function $\Phi=\Phi(x_1,x_2,x_3)$,

$$\partial_{33}\Phi = -\partial_{11}\Phi - \partial_{22}\Phi. \tag{7}$$

These equations are valid for the layer as well as for the half-space, but in the latter case the moduli G and κ should by replaced by G_0 and κ_0 respectively. As evident from the system it can be decomposed into two independent groups of equations. Eqns (1) and (2) form the first group for the determination of the function χ and four eqns (eqns (3)–(6)) form the second group for the determination of ϕ and ψ.

2.2 Fourier transform of 3D harmonic functions

We further use the double Fourier transform applied with respect to x_1 and x_2 in the form

$$\begin{aligned}\mathbf{DF}\big(Y(x_1,x_2)\big) &\equiv \hat{Y}(s_1,s_2) = \int_{-\infty}^{\infty}\int_{-\infty}^{\infty} Y(x_1,x_2)e^{-is_1x_1}e^{-is_2x_2}\,dx_1dx_2, \\ \mathbf{DF}^{-1}\big(\hat{Y}(s_1,s_2)\big) &\equiv Y(x_1,x_2) = \frac{1}{(2\pi)^2}\int_{-\infty}^{\infty}\int_{-\infty}^{\infty} \hat{Y}(s_1,s_2)e^{-is_1x_1}e^{-is_2x_2}\,ds_1ds_2.\end{aligned} \tag{8}$$

By applying eqn (8) to eqn (7), one finds

$$\partial_{33}\hat{\Phi}(s_1,s_2,x_3) - \lambda^2(s_1,s_2)\hat{\Phi}(s_1,s_2,x_3) = 0, \quad \lambda(s_1,s_2) = \sqrt{s_1^2+s_2^2}. \tag{9}$$

The solution for $\hat{\Phi}(s_1, s_2, x_3)$ and its normal derivative can be presented in the form

$$\hat{\Phi}(s_1,s_2,x_3) = \hat{\Phi}_1(s_1,s_2)e^{-\lambda(s_1,s_2)x_3} + \hat{\Phi}_2(s_1,s_2)e^{\lambda(s_1,s_2)(x_3-H)}, \tag{10}$$

$$\frac{\partial_3\hat{\Phi}(s_1,s_2,x_3)}{\lambda(s_1,s_2)} = -\hat{\Phi}_1(s_1,s_2)e^{-\lambda(s_1,s_2)x_3} + \hat{\Phi}_2(s_1,s_2)e^{\lambda(s_1,s_2)(x_3-H)}, \tag{11}$$

for the layer $0 \le x_3 \le H$ and

$$\hat{\Phi}(s_1,s_2,x_3) = \hat{\Phi}_0(s_1,s_2)e^{\lambda(s_1,s_2)x_3}, \tag{12}$$

$$\partial_3\hat{\Phi}(s_1,s_2,x_3) = \hat{\Phi}_0(s_1,s_2)\lambda(s_1,s_2)e^{\lambda(s_1,s_2)x_3}, \tag{13}$$

for the half-space $x_3 \le 0$.

It is evident from eqns (12) and (13) that one cannot simultaneously impose two independent boundary conditions of the Dirichlet and Neumann type on the boundary of half-space as the problem becomes overspecified. This fact will be used further on for outlining a procedure for crack detection.

2.3 Solution for the layer in Fourier space

Let us consider the Cauchy IVP for the layer assuming that the boundary values of both the function and its normal derivatives are given on the boundary x_3=H. It follows from eqns (10) and (11) that for $0 \le x_3 \le H$

$$\hat{\Phi}(s_1,s_2,x_3) + \frac{\partial_3\hat{\Phi}(s_1,s_2,x_3)}{\lambda(s_1,s_2)} = 2\hat{\Phi}_2(s_1,s_2)e^{\lambda(s_1,s_2)(x_3-H)}, \tag{14}$$

$$\hat{\Phi}(s_1,s_2,x_3) - \frac{\partial_3\hat{\Phi}(s_1,s_2,x_3)}{\lambda(s_1,s_2)} = 2\hat{\Phi}_1(s_1,s_2)e^{-\lambda(s_1,s_2)x_3}. \tag{15}$$

On the boundary x_3=H

$$\hat{\Phi}(s_1,s_2,H) + \frac{\partial_3\hat{\Phi}(s_1,s_2,H)}{\lambda(s_1,s_2)} = 2\hat{\Phi}_2(s_1,s_2), \tag{16}$$

$$\hat{\Phi}(s_1,s_2,H) - \frac{\partial_3\hat{\Phi}(s_1,s_2,H)}{\lambda(s_1,s_2)} = 2\hat{\Phi}_1(s_1,s_2)e^{-\lambda(s_1,s_2)H}. \tag{17}$$

On the interface x_3=0

$$\hat{\Phi}(s_1,s_2,0) + \frac{\partial_3\hat{\Phi}(s_1,s_2,0)}{\lambda(s_1,s_2)} = 2\hat{\Phi}_2(s_1,s_2)e^{-\lambda(s_1,s_2)H}, \tag{18}$$

$$\hat{\Phi}(s_1,s_2,0) - \frac{\partial_3\hat{\Phi}(s_1,s_2,0)}{\lambda(s_1,s_2)} = 2\hat{\Phi}_1(s_1,s_2). \tag{19}$$

It is evident from eqns (16) to (19) that the initial value problem for the harmonic function is readily found by known values of the function and its normal derivative at either end (x_3=0 or x_3=H). Thus, if the initial conditions are specified for x_3=H, then

$$2\hat{\Phi}(s_1,s_2,x_3)=\left(\hat{\Phi}(s_1,s_2,H)-\frac{\partial_3\hat{\Phi}(s_1,s_2,H)}{\lambda(s_1,s_2)}\right)e^{-\lambda(s_1,s_2)(x_3-H)}+ \\ +\left(\hat{\Phi}(s_1,s_2,H)+\frac{\partial_3\hat{\Phi}(s_1,s_2,H)}{\lambda(s_1,s_2)}\right)e^{\lambda(s_1,s_2)(x_3-H)},\quad 0\le x_3\le H. \tag{20}$$

It is evident, however, that the application of the inverse Fourier transform cannot be performed directly due to positive power of the first exponent in the right hand of eqn (20). This fact also indicates that the solution of the initial value problem for harmonic functions is unstable.

Let us find the relationships between the Fourier transforms on the boundaries. It follows from eqns (16) and (18) that

$$\hat{\Phi}(s_1,s_2,0)+\frac{\partial_3\hat{\Phi}(s_1,s_2,0)}{\lambda(s_1,s_2)}=\left(\hat{\Phi}(s_1,s_2,H)+\frac{\partial_3\hat{\Phi}(s_1,s_2,H)}{\lambda(s_1,s_2)}\right)e^{-\lambda(s_1,s_2)H}. \tag{21}$$

From eqns (17) and (19), one finds

$$\hat{\Phi}(s_1,s_2,H)-\frac{\partial_3\hat{\Phi}(s_1,s_2,H)}{\lambda(s_1,s_2)}=\left(\hat{\Phi}(s_1,s_2,0)-\frac{\partial_3\hat{\Phi}(s_1,s_2,0)}{\lambda(s_1,s_2)}\right)e^{-\lambda(s_1,s_2)H}. \tag{22}$$

These relationships are used further on for derivation of a system of integral equations.

2.4 Reduction to integral equations

Let us apply the inverse Fourier transform to eqns (21) and (22) and use the convolution theorem, which leads to the following system of integral equations of the Fredholm type of the first kind:

$$\Phi(x_1,x_2,0)+\frac{1}{2\pi}\int_{-\infty}^{\infty}\int_{-\infty}^{\infty}\frac{\partial_3\Phi(\xi,\eta,0)}{\rho(\xi-x_1,\eta-x_2,0)}d\xi d\eta= \\ =\frac{1}{2\pi}\int_{-\infty}^{\infty}\int_{-\infty}^{\infty}\left(\frac{H\Phi(\xi,\eta,H)}{\rho^3(\xi-x_1,\eta-x_2,H)}+\frac{\partial_3\Phi(\xi,\eta,H)}{\rho(\xi-x_1,\eta-x_2,H)}\right)d\xi d\eta,\quad -\infty<x_1,x_2<\infty, \tag{23}$$

$$\frac{1}{2\pi}\int_{-\infty}^{\infty}\int_{-\infty}^{\infty}\left(\frac{H\Phi(\xi,\eta,0)}{\rho^3(\xi-x_1,\eta-x_2,H)}-\frac{\partial_3\Phi(\xi,\eta,0)}{\rho(\xi-x_1,\eta-x_2,H)}\right)d\xi d\eta= \\ =\Phi(x_1,x_2,H)-\frac{1}{2\pi}\int_{-\infty}^{\infty}\int_{-\infty}^{\infty}\frac{\partial_3\Phi(\xi,\eta,H)}{\rho(\xi-x_1,\eta-x_2,0)}d\xi d\eta,\quad -\infty<x_1,x_2<\infty. \tag{24}$$

Here,

$$\rho(x_1,x_2,x_3)=\sqrt{x_1^2+x_2^2+x_3^2}\,, \tag{25}$$

whose reciprocal ($1/\rho$) is a harmonic function and the following integrals [13] have been used taking into account that ρ is even with respect to all variables:

$$\frac{2}{\pi}\int_0^\infty\int_0^\infty\frac{1}{\lambda(\alpha,\beta)}e^{-\lambda(\alpha,\beta)z}\cos(\alpha x)\cos(\beta y)d\alpha d\beta=\frac{1}{\rho(x,y,z)}\,, \tag{26}$$

$$\frac{2}{\pi}\int_0^\infty\int_0^\infty e^{-\lambda(\alpha,\beta)z}\cos(\alpha x)\cos(\beta y)d\alpha d\beta=\frac{z}{\rho^3(x,y,z)},\quad z>0\,. \tag{27}$$

Note that the integral in eqn (27) can be obtained by differentiation of eqn (26) with respect to x_3.

The system eqns (23) and (24) is sought system of governing integral equations for the elastic layer.

3 THE CASE OF NORMAL LOADING

3.1 Solution for the layer

Let us consider a special case of normal loading of the upper boundary ($x_3=H>0$) of the layer, which is most interesting from practical point of view. Assume $\sigma_{13}(x_1,x_2,x_3)=0$ and $\sigma_{23}(x_1,x_2,x_3)=0$ in the system eqns (1)–(6) and introduce the following notations for known (measured) combinations of the displacements on $x_3=H$:

$$\begin{aligned}
&2(1-\nu)U(x_1,x_2)=u_3(x_1,x_2,H),\\
&2(1-\nu)R(x_1,x_2)=\partial_1u_2(x_1,x_2,H)-\partial_2u_1(x_1,x_2,H),\\
&2(1-\nu)S(x_1,x_2)=\partial_1u_1(x_1,x_2,H)+\partial_2u_2(x_1,x_2,H).
\end{aligned} \tag{28}$$

The first group of eqns (1) and (2) takes the form

$$\chi(x_1,x_2,H)=R(x_1,x_2),\quad \partial_3\chi(x_1,x_2,H)=0\,. \tag{29}$$

The second group is obtained by summing eqns (4) and (6) and by taking into account that the last two terms in eqn (5) can be neglected due to homogeneity of eqn (3) in the case of normal load. Therefore, we arrive at the problem of finding the sum of two unknown harmonic functions as follows:

$$u_3=-2(1-\nu)(\varphi+\psi)\,, \tag{30}$$

$$\partial_1u_1+\partial_2u_2+\frac{\sigma_{33}}{2G}=-2(1-\nu)\partial_3(\varphi+\psi)\,. \tag{31}$$

As soon as the sum $\varphi+\psi$ is determined one can derive the solutions for φ and ψ and their normal derivatives by using eqns (3)–(5). By introducing the auxiliary harmonic function $\omega=-\varphi-\psi$ and employing eqn (28), one obtains the following initial value problem:

$$\omega(x_1,x_2,H)=U(x_1,x_2),\quad \partial_3\omega(x_1,x_2,H)=S(x_1,x_2)+P(x_1,x_2), \tag{32}$$

where the function $P(x_1,x_2)=\sigma_{33}(x_1,x_2,H)(2D)^{-1}$ is known exactly in contrast to the other terms on the right-hand sides of eqns (29) and (32) that contain measurement errors.

It is physically more realistic to measure displacements (strains) on the stress-free surfaces. In this case the dimensions of the zones of load application should be relatively small compared with the areas of displacement monitoring. This is valid in case of concentrated loads, which ensures that displacements are known over almost the whole boundary except for a finite number of points and thus all BC1 are satisfied. For simplicity it is assumed that all the loads acting on the surface $x_3=H$ are concentrated forces applied at different locations. For a normal concentrated load of intensity P_0 applied at the origin the function $P(x_1,x_2)=P_0\delta(x_1,x_2)$, where $\delta(x_1,x_2)$ is the Delta function. Its Fourier transform is constant P_0. In order to keep the layer in equilibrium the resultant normal stress acting on the opposite side of the layer should have the same value P_0 regardless of the its distribution over the plane $x_3=0$. If the load on the opposite side is axisymmetric then all the conditions of global equilibrium are satisfied. This load can be selected, for example, as $P_0\exp(-x_1^2-x_2^2)/\pi$, which emphasizes the fact that the stress and displacement vectors in the layer can possess polar symmetry. Let ω_0 be a harmonic function that presents the solution of the boundary value problem in normal stresses for the elastic layer. Then we can seek for a solution of the initial value problem with modified conditions eqn (32) as follows:

$$\omega(x_1,x_2,H)=U(x_1,x_2)-U_0(x_1,x_2),\quad \partial_3\omega(x_1,x_2,H)=S(x_1,x_2)-S_0(x_1,x_2), \tag{33}$$

where U_0 and S_0 are the combinations of the displacement in eqn (28) that correspond to the solution of the BVP in terms of stresses. In the case of single concentrated load these functions possess polar symmetry, i.e. they depend only on the polar radius. In the case of several concentrated forces applied on $x_3=H$ one should use superposition of the particular solutions for every force and thus the polar symmetry does not take place anymore. We do not use symmetry further.

3.1.1 System for χ

In this case, we have to use eqn (29). Then the system eqns (23) and (24) becomes $(-\infty<x_1,x_2<\infty)$:

$$\begin{cases} \chi(x_1,x_2,0)+\dfrac{1}{2\pi}\displaystyle\int_{-\infty}^{\infty}\int_{-\infty}^{\infty}\frac{\partial_3\chi(\xi,\eta,0)}{\rho(\xi-x_1,\eta-x_2,0)}d\xi d\eta=\frac{H}{2\pi}\int_{-\infty}^{\infty}\int_{-\infty}^{\infty}\frac{R(\xi,\eta)}{\rho^3(\xi-x_1,\eta-x_2,H)}d\xi d\eta, \\ \dfrac{1}{2\pi}\displaystyle\int_{-\infty}^{\infty}\int_{-\infty}^{\infty}\left(\frac{H\chi(\xi,\eta,0)}{\rho^3(\xi-x_1,\eta-x_2,H)}-\frac{\partial_3\chi(\xi,\eta,0)}{\rho(\xi-x_1,\eta-x_2,H)}\right)d\xi d\eta=R(x_1,x_2). \end{cases} \tag{34}$$

It follows from eqns (21) and (22) with the use of eqn (29) that this system is consistent only when the following relationship between $\chi(x_1,x_2,0)$ and $\partial_3\chi(x_1,x_2,0)$ holds:

$$\chi(x_1,x_2,0)+\frac{1}{2\pi}\int_{-\infty}^{\infty}\int_{-\infty}^{\infty}\frac{\partial_3\chi(\xi,\eta,0)}{\rho(\xi-x_1,\eta-x_2,0)}d\xi d\eta=$$

$$=\frac{1}{2\pi}\int_{-\infty}^{\infty}\int_{-\infty}^{\infty}\left(\frac{2H\chi(\xi,\eta,0)}{\rho^3(\xi-x_1,\eta-x_2,2H)}-\frac{\partial_3\chi(\xi,\eta,0)}{\rho(\xi-x_1,\eta-x_2,2H)}\right)d\xi d\eta. \quad (35)$$

3.1.2 System for ω

Substitution of eqn (33) into eqns (23)–(24) results in ($-\infty < x_1, x_2 < \infty$):

$$\begin{cases}\omega(x_1,x_2,0)+\dfrac{1}{2\pi}\displaystyle\int_{-\infty}^{\infty}\int_{-\infty}^{\infty}\frac{\partial_3\omega(\xi,\eta,0)}{\rho(\xi-x_1,\eta-x_2,0)}d\xi d\eta=F_1(x_1,x_2,H),\\ \dfrac{1}{2\pi}\displaystyle\int_{-\infty}^{\infty}\int_{-\infty}^{\infty}\left(\frac{H\omega(\xi,\eta,0)}{\rho^3(\xi-x_1,\eta-x_2,H)}-\frac{\partial_3\omega(\xi,\eta,0)}{\rho(\xi-x_1,\eta-x_2,H)}\right)d\xi d\eta=F_2(x_1,x_2).\end{cases} \quad (36)$$

Here the right-hand side of the system is as follows:

$$F_1(x_1,x_2,H)=\frac{1}{2\pi}\int_{-\infty}^{\infty}\int_{-\infty}^{\infty}\left(\frac{H(U(\xi,\eta)-U_0(\xi,\eta))}{\rho^3(\xi-x_1,\eta-x_2,H)}+\frac{S(\xi,\eta)-S_0(\xi,\eta)}{\rho(\xi-x_1,\eta-x_2,H)}\right)d\xi d\eta, \quad (37)$$

$$F_2(x_1,x_2)=U(x_1,x_2)-U_0(x_1,x_2)-\frac{1}{2\pi}\int_{-\infty}^{\infty}\int_{-\infty}^{\infty}\frac{S(\xi,\eta)-S_0(\xi,\eta)}{\rho(\xi-x_1,\eta-x_2,0)}d\xi d\eta. \quad (38)$$

As soon as the solution of eqn (36) is found one can find the total solution for the case of concentrated loads by summing ω with ω_0. As a result, we determine the combination $\partial_1 u_1+\partial_2 u_2+\sigma_{33}(2G)^{-1}$ and the normal displacement u_3. The normal stresses on the interface $x_3=0$ is found by differentiating eqn (5) with respect to x_3 and summing with eqn (6), which results in

$$\frac{\sigma_{33}(x_1,x_2,0)}{2G}=(1-2\nu)\partial_3\omega(x_1,x_2,0)+\partial_3 u_3(x_1,x_2,0). \quad (39)$$

3.2 Numerical approach

Several standard methods can be employed for solving the system of integral eqn (36) taking into account that the second integral equation has regular kernels while the first one has a weak singularity and therefore the second term in its left hand side should be treated as a convergent improper integral.

Therefore, eqn (34) can be referred to as a system of Fredholm integral equations of the first kind, which usually possess unstable solutions. Alternatively, one can use the Trefftz-type approach that demonstrated satisfactory performance for solving unstable Fredholm equations when complemented by the SVD regularisation (see Galybin [15] for detail).

Let us seek the function ω and its derivative with respect to x_3 (that is also harmonic) as two linear combinations of the radial basis functions $1/\rho$ with the nodes (ξ_k, η_k, ζ_k) lying outside the layer

$$\omega(x_1,x_2,x_3)=\sum_{k=1}^{N}\frac{a_k}{\rho(x_1-\xi_k,x_2-\eta_k,x_3-\xi_k)}, \tag{40}$$

$$\partial_3\omega(x_1,x_2,x_3)=\sum_{k-1}^{N}\frac{b_k}{\rho(x_1-\xi_k,x_2-\eta_k,x_3-\xi_k)}. \tag{41}$$

Then the following system of ($3M$ by $2N$) linear algebraic equations can be formed for determination of unknown coefficient a_k and b_k:

$$\begin{cases}\sum_{k=1}^{N}\frac{a_k}{\rho(x_1^m-\xi_k,x_2^m-\eta_k,H-\xi_k)}=U(x_1^m,x_2^m)-U_0(x_1^m,x_2^m), \quad m=1\ldots M,\\ \sum_{k=1}^{N}\frac{b_k}{\rho(x_1^m-\xi_k,x_2^m-\eta_k,H-\xi_k)}=S(x_1^m,x_2^m)-S_0(x_1^m,x_2^m), \quad m=1\ldots M,\\ \sum_{k=1}^{N}\frac{(H-\xi_k)a_k-\rho^2(x_1^m-\xi_k,x_2^m-\eta_k,H-\xi_k)b_k}{\rho^3(x_1^m-\xi_k,x_2^m-\eta_k,H-\xi_k)}=0, \quad m=1\ldots M,\end{cases} \tag{42}$$

where x_1^m, x_2^m are collocation points m=1…M.

This system is overdetermined if $3M > 2N$ and therefore its approximate solution can be found by the least square method. It should be mentioned that a regularisation procedure is necessary in the case when the condition number of the matrix is large. As shown in Galybin [15] the SVD regularisation is efficient for medium size matrices while for large matrices one can use Tikhonov's regularisation.

3.3 A procedure for interface delamination detection

The solution for the layer can be used to suggest a procedure for checking if the layer and the half-space are fully bonded or delamination has occurred on the interface x_3=0. In the case of full bonding between the layer and the half-space the continuity of the stress and displacement vectors on the interface x_3=0 should be satisfied. Otherwise they are violated. As follows from the solution for the lower half-space:

$$\partial_3\hat{\Phi}(s_1,s_2,0)=\lambda(s_1,s_2)\hat{\Phi}(s_1,s_2,0), \tag{43}$$

Therefore,

$$\lambda(s_1,s_2)\hat{u}_3(s_1,s_2,0)=\partial_1\hat{u}_2(s_1,s_2,0)+\partial_1\hat{u}_2(s_1,s_2,0)+\frac{1}{2G_0}\hat{\sigma}_{33}(s_1,s_2,0). \tag{44}$$

After application of the inverse Fourier transform to eqn (44), it can be presented in one of the following equivalent forms (we assume that the solutions are smooth enough):

$$u_3(x_1,x_2,0)=\frac{1}{2\pi}\int_{-\infty}^{\infty}\int_{-\infty}^{\infty}\frac{\partial_1 u_2(\xi,\eta,0)+\partial_1 u_2(\xi,\eta,0)+\frac{1}{2G_0}\sigma_{33}(\xi,\eta,0)}{\rho(\xi-x_1,\eta-x_2,0)}d\xi d\eta\,, \quad (45)$$

or

$$\frac{-1}{2\pi}\int_{-\infty}^{\infty}\int_{-\infty}^{\infty}\frac{\left(\partial_{11}+\partial_{22}\right)u_3(\xi,\eta,0)}{\rho(\xi-x_1,\eta-x_2,0)}d\xi d\eta=$$

$$=\partial_1 u_2(x_1,x_2,0)+\partial_1 u_2(x_1,x_2,0)+\frac{1}{2G_0}\sigma_{33}(x_1,x_2,0). \quad (46)$$

The solution of the Cauchy IVP for the layer determines two harmonic functions, one for the combination $\partial_1 u_1+\partial_2 u_2+\sigma_{33}(2G)^{-1}$ and another for the normal displacement u_3 in the layer as well as their boundary values. The normal stress is also known from eqn (39). Therefore, one can substitute these solutions into either eqn (45) or eqn (46) to check if the initial conditions given by eqn (28) satisfy them. Given that the strain/displacement measurements on the boundary $x_3=H$ are subjected to errors, eqns (45) and (46) should be satisfied approximately. Discussion of the choice of a proper level of approximation is beyond the scope of this paper. We should assume that if both eqns (45) and (46) are not fulfilled, then we expect the presence of a delamination on the interface. It can also be located by calculating the normal stresses that have to be close to zero on the surface of delamination.

4 CONCLUSIONS

This study concerns the 3D elastic problem for the coating bonded to a substrate from a dissimilar elastic material. We presented the system of integral equations for the Cauchy initial value problem for an elastic layer that followed from the analytical solution in the Fourier space. It is shown that the solution of the system is unstable because the system consists of the Fredholm equations of the first kind. A numerical procedure is suggested that can hamper instability. We have also presented the relationships that have to be imposed on the initial conditions that satisfy continuity of the stress and displacement vectors on the interface between the layer and the half-space.

ACKNOWLEDGEMENT

This work was supported by the grant of the Government of the Russian Federation No. 14.Z50.31.0046.

REFERENCES

[1] Galybin, A.N. & Rogerson, G.A., Ill-posed problem of the Cauchy type for an elastic strip. *Mathematics and Mechanics of Solids*, 2019. (E-pub ahead of print). DOI: 10.1177/1081286519826395.

[2] Shvab, A.A., Incorrectly posed static problems of elasticity. *Solid Mechanics*, **24**, pp. 98–106, 1989.

[3] Bonnet, M. & Constantinescu, A., Inverse problems in elasticity. *Inverse Problems*, **21**, pp. R1–R50, 2005.

[4] Marin, L., Elliott, L., Ingham, D.B. & Lesnic, D., Boundary element method for the Cauchy problem in linear elasticity. *Engineering Analysis with Boundary Elements*, **25**(9), pp. 783–793, 2001.

[5] Marin, L. & Lesnic, D., Boundary element solution for the Cauchy problem in linear elasticity using singular value decomposition. *Computer Methods in Applied Mechanics and Engineering*, **191**(6), pp. 3257–3270, 2002.

[6] Marin, L. & Lesnic, D., The method of fundamental solutions for inverse boundary value problems associated with the two-dimensional biharmonic equation. *Mathematical and Computer Modelling*, **42**(3–4), pp. 261–278, 2005.

[7] Galybin, A.N., A method for determination of stress distributions in the process zone ahead of a 2D crack. *"Moving Boundaries VI" Computational Modelling of Free and Moving Boundary Problems*, eds B. Sarler & C.A. Brebbia, WIT Press: Southampton, pp. 243–252, 2001.

[8] Schwab, A.A., The inverse problem of elasticity theory: Application of the boundary integral equation for the holomorphic vector. *Physics of the Solid Earth*, **30**(4), pp. 342–348, 1994.

[9] Moisil, G.C. & Theodoresco, N., Fonctions holomorphes dans l'espace. *Mathematica*, **5**, pp. 141–153, 1931.

[10] Galybin, A.N. & Irša, J., On reconstruction of 3D harmonic functions from discrete data. *Proceedings of Royal Society A*, **466**(2119), pp. 1935–1955, 2010.

[11] Paige, C.C. & Saunders, M.A., LSQR: An algorithm for sparse linear equations and sparse least squares. *ACM Transactions on Mathematical Software*, **8**, pp. 43–71, 1982.

[12] Shevljakov, Y.A., *Matrix Algorithms in the Theory of Elasticity of Inhomogeneous Media*, Vischa Schola: Kiev and Odessa, p. 216, 1977. (In Russian.)

[13] Vafa, J.P. & Fariborz, S.J., Analysis of rectangular cracks in elastic bodies. *Theoretical and Applied Fracture Mechanics*, **87**, pp. 78–90, 2017.

[14] Sih, G.C. & Liebowitz, H., Mathematical theories on brittle fracture. *Fracture: An Advanced Treatise*, vol. 2, ed. H. Liebowitz, Academic Press: New York, 1968.

[15] Galybin, A.N., Contact inverse problem for an elastic half-space. *Engineering Analysis with Boundary Elements*, **68**, pp. 35–41, 2016.

[16] Galybin, A.N., Integral equations and Gauss–Chebyshev quadratures for planar rectangular cracks. *International Journal of Mechanical Sciences*, **146–147**, pp. 272–279, 2018.

Author index

Computational and Experimental Methods in Multiphase and Complex Flow X

Edited by: ***S. HERNÁNDEZ****, University of A Coruña, Spain and* ***P. VOROBIEFF****, University of New Mexico, USA*

Composed of papers presented at the 10th conference on Multiphase flow this book presents the latest research on the subject. The research included in this volume focuses on using synergies between experimental and computational techniques to gain a better understanding of all classes of multiphase and complex flow.

The presented papers illustrate the close interaction between numerical modellers and researchers working to gradually resolve the many outstanding issues in our understanding of multiphase flow.

Recently multiphase fluid dynamics have generated a great deal of attention, leading to many notable advances in experimental, analytical and numerical studies. Progress in numerical methods has permitted the solution of many practical problems, helping to improve our understanding of the physics involved.

Multiphase flows are found in all areas of technology and the range of related problems of interest is vast, including astrophysics, biology, geophysics, atmospheric process, and many areas of engineering.

The papers in the book cover a number of topics, including: Experimental measurements; Numerical methods; Multiphase flows and Flow in porous media.

ISBN: 978-1-78466-329-2 **eISBN: 978-1-78466-330-8**
Published 2019 / 250pp

www.ingramcontent.com/pod-product-compliance
Lightning Source LLC
LaVergne TN
LVHW062223200926
848484LV00006B/1460